Coastal Monitoring Through Partnerships

COASTAL MONITORING THROUGH PARTNERSHIPS

Proceedings of the Fifth Symposium on
the Environmental Monitoring and Assessment Program (EMAP)
Pensacola Beach, FL, U.S.A., April 24–27, 2001

Scientific Editors:

Brian D. Melzian
U.S. Environmental Protection Agency
National Health and Environmental Effects Research Laboratory
Atlantic Ecology Division
Narrangansett, Rhode Island, U.S.A.

Virginia Engle
U.S. Environmental Protection Agency
National Health and Environmental Effects Research Laboratory
Gulf Ecology Divison
Gulf Breeze, Florida, U.S.A.

Malissa McAlister
The Council of State Governments
Lexington, Kentucky, U.S.A.

Shabeg Sandhu
U.S. Environmental Protection Agency
National Health and Environmental Effects Research Laboratory
Research Triangle Park, North Carolina, U.S.A.

Technical Editor

Lisa Kay Eads
The Council of State Governments
Lexington, Kentucky, U.S.A.

Reprinted from *Environmental Monitoring and Assessment*, Volume 81, Nos. 1–3, 2003

KLUWER ACADEMIC PUBLISHERS
Dordrecht/Boston/London

A C.I.P. Catalogue record for this book is available from the Library of Congress.

ISBN 1-4020-1089-3

Published by Kluwer Academic Publishers
P.O. Box 17, 3300 AA Dordrecht, The Netherlands

Sold and distributed in North, Central and Latin America
by Kluwer Academic Publishers
101 Philip Drive, Norwell, MA 02061, U.S.A.

In all other countries, sold and distributed
by Kluwer Academic Publishers
P.O. Box 322, 3300 AH Dordrecht, The Netherlands

Printed on acid-free paper

Printed in the Netherlands

TABLE OF CONTENTS

COASTAL MONITORING THROUGH PARTNERSHIPS

Proceedings of the Fifth Symposium on
the Environmental Monitoring and Assessment Program (EMAP)
Pensacola Beach, FL, U.S.A., April 24–27, 2001

A. Preface

B. Regional and National Coastal Monitoring Partnership Programs

C. Monitoring Approaches, Modeling, and Data Management

D. Benthic Communities Monitoring and Assessment

E. Biological Indicators & Interlaboratory Sediment Comparisons

F. Microbiological Modeling, Indicators, and Monitoring

G. Monitoring and Assessment of Phytoplankton and Submerged Aquatic Vegetation Communities

U.S. EPA funds placed into the "Promoting a Science-Based Framework for Ecosystem Protection Program's" Cooperative Agreement (CR-82658401) with The Council of State Governments were used to support the EMAP 2001 "Coastal Monitoring Through Partnerships" Symposium; and the production of these Proceedings.

PREFACE

Coastal areas in the United States are growing rapidly, and currently about 45% of the nation's human population lives in these areas. As the coastal population increases, there will likely be increasing environmental and socioeconomic pressures on our estuaries and coastal environments. Establishing the condition of the nation's estuarine and coastal environments is one key to protecting and restoring our estuaries and coastal environments. Monitoring the condition of all our nation's estuarine and coastal ecosystems over the long term is more than any one program can accomplish. Therefore, it is crucial that monitoring programs at all levels (local, state, and federal) cooperate in the collection and sharing of environmental data. To this end, EPA's Office of Research and Development's Environmental Monitoring and Assessment Program's (EMAP) 2001 Symposium entitled "**Coastal Monitoring Through Partnerships**" provided a forum to present and discuss the results of successful partnerships among federal, state, tribal, and academic scientists and managers to advance the science of monitoring and assessment of estuarine and coastal ecosystems.

The National Coastal Assessment (NCA) is one model for a successful federal, state, tribal, and academic partnership program. The NCA is a national demonstration of EMAP's integrated probabilistic monitoring approach to produce comprehensive assessments of the condition of all the nation's estuarine and coastal environments in partnership with the coastal states, the National Oceanic and Atmospheric Administration (NOAA), and the United States Geological Survey (USGS). In the NCA, we have successfully partnered with all 24 marine coastal states (including Alaska and Hawaii) and Puerto Rico in the use of a compatible, probabilistic monitoring design. With a minimum of 50 sampling locations and a common set of core indicators, each state will be able to independently assess conditions of their estuarine and coastal environments. Subsequently, these estimates will be aggregated to assess conditions at the EPA Regional, biogeographical, and national levels.

Rather than assess estuarine conditions on a chemical by chemical basis, the NCA uses biological indicators to integrate the various stressors acting on the community, and to characterize the condition of the estuarine and coastal environments. However, chemical and physical parameters are also collected to interpret and rank the likely stressors associated with impaired estuarine and coastal conditions in each state. The initial outcome of the NCA will be to provide the public and decision-makers with a reliable baseline for the current condition of the nation's estuaries and coastal environments. As the states adopt the approach used in the NCA, and as they continue to assess estuarine and coastal ecological conditions through time, the trends established will provide unbiased, quantitative evidence for the effectiveness of our national, regional and state environmental protection and restoration policies and programs.

There are many components involved in determining the overall impacts of anthropogenic stressors on estuarine and coastal waters. It will take strong "partnerships" like those described during the EMAP 2001 Symposium, to ensure that we have healthy, sustainable estuaries and coastal environments, now and in the future.

Michael E. McDonald
Director, Environmental Monitoring and Assessment Program
National Health and Environmental Effects Research Laboratory
Office of Research and Development
U.S. Environmental Protection Agency

Environmental Monitoring and Assessment **81**: 1, 2003.

SOUTHERN CALIFORNIA'S MARINE MONITORING SYSTEM TEN YEARS AFTER THE NATIONAL RESEARCH COUNCIL EVALUATION

BROCK B. BERNSTEIN[1] and STEPHEN B. WEISBERG[2*]

[1]*Independent Consultant, Ojai, CA, U.S.A.;*
[2]*Southern California Coastal Water Research Project, Westminster, CA, U.S.A.*
*(author for correspondence, e-mail: stevew@sccwrp.org)

Abstract. In 1990, the National Research Council (NRC) published two in-depth assessments of marine environmental monitoring effectiveness. The first of these, *Managing Troubled Waters: The Role of Marine Environmental Monitoring*, provided a national perspective and the second, *Monitoring Southern California's Coastal Waters*, examined the specifics of monitoring design and implementation in a densely populated, highly urbanized coastal region. The reports include explicit recommendations about the need for greater regionalization of monitoring efforts, supported by greater standardization of field, laboratory, and data analysis methods. They also identified the need for centralized data management and for greater flexibility in the language of standard discharge permits, flexibility that would permit discharge agencies to more readily participate in regional monitoring and research programs. Other recommendations identified a need for EPA and NOAA to focus on creating a national monitoring program structured as a network of coordinated local and regional efforts. Finally, the NRC emphasized the need for better reporting and for periodic review of monitoring's relevance to management concerns. In this paper, we use southern California as a test case to assess progress made in implementing the NRC's recommendations. We review progress made on each recommendation and discuss the features of the regulatory and management climate that contributed to or impeded this progress. We also consider whether, and to what extent, the NRC's recommendations remain relevant in the present context.

Keywords: regional monitoring, southern California, monitoring design, coastal zone management

1. Introduction

Ten years ago, the National Research Council (NRC) prepared two documents reviewing the effectiveness of marine monitoring programs, a national assessment (NRC, 1990a) and a more detailed look at programs in southern California (NRC, 1990b). Both documents found numerous deficiencies in overall monitoring systems, as well as in the design and implementation of individual programs, and both documents made recommendations for improvement. In particular, the southern California assessment found that there was a minimum of \$17 million per year of ongoing effort in the late 1980s, a number which was refined and updated by Schiff *et al.* (2002a) to \$31 million per year in 1997. The bulk of monitoring is carried out by dischargers, most notably municipal wastewater dischargers. The NRC review found that despite the large expenditure, marine monitoring was spatially restricted, with 70% allocated to localized discharge permit monitoring and much of the remainder to selected historic trend sites. As a result, less than 5% of the area of the Southern California Bight (the coastal region between Point Conception in the north and the U.S.-Mexico border in the south) was monitored. Effort was incon-

Environmental Monitoring and Assessment **81**: 3–14, 2003.

sistent, using different indicators and field and laboratory methods, and with data stored in incompatible data formats. Because there was no institutional mechanism for integrating or reporting findings, these deficiencies prevented a coherent assessment of the state of the Bight as a whole. They also made it extremely difficult to compare localized patterns and trends to the regional background. Together, these deficiencies severely limited the ability to use marine monitoring data in management decision making.

To address these shortcomings, the NRC reports made a series of recommendations. In this paper, we use southern California as a test case to examine the extent to which these recommendations have been addressed ten years after they were published. We then discuss factors underlying the monitoring system's successful response to these recommendations and identify next steps in the system's ongoing evolution, which has included a shift from a primary emphasis on single large discharge points to more of an ecosystem perspective.

2. Recommendations and Progress

2.1 RECOMMENDATION #1: ESTABLISH A REGIONAL MONITORING PROGRAM

The NRC studies identified that a large amount of monitoring was being conducted in southern California, but that most of it was local and disconnected. It was impossible to make regional assessments and to place local patterns and trends in a regional context. A major recommendation was to establish a regional monitoring program, with standardized field, laboratory, and data management methods, and with a process for analyzing and integrating results on a regional scale. The recommendation focused on three major areas: 1) sediment condition; 2) water quality and public health at swimming beaches; and 3) contaminant input sources. It is worth noting that a later NRC report on coastal wastewater management (NRC, 1993) also stressed the value of integrating regional monitoring more fully into the management process.

Southern California's monitoring entities have been especially responsive to this recommendation, with the latest and most direct aspects of an evolving regional monitoring program being two regional surveys facilitated by the Southern California Coastal Water Research Project (SCCWRP). The first survey was conducted in 1994, involved 12 organizations and focused on fish and sediment quality (Hashimoto and Weisberg, 1998). The second (Bight'98) was conducted in 1998, involved 62 organizations and extended beyond the first survey in three ways: 1) it extended inshore to include an assessment of bays, harbors, and port areas, 2) it extended southward to include Mexican waters between the U.S.-Mexico border and Ensenada, and 3) it added a shoreline microbiology component. Both surveys focused on estimating the spatial extent of perturbation and used stratified random sampling designs to achieve this objective.

The results of these studies (Schiff, 2000; Schiff and Allen, 2000; Bergen *et al.*, 2000; Noble *et al.*, 2000) enabled scientists and managers, more completely than ever before, to map the extent of perturbations in a variety of indicators. These efforts were unique to the region in their use of common metrics and data collected from the entire region, not just from around individual discharge points. This enabled managers to assess relative degrees of perturbation at different locations and to begin tracking trends against an accurate description of the regional background.

These regional efforts focused on an assessment objective. However, success in that effort was entirely dependent on the achievement of regional methods standardization, which was accomplished in several ways and built on earlier efforts to standardize benthic sampling gear and invertebrate taxonomy. First, methods manuals were developed during the regional program and have since been adopted as required procedures for facility-specific monitoring. Second, training exercises were conducted. Third, quality assurance protocols were implemented in which each participating organization had to demonstrate competency prior to collecting data for the regional survey. Quality assurance involved field audits of onboard activities, as well as blind samples to assess competency in benthic infaunal identification (Ranasinghe *et al.*, This Volume), chemistry (Gossett *et al.*, This Volume), toxicology (Bay *et al.*, This Volume), microbiology (McGee *et al.*, 1999), and fish identification. While these exercises demonstrated the considerable differences in laboratory performance that were anticipated by the NRC review, the intercalibration exercises provided a means to resolve these differences.

Standardization extended beyond data collection to include data management, as the project required data sharing among the more than 40 organizations that collected or processed samples. This was accomplished through a set of standardized data transfer protocols. While successful on a project basis, this didn't meet the larger NRC goal of creating centralized data bases that make data available from all of the ongoing programs. The State Water Resources Control Board has attempted to build from the standardized data transfer protocols to create their SWIM (Surface Waters Information Management) data base.

While Bight '98 represents the most visible regional monitoring advance since the NRC report, there have been several other movements towards regionalization. Within San Diego County, more than ten dischargers are contributing funds toward a set of quarterly regional kelp monitoring overflights. The Minerals Management Service has organized a Bight-wide network of organizations that monitor the rocky intertidal zone. It has developed consistency in data collection methods among programs, identified spatial gaps that must be filled to complete the regional picture, and created a regional data sharing scheme. The Southern California Wetlands Recovery Program was formed as a cooperative effort of 17 state and federal organizations to integrate wetlands acquisition and restoration activities. It is now developing an integrated regional monitoring program to assess their progress towards specific goals. While these other efforts are still in their early stages and have not yet reached their full potential, they represent substantial progress toward fulfilling the NRC recommendation.

2.2 RECOMMENDATION #2: INCREASE THE FLEXIBILITY OF MONITORING PROGRAMS

The NRC evaluation pointed out the segregation in monitoring efforts among government, academic, and discharger groups and argued that the perceived dichotomy between research and monitoring was artificial. This dichotomy is based on the assumptions that monitoring is a routine and relatively uncreative activity best performed by discharge agencies, while research is best carried out by academic scientists at research institutions. Program designs based on these assumptions, and on the consequent division of labor among government, academic, and discharger entities, reduces needed flexibility and hampers cost-effective planning and utilization of monitoring resources. It impairs the ability of scientists at discharge agencies, who often know the most about the affected environment, to pose and address important new questions. It also ignores the fact that monitoring is not always a straightforward, routine activity but often requires input from specialists not found in discharge agencies (e.g., physical oceanographers, toxicologists, statisticians). The NRC evaluation recognized that many questions can best be addressed by a combination of focused research and monitoring. This recommendation focused specifically on improving the ability of monitoring organizations to participate in regional monitoring and research activities and on better integrating the efforts of government, academic, and traditional monitoring entities.

There has been considerable progress in responding to the spirit of this recommendation, although not necessarily in the way the NRC panel envisioned at the time. The mechanisms of resource exchange and more broadly written NPDES permits enable resources and expertise to be accessed and combined in many more ways than previously. In addition, the new NPDES permits (described more fully below) enable the discharge monitoring programs to be used as focal points for bringing expertise from a variety of sources to bear on questions that encompass traditional elements of both research and monitoring. Improved information about spatial and temporal patterns of impact and about the relative contribution of different sources of perturbation and contaminant input are essential for the Integrated Coastal Management approach described in NRC (1993).

The most direct advance has been the concept of "resource exchange," in which regulatory agencies that issue discharge permits allow dischargers to trade a portion of their ongoing ambient site-specific monitoring requirements for an equivalent amount of monitoring in the Bight-wide regional surveys facilitated by SCCWRP. This was the primary resource used to "fund" these regional surveys. The regulatory agencies' intent in allowing this flexibility was to allow participation at minimal incremental cost (i.e., cost-neutrality).

While largely successful, there were unanticipated costs that rendered the resource exchange less than cost-neutral for dischargers. Most of the additional costs were associated with quality assurance and intercalibration exercises necessary to develop comparability among participating organizations. For several dischargers, particularly with regard to processing chemistry samples, the cost of meeting quality assurance goals exceeded the cost of sample processing in the study itself. There were also considerable unrecovered costs associated with planning, data analysis, and interpretation and synthesis, which are always more time consuming when conducted as part of a cooperative, multi-organizational effort. Most participants, however, believed that the increased knowledge and staff education gained through participation in all these activities offset the extra cost of the time invested.

While resource exchange to accomplish regional surveys is a good example of increased regulatory flexibility that yields more effective monitoring, the regional surveys are static, intermittent events. Even greater progress towards the NRC recommendation is exhibited in the monitoring permit issued to the Orange County Sanitation District (OCSD) in 1998, which divides their monitoring obligations into three components: (1) core monitoring; (2) regional monitoring; and (3) special studies. Core monitoring consists of the basic site-specific monitoring necessary to address individual discharger limits and impacts. It is mostly conducted in the immediate vicinity of the discharge and focuses on small-scale spatial effects and trends analysis. Regional monitoring provides the information necessary to make assessments over large areas and represents institutionalization of the resource exchange that was conducted in an ad hoc manner for the first two regional surveys. Special studies are focused short-term research efforts intended to help managers understand core or regional monitoring results, or to address unique issues of local importance. It represents an opportunity for the discharger to adapt its program, cooperatively with the regulators, on an annual basis to address issues of concern. Collectively, these represent a significant advance from the typical historic permit, in which a comparable amount of effort was focused entirely on the core program, without the flexibility to address regional or shorter-term research questions.

Orange County Sanitation District is the only organization to have yet received this kind of progressive permit, but there is movement toward adopting this framework throughout southern California. SCCWRP has been tasked by its Commission, which includes all of the permit-granting water quality regulatory agencies in southern California, to develop a model monitoring program for ocean dischargers. Their recommendations mimic the three-pronged permit issued to OCSD, with the idea that such a permit approach is responsive to management questions about the processes leading to impacts and adaptive to the results of both research and monitoring (Schiff *et al.*, 2002b). Ongoing permit negotiations for other dischargers are beginning to follow this model.

2.3 RECOMMENDATION #3: EPA AND NOAA SHOULD PROVIDE GUIDANCE

The NRC evaluation recognized that marine monitoring in southern California and throughout the country was excessively dominated by primarily local (i.e., in the vicinity of the discharge) concerns. In contrast, both U.S. EPA and NOAA had direct experience with conceiving, planning, and implementing larger monitoring programs of both regional and national scope. The NRC therefore recommended that these agencies provide guidance and direct assistance in developing regional monitoring programs, and perhaps a national network of such regional programs.

The U.S. EPA was an effective partner in helping to establish southern California's regional monitoring program. EPA's national office, primarily through their Environmental Monitoring and Assessment Program (EMAP), provided considerable assistance in creating the program's technical underpinnings. The primary assistance was in helping to focus the local questions around which to design the program and then in developing a design that would meet these objectives. This turned out to be a successful two-way interaction, as the local needs for enhanced stratification and nested levels of stratification for different indicators caused EMAP to reconsider their national design approach (Stevens *et al.*, 1997). EMAP also provided assistance in developing field and data management manuals. While most of the organizations participating in the regional surveys had their own manuals and procedures in these areas, EMAP's experience in merging efforts across programs proved invaluable.

EMAP also provided seed money for the first regional survey. While local organizations outspent EMAP almost 8 to 1, the EMAP contribution proved valuable when and where individual organizations were incapable of certain types of sampling or analysis and contractor support was needed to fill the gaps. Interestingly, EMAP provided no financial assistance for the 1998 survey, but by that point such external support was no longer necessary. The proof of concept that EMAP invested in during the first survey was sufficient to generate additional and ongoing local support for subsequent efforts. This investment turned out to be a valuable one in return for EPA, as the skills that developed in southern California served as a launching point for EPA's subsequent west coast monitoring initiative in 1999.

In contrast, NOAA played no significant role in developing southern California's regional program. NOAA made some cash contributions to California's Bay Protection and Cleanup Program to conduct targeted sampling of Los Angeles Harbor and San Diego Bay during the last decade, but this contribution contrasted with EPA's approach in two ways. First, EPA allowed the local programs to define questions of interest. Second, EPA placed priority on investing in developing local expertise needed for program continuity. NOAA's interaction seemed more motivated by the opportunity for cost-effective acquisition of data that NOAA needed for their own assessment purposes, rather than on providing guidance and developing the lasting relationship advocated by the NRC. As a result, the parameters and methods chosen for the regional surveys differ slightly from those used by NOAA nationally. Furthermore, NOAA's National Ocean Data Center data bases do not yet include the extensive regional survey data, or that from the routine surveys conducted by most of the regional survey participants. The principal NOAA contribution to the success of the regional programs was the chemistry intercalibration exercises they co-sponsor with the National Institute of Standards and Technology. Although subsequent local intercalibration exercises were necessary to enhance comparability (Gossett *et al.*, This Volume), the procedures developed by NOAA served as a guide to the local group.

This disparity in the two agency's contributions to regional monitoring in southern California during the mid to late 1990s is in contrast to their respective contributions to other similar efforts at other times and places. For example, EMAP's Estuaries component, in one of the program's first efforts, tried to impose a standardized design in the mid-Atlantic region, without focusing on local questions that could have encouraged support and participation by local and regional participants. This resulted in more resistance and a much less collaborative atmosphere than characterized the southern California experience. In contrast, NOAA has a long history of local involvement in studies in east coast estuaries. Their participation in the CalCOFI Program helped create and sustain a world-class large-scale research and monitoring program, and data from NOAA's Status and Trends Program and data syntheses prepared by the agency's National Ocean Service helped provide an important basis for understanding contaminant patterns and impacts in southern California (e.g., Mearns *et al.*, 1991). In understanding the two agency's respective contributions to regional monitoring in southern California, it is important to understand, first, that the EMAP program had matured to the point where it realized the long-term benefits of collaboration with local parties and, second, that EMAP expenditures increased after 1990 while those for NOAA's contaminant monitoring efforts decreased.

2.4 RECOMMENDATION #4: ENHANCE COMMUNICATION WITH USERS

The NRC evaluation recognized that even superlative monitoring serves no useful purpose unless and until the results are communicated effectively to a range of user communities. These include primarily managers, scientists, and the general public. The NRC therefore recommended that a variety of methods be used to summarize results and apply them directly to decision contexts.

Most historic monitoring data have been used primarily for site-specific decisions, such as for 301h waivers. This regulatory framework has a limited audience. In contrast, the more public-friendly nature of the questions asked in regional monitoring, particularly in the shoreline microbiology component (e.g., How safe is it to swim in the ocean?), led to extensive media coverage and requests for information by several of California's legislative committees. As a result, monitoring data are having a greater impact on key management decisions, such as greater focus on urban runoff as a source of shoreline bacteria, and have contributed to increased funding for initiatives to resolve such problems.

Regional monitoring also led to improved dialogue between regulators and dischargers on the goals of monitoring, the methods used to achieve these goals, and the ways in which monitoring data should be interpreted. When these kinds of issues are discussed in the context of regulators asking dischargers how to cooperatively build a regional program, the discussions take on a more positive tone than when dischargers initiate the discussion by disagreeing about facility-specific issues or requirements.

These discussions also included a broader array of organizations, including several non-profit environmental advocacy organizations that participated in Bight'98 by collecting, and in some cases, processing samples. As full partners in the process, these groups participated in the discussions about analysis and presentation of results. The honest discussions about data analysis and interpretation alternatives served to build a degree of trust among participants that has transcended the intermittent surveys. One positive example of this interaction is the greater sharing of data among public health agencies and non-profit environmental organizations that publish beach report cards, a collaboration that was based on a better understanding that both groups were serious about doing their jobs well.

The NRC recommendation also focused on breaking down barriers between academic researchers and routine monitoring entities. The regional surveys provide a tremendous opportunity for value-added participation by academics as the cost for sample collection and processing of routine variables is already covered. Some researchers took advantage by, for example, adding viral measurements to the measurement of routine bacterial indicators (Jiang *et al.*, 2001) and testing new biomarkers at the same time the large monitoring entities were collecting routine fish tissue chemistry measures. This still falls short of the NRC goal of continuing interaction outside of these one-time events. It is not clear whether continuing interactions outside of regional surveys will ever be the norm, given the different missions of the respective organizations. However, these fruitful experiences, combined with the increasing addition of special studies to routine monitoring permits, provide a solid basis for expanding such value-added cooperation in the future.

3. Factors Underlying the Effectiveness of the Southern California Response

In the context of regulatory and management policy making, ten years is not a long time. Yet the monitoring and management systems in southern California have undergone significant changes in that time and their evolution is continuing. While, in hindsight, the NRC recommendations and the past decade's changes seem obvious and quite reasonable, the reality is that it is often very difficult to achieve fundamental change in long-standing compliance monitoring programs (Bernstein *et al.*, 1999). Fear of undermining well-established compliance procedures, issues of organizational inertia, and the costs of coordination can also weaken tendencies toward regionalization (Bernstein *et al.*, 1999; NRC, 1997).

The single most important reason for the changes in southern California was the NRC evaluation and the reports it produced. It alerted managers to inefficiencies and focused scientists towards improvement. It provided a neutral forum for participants and observers of the monitoring system to speak openly about its shortcomings. Newton's Law that a body in motion tends to stay in motion seems to apply equally well to permitted monitoring requirements, which tend to remain unchanged unless a catalyst such as the NRC reports serve to deflect their trajectory.

The second contributing factor was EMAP, which proved through demonstration projects on the east coast of the U.S. that large scale regional programs were logistically feasible and could yield effective management information (Weisberg *et al.*, 1993). EMAP also provided technical guidance in areas such as sampling design, field procedures, and data structures, based on their demonstration project experience. Finally, EMAP provided seed money which challenged the local monitoring entities to expend their own resources.

Third, there was a neutral local party to facilitate and coordinate regional monitoring. The Southern California Coastal Water Research Project (SCCWRP) is a non-profit, local, marine research agency that is jointly administered by the four largest NPDES dischargers in the Southern California Bight and the five regulatory agencies that oversee those dischargers (Mearns *et al.*, 2001). Cooperation of the regulator and discharger communities was forged through their mutual participation in SCCWRP. The SCCWRP Commission, composed of senior management representatives from each participating agency, became a formal organizational body to receive, review, and respond to the results of the monitoring plans. A major strength of the SCCWRP Commission in this role is that the recipients of that information have the direct authority to implement management actions in response to the project results. SCCWRP staff also had the scientific skills to ensure that the project was completed. When other participating organizations could not undertake selected activities within their available resources, particularly in areas such as statistical design, database management, and report preparation, SCCWRP staff provided the technical expertise and personnel resources to conduct such tasks. Since SCCWRP is jointly administered by regulators and dischargers, their staff also provided non-partisan credibility in project development and interpretation of results.

Fourth, regulatory agencies provided the flexibility to allow discharger participation at minimal incremental cost. This flexibility should be commended because it runs contrary to typical regulator oversight behavior. Placing decisions about how to monitor largely in the hands of a regional planning group that included the dischargers themselves dramatically improved the development and acceptance of improved and standardized approaches. Concern was expressed that merging of programs would lead to a downward drift in quality to accommodate the least common denominator, with the least precise or least expensive approach being used.

However, the opposite occurred. Many of the methods required in existing permits were outdated (e.g., requiring PCB aroclors rather than PCB congeners) and the cooperative nature of the interactions provided the dischargers a forum for proactively improving their methods. Additionally, the mutual participation of dischargers and regulators in SCCWRP, as well as the trust exhibited by all parties in SCCWRP to ensure the technical adequacy and relevancy of the regional monitoring, added a comfort level to the regulator's decision to add flexibility.

Lastly, southern California had a series of previously developed networks that facilitated the additional interactions necessary to accomplish the regional survey (Bernstein *et al.*, 1999). These long-term relationships among participants contributed to trust in each other's motives and to confidence in their joint problem-solving abilities. One example is the Southern California Association of Marine Invertebrate Taxonomists (SCAMIT) that was formed more than 10 years prior to enhance the quality of benthic infaunal analysis. Another example is that there are a limited number of academic institutions at which many of the environmental professionals in southern California have been trained. Thus, many participants have ties that predate their agency employment. As a result of these and other long-standing interactions, agency staff have parallel views of themselves as belonging to multiple communities, most notably ones focused on common problems and regional collaboration (Bernstein *et al.*, 1999). These other identities served to soften boundaries when people from institutions with different missions were called upon to work cooperatively.

4. Next Steps

The regional surveys have effectively responded to many of the NRC recommendations and have improved the efficiency and usefulness for managers of monitoring data. For example, the expansion of regional monitoring to include a broader range of habitats such as bays and estuaries, as well as a more extensive set of participants (e.g., from urban stormwater management programs) has laid the groundwork to address questions about a much wider set of sources and impacts. However, two shortcomings remain as the next roadblocks to the effective use of monitoring in a management context. The first is the need to address trends. The present regional surveys focus on status across a broad area at single points in time and the remaining annual monitoring is too site-specific to be the basis for ongoing regional planning. Managers have made it clear that they also need to know the trends in environmental condition before taking actions because a resource in poor but improving condition may not require further management action, whereas a resource in good but declining condition may require more immediate action. While repeated implementation of the regional survey eventually might provide such answers, managers cannot wait the extended period necessary to establish trends through this means.

Southern California has embarked upon two alternative approaches for meeting this need for more timely trend information. The first is to not rely entirely on probabilistic sampling for the regional effort. In Bight '98, thirteen strategically selected sites that had been sampled periodically since 1977 as reference sites were included as sample sites. A second approach has been the use of coring at a subset of probability sites to assess trends in selected indicators through radiodated sampling strata (Zeng *et al.*, 2001). While very expensive, this approach allows for two kinds of information that cannot be gathered through repeated sampling of selected historical sites: 1) spatially-based assessment of trends (i.e., what percent of the sediment in southern California has declining chemical concentrations, which is best done through probability-based designs), and 2) extending the analysis to preindustrial periods, rather than the shorter period from which the 13 historical sites were first measured.

Managers have also identified that they need to know the causes of problems in order to develop appropriate solutions. Cause and effect assessments will most likely have to be addressed through the special studies portion of the monitoring requirements or through cooperative relationships with academic researchers, but there have also been some interesting additions to the regional program that begin to address this issue. One of the additions has been the use of chemical markers that are indicative of contributions from specific sources, such as linear alkyl benzenes (detergent byproducts) which are markers for sewage sources (Zeng *et al.*, 1998). Another strategy has been the use of multivariate pattern analysis to look for chemical signatures of plume tracks. While insufficient by themselves, these approaches begin to address the next management question and bring the monitoring community closer to addressing the management needs.

References

Bay, S.M., Jirik, A. and Asato, S.: 2002, 'Interlaboratory variability of amphipod sediment toxicity tests in a cooperative regional monitoring program', This Volume.

Bergen, M., Cadien, D., Dalkey, A., Montagne, D., Smith, R.W., Stull, J., Velarde, R.G. and Weisberg, S.B.: 2000, 'Assessment of benthic infaunal condition on the mainland shelf of southern California,' *Environmental Monitoring and Assessment* **64**, 421–434.

Bernstein, B. B., Allen, M. J., Dorsey, J., Gold, M. Lyons, M. J., Pollock, G. A., Smith, D., Stull, J. K., and Wang, G. Y.: 1999, 'Compliance monitoring in a regional context: revising seafood tissue monitoring for risk assessment,' *Ocean and Coastal Management* **42**, 399–418.

Gossett, R., Baird, R., Christensen, K. and Weisberg, S.B.: 2002, 'Making performance-based chemistry work: How we created comparable data among 10 laboratories as part of a southern California regional assessment', This Volume.

Hashimoto, J. and Weisberg, S.B.: 1998, 'Coordinating site-specific NPDES monitoring to achieve regional monitoring in southern California' pp 41–48 in Monitoring: Critical Foundations to Protect Our Waters, Proceedings of the National Water Quality Monitoring Council Conference, Reno NV. Environmental Protection Agency, Washington, DC.

Jiang, S., Noble, R. and Chu, W.: 2001, 'Human adenoviruses and coliphages in urban runoff-impacted coastal waters of southern California', *Applied and Environmental Microbiology* **67**, 179–184.

McGee, C.D., Leecaster, M.K., Vainik, P.M., Noble, R.T., Walker K.O. and Weisberg, S.B.: 1999, 'Comparison of bacterial indicator measurements among southern California marine monitoring' pp. 1187–198 in S.B. Weisberg and D. Hallock (eds). Southern California Coastal Water Research Project 1997–98 Annual Report. Westminster, CA.

Mearns, A.J. Matta, M., Shigenaka, G., MacDonald, D., Buchman, M., Harris, H., Golas, J., and Lauenstein, G.: 2001, 'Contaminant trends in the Southern California Bight: inventory and assessment', NOAA Technical Memorandum NOS ORCA 62.

Mearns, A.J., Allen, M.J. and Moore, M.D.: 2001, 'The Southern California Coastal Water Research Project—30 years of environmental research in the Southern California Bight' pp. 2–27 in S.B. Weisberg and D. Elmore (eds). Southern California Coastal Water Research Project 1999–2000 Annual Report. Westminster, CA.

National Research Council (NRC): 1990a, *Managing Troubled Waters: The Role of Marine Environmental Monitoring*, National Academy Press, Washington, D.C., 125 pp.

National Research Council (NRC): 1990b, *Monitoring Southern California's Coastal Waters*, National Academy Press, Washington, D.C., 154 pp.

National Research Council (NRC): 1993, Managing Wastewater in Coastal Urban Areas, Washington, D.C., 477 pp.

National Research Council (NRC): 1997, Striking a Balance: Improving Stewardship of Marine Areas, Washington, D.C., 177 pp.

Noble, R.T., Dorsey, J.H., Leecaster, M., Orozco-Borbon, V., Reid, D., Schiff, K.C. and Weisberg S.B.: 2000, 'A regional survey of the microbiological water quality along the shoreline of the Southern California Bight' *Environmental Monitoring and Assessment* **64**, 435–447.

Ranasinghe, J.A., Montagne, D.E., Weisberg, S.B., Bergen, M. and Velarde, R.G.: 2002, 'Variability in the identification and enumeration of marine benthic invertebrate samples and its effect on benthic assessment measures', This Volume.

Schiff, K.C.: 2000, 'Sediment chemistry on the mainland shelf of the southern California Bight', *Marine Pollution Bulletin* **40**, 268–276.

Schiff, K.C. and Allen, M.J.: 2000, 'Chlorinated hydrocarbons in flatfishes from the southern California, USA, Bight' *Environmental Toxicology and Chemistry* **9**, 1559–1565.

Schiff, K.C., Weisberg, S.B. and Raco-Rands, V.E.: 2002a, 'Inventory of ocean monitoring in the Southern California Bight', *Environmental Management* **29**, 871–876.

Schiff, K.C., Brown, J., and Weisberg, S.B.: 2002b, 'Model monitoring program for large ocean dischargers in southern California', Southern California Coastal Water Research Project, Westminster, CA., 101 pp.

Stevens, D.L.: 1997, 'Variable density grid-based sampling designs for continuous spatial populations', *Environmetrics* **8**, 167–195.

Weisberg, S.B., Frithsen, J.B., Holland, A.F., Paul, J., Scott, K.J., Summers, J.K., Wilson, H., Valente, R., Heimbuch, D., Gerritsen, J., Schimmel, S. and Latimer, R.: 1993. EMAP—Estuaries Virginian Province 1990 Demonstration Project Report. EPA Report 600/R92/100. U.S. Environmental Protection Agency, Environmental Research Laboratory, Narragansett, RI.

Zeng, E.Y., Cheng, D., Khan, A.R. and Vista, C.L.: 1998, 'Validity of using linear alkylbenzenes as markers of sewage contamination with interference from tetrapropylene-based alkylbenzenes' *Environmental Toxicology and Chemistry* **17**, 394–397.

Zeng, E.Y., Bay, S.M., Tran, K., and Alexander, C.: 2001, 'Temporal and spatial distributions of contaminants in sediments of Santa Monica Bay, California' pp. 96–113 in S.B. Weisberg and D. Elmore (eds). Southern California Coastal Water Research Project 1999–2000 Annual Report. Westminster, CA.

EFFECTIVE APPLICATION OF MONITORING INFORMATION: THE CASE OF SAN FRANCISCO BAY

RAINER HOENICKE[1*], JAY A. DAVIS[2], ANDREW GUNTHER[3], THOMAS E. MUMLEY[4], KHALIL ABU-SABA[4], AND KAREN TABERSKI[4]

[1]*The Resources Agency, 1416 Ninth St., Suite 1311, Sacramento, CA 95814;*
[2]*San Francisco Estuary Institute, 7770 Pardee Lane, Oakland, CA 94621;*
[3]*Applied Marine Sciences, 4235 Piedmont Avenue, Oakland, CA 94611;*
[4]*San Francisco Bay Regional Water Quality Control Board, 1515 Clay Street, Oakland, CA 94612*
[*]*(author for correspondence, e-mail: rainer.hoenicke@resources.ca.gov)*

Abstract. The San Francisco Estuary Regional Monitoring Program (RMP) for Trace Substances is an innovative partnership among a regulatory agency, more than 70 regulated entities, and an independent scientific organization. The institutional arrangement behind the RMP has made the regulatory system increasingly responsive to emerging management needs, particularly with regard to the development of total maximum daily loads and ecosystem impairment assessment. Through multi-agency partnerships within and outside the RMP institutional structure, major information gaps for several pollutants of concern have been narrowed, resulting in a successful consensus-based regulatory approach to managing copper and nickel mass inputs into the Estuary. Short-term research efforts, based upon monitoring results, helped identify the most cost-effective control and remediation options for various bioaccumulative substances. Additionally, adaptive changes to the monitoring program documented the existence of widespread aquatic toxicity in the Estuary that is apparently due to pesticide runoff from agricultural and urban areas. One of the most important contributions of this collaborative monitoring program is the deliberate and systematic adjustment of management and research questions that serve to influence and add relevance to the overall research agenda related to San Francisco Estuary ecosystem assessment.

Keywords: monitoring and decision-making, information integration, data vs. information, toxicity, legacy pollutants, mass loads, remediation.

1. Introduction

Numerous attempts have been made to implement the National Research Council's (NRC, 1990a,b) recommendations for effective environmental monitoring of near-coastal waters and other ecosystems (Bernstein *et al.*, 1997; Eichbaum and Bernstein, 1990; Thompson *et al.*, 2000; U.S. Forest Service, 1999). Many of these attempts have resulted in major improvements in management responses to newly developed scientific information (SFEI, 2001). These attempts have been characterized by an "adaptive management" process, in which monitoring results are used to guide management decisions and to refine the monitoring program (Healey and Hennessey, 1994; Ludwig *et al.*, 1993). The National Research Council termed this approach "Integrated Coastal Management" (NRC, 1993) and has been effective in promoting continuous feedback loops between management decisions and monitoring and research.

An almost equally important shift in the mind-set of decision-makers has occurred with regard to the relative allocation of resources to data collection and data interpretation and dissemination. Monitoring does not assist environmental managers until rigor-

Environmental Monitoring and Assessment **81**: 15–25, 2003.

ously interpreted and relevant data are available upon which management decisions can be based. Historically, data have been collected faster than they can be analyzed using sophisticated statistical procedures or simple numerical models. As a consequence, monitoring data frequently did not contribute to critical processes such as the development and revision of conceptual models or the formulation of hypotheses for testing with subsequent research. Yet it is precisely these processes that connect monitoring to environmental decision making that allows managers to determine if their decisions are likely to result in environmental benefits to society that outweigh their costs to the tax- or rate-payer (NRC, 1993).

Additionally, scientists and managers alike have become increasingly aware of the need to communicate to a variety of audiences the value of a monitoring or assessment program, the kind of information it is expected to provide, and how that information is linked to decision-making (Olivieri *et al.*, 2000). Unlike scientists in the health fields, those in the environmental sciences have only recently come to appreciate that the audiences of their monitoring and assessment program results are not restricted to their scientific and technical peers and environmental managers at the agency level, but that they are also housed in state and federal legislatures.

This paper describes how an adaptive management process has been used to improve the linkage between monitoring and environmental decision-making in the San Francisco Estuary. Management questions drive: (1) the design and scoping efforts for trend monitoring (including pilot studies of promising techniques), (2) evaluation of environmental indicators with regard to regulatory guidelines or benchmarks, as well as pollutant trends over time, and (3) special studies designed to test specific hypotheses. We review how the information generated by the Regional Monitoring Program (RMP) for Trace Substances has been used in various decisions and how information dissemination could be expanded to reach audiences that are likely to have the greatest impact with regard to translating scientific information into environmental policy improvements.

2. Evolution of the Monitoring Program

The Regional Monitoring Program (RMP) for Trace Substances was created in 1993 through discharge or use-permit conditions by the San Francisco Bay Regional Water Quality Control Board (Regional Board), the state's regulatory agency implementing the federal Clean Water Act provisions and the California Water Code. An overview of the history, institutional framework, monitoring objectives, and results of the monitoring program can be found in Thompson *et al.* (2000), in SFEI (2001), and through the Internet at www.sfei.org.

For the first five years (1993–1997), preceding a program review (Bernstein and O'Connor, 1997), the monitoring program provided baseline data on contaminants of critical regulatory concern. These contaminants included a list of ten trace metals and metalloids, selected chlorinated pesticides, polychlorinated biphenyls

(PCBs), polycyclic aromatic hydrocarbons (PAHs), and alkanes. The list of analytes has undergone slight modifications throughout the years, varies between water, sediment, and tissue samples, and will continue to be adjusted based on information generated by the RMP and other assessment efforts. As a result of the program review, the RMP has expanded its objectives beyond mere pollutant characterization to include evaluation of loads to the Estuary, pollutant pathways, and effects, all of which have become critical components in assisting the Regional Board in its development of Total Maximum Daily Loads (TMDLs) for those pollutants that are on the so-called Impaired Waters List. The parties involved in the RMP developed management questions based on the expanded objectives and have since continually used the feedback mechanisms between management questions and monitoring information to adjust the program.

It became apparent that comparisons of ambient concentrations of pollutants to water quality objectives alone were not always appropriate to evaluate if "designated uses" of the Estuary were met. Therefore, fish tissue monitoring became part of the status and trends monitoring component in 1999, as well as characterizing the duration and spatial extent of episodic aquatic toxicity. Numerous scoping efforts, involving experts from within and outside the San Francisco Bay Area, are currently being conducted to further adjust the monitoring program in response to a second generation of management questions, demonstrating the application of the methodology outlined in the NRC reports (NRC, 1990a,b). These efforts are extended beyond Estuary monitoring activities to include tidal wetlands and watersheds.

More recently, Regional Board staff recognized that the current deterministic sampling design, while suitable for characterization of "background" conditions, does not serve to answer questions related to how well each Estuary segment meets pollutant guidelines for water, sediment, or tissue concentrations. A shift toward a probabilistic sampling design for the purposes of assessing pollutant trends and determining if guidelines are being met is currently underway.

3. Overview of Recent Findings

The RMP database, spanning nine consecutive years, shows that several legacy pollutants continue to impair ecosystem integrity and threaten human health. Among these, mercury and PCBs have the highest regulatory priority. Table 1 summarizes the percent exceedances for contaminants for which environmental benchmarks in water exist. Another measure of impairment is based on human health-related screening values for PCBs, DDT, and mercury applied to fish tissue (Figure 1). Most metals are no longer of concern, and related monitoring efforts will focus primarily on using results as triggers for additional management actions should metal concentrations exceed targets (Olivieri *et al.*, 2000).

Several pollutants that do not fall into the "legacy" category and are still being emitted, entering the Estuary via tributaries, outfalls, or air deposition, have been

Table 1. Water quality criteria and the percentage of RMP samples collected 1993–1999 that exceeded those criteria.

Parameter	*Criterion (µg/L)**	*% Exceeded*
Ag	1.9 (salt), 3.4 (fresh)	0.0
As	36 (salt), 150 (fresh)	0.0
Cd	9.3 (salt), 2.2 (fresh)	0.0
Cr	50 (salt), 11 (fresh)	0.0
Cu	3.1 (salt), 9.0 (fresh)	9.4
Hg	0.025 (salt), 0.012 (fresh)	37
Ni	8.2 9salt), 52 (fresh)	1.4
Pb	8.1 (salt), 2.5 (fresh)	0.0
Se	5	0.8
Zn	81 (salt), 120 (fresh)	0.0
Total PCBs	0.00017	85
Total DDTs	0.00059	43
p,p´-DDD	0.00084	4.5
p,p´-DDE	0.00059	21
p,p´-DDT	0.00059	3.5
Total Chlordanes	0.00059	11
Heptachlor	0.00021	0.0
Heptachlor Epoxide	0.00011	12
alpha-HCH	0.013	0.0
beta-HCH	0.046	0.0
gamma-HCH	0.063	0.0
Dieldrin	0.00014	20
Endrin	0.81	0.0
Mirex	0.001	0.0
Total PAHs	0.031	44
Acenaphthene	2700	0.0
Anthracene	110000	0.0
Fluorene	14000	0.0
Benz(a)anthracene	0.049	0.7
Chrysene	0.049	0.0
Pyrene	11000	0.0
Benzo(a)pyrene	0.049	0.7
Benzo(b)fluoranthene	0.049	1.3
Benzo(k)fluoranthene	0.049	0.3
Dibenz(a,h)anthracene	0.049	0.0
Fluoranthene	370	0.0
Indeno(1,2,3-cd)pyrene	0.049	1.3

*California Toxics Rule criteria, except for Hg and Se. All trace elements except Hg and Se are compared to dissolved criteria.

found to exceed various benchmarks. Several of them have been included on the impaired waters list or are being considered for listing. Among these are the registered organophosphate pesticides diazinon and chlorpyrifos, as well as PAHs.

During the fifth year of monitoring (1997), aquatic toxicity test results indicated that toxic water masses originating from the watersheds of the Sacramento and San Joaquin Rivers may remain toxic for long enough to impact the northern and eastern segments of the San Francisco Estuary. As a result, in 1998 the RMP implemented a monitoring element to investigate the duration and frequency of occur-

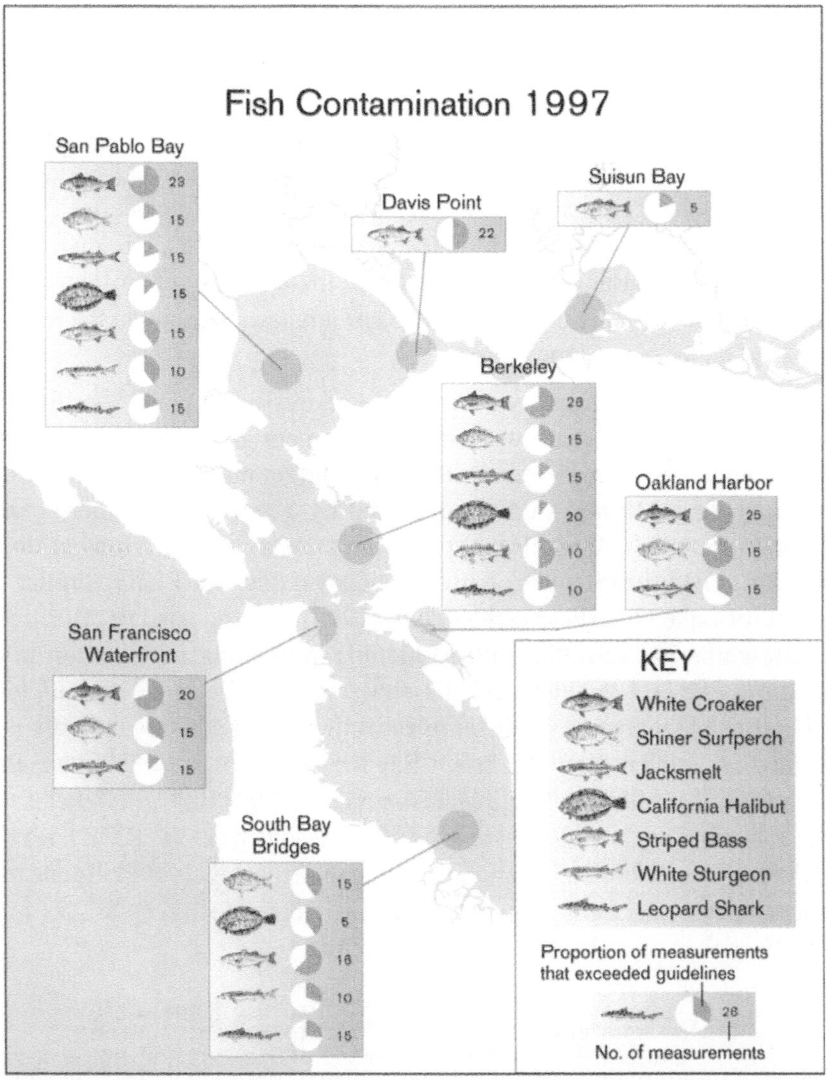

Figure 1. Percent of fish tissue concentrations exceeding human health screening values.

rence of toxic runoff episodes. This monitoring has documented extended periods of toxicity to the mysid shrimp *Americamysis bahia*, a standard, non-resident EPA test organism (Ogle *et al.*, 2001; Thompson *et al.*, 2000). Frequently, this toxicity has been associated with concentrations of organophosphate pesticides above acute toxicity thresholds. Since then, runoff event-based sampling has been done in the northern segment of the Estuary, closest to the Sacramento-San Joaquin River Delta and in the smaller tributaries of Pacheco Slough and the Napa River. This sampling approach has tracked the occurrence of toxicity and expanded the suite of indicators to include juvenile *Menidia beryllina*, a fish species believed to be more sensitive to pyrethroid pesticides than the crustacean *A. bahia*.

Details of this monitoring element are described in Ogle *et al.* (2001). It is too early to tell if decreasing frequencies and duration of toxic episodes since 1998 (Figure 2) can be credited to successful implementation of voluntary agricultural management practices designed to reduce effects of pesticides on non-target species. Alternative explanations for this apparent downward trend may be a shift to a newer generation of pesticides, such as pyrethroids, which are known to affect different indicator organisms than organophosphate pesticides, or merely weather-imposed reductions in pesticide applications. In recent years, toxicity identification evaluations (TIEs) have not been able to point out the causative agents of observed toxicity, and very resource-intensive efforts are underway to investigate the causes of unknown toxicity in the Sacramento and San Joaquin Rivers and the Estuary.

Other pollutants of regulatory concern are addressed in close coordination between the RMP and upstream watershed assessment activities. For example, the use of conceptual and simple numerical mass budget models populated with data generated by the RMP and studies supported by a variety of partner agencies is beginning to show where additional pollution prevention and remediation activities may be most cost-effective. A mass budget model for PCBs, similar to that developed for Lake Ontario (Mackay *et al.*, 1994) is being used to prioritize additional data-gathering activities and to identify remaining remediation priorities (Davis *et al.*, 2002). Congener-specific analysis of PCBs in water samples from two tributaries compared to water samples from Estuary stations not only revealed a concentration gradient from the South-Bay tributaries toward the Central Bay, but that the congener profiles in the tributaries are reflecting a different Aroclor mixture than those in the Estuary (Johnson *et al.*, 1999). Results from the RMP have demonstrated that, without additional intervention steps, external inputs of PCBs and mercury from surrounding watersheds may be of sufficient magnitude to keep fish consumption health advisories in place for the foreseeable future.

4. Management Responses to Monitoring Information

The five-year program review changed the course of thinking and action and caused the monitoring program to shift from a pollutant characterization phase to narrow-

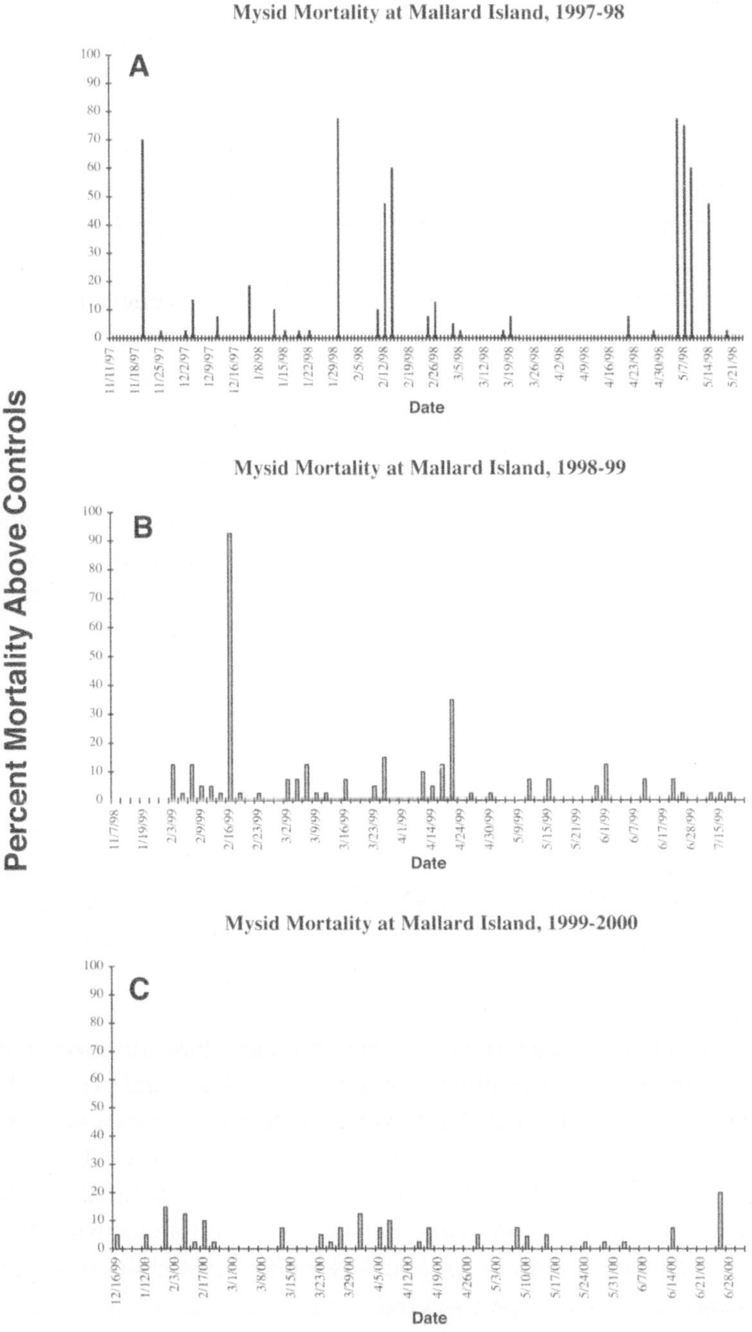

Figure 2. Toxicity of ambient water samples collected in the western part of the Delta at Mallard Island to *Americamysis bahia*. A. 1997–98; B. 1998-99; C. 1999–2000. Percent survival data have been normalized to the control responses. From: Ogle *et al.*, 2001.

ing critical data gaps. The review allowed the Regional Board to narrow their regulatory focus on a few priority issues. While in the past, the Estuary was listed as impaired due to metals, the current 303(d) list developed in 1998 focuses specifically on mercury, selenium, copper, and nickel. Due in large part to the monitoring database accumulated by the RMP, other trace elements no longer require the same amount of regulatory and management attention as prior to 1998. Concurrent efforts to determine if copper and nickel concentrations in the Estuary are causing adverse effects to biota (or impairment to certain "designated uses") have been successful in demonstrating through an extensive stakeholder-driven fact-finding process that impairment due to copper and nickel is "unlikely" (Olivieri *et al.*, 2000). The stakeholders recognized that actions could proceed despite scientific uncertainties. The use of conceptual models as educational and planning tools helped convey complex scientific information to a broad audience. Combined with expert scientific review, this has primarily led to the recommendation by the stakeholders to develop site-specific water quality objectives for copper and nickel and the subsequent decision by the Regional Board to implement this recommendation. This accomplished two important results: first, it avoided large expenditures associated with full implementation of mass load reductions; and second, it showed that consensus-based rule making is possible under certain circumstances, particularly if decisions are reversible, if they take uncertainties into account, and if monitoring and assessment results automatically trigger policy and management adjustments.

The accumulated RMP database on mercury concentrations in water, sediment, and fish tissue demonstrated that pollutant concentrations in water compared to water quality criteria are not always good indicators of impairment. The water quality criterion exceedances of mercury do not reflect the degree of beneficial use impairment as measured by fish tissue contamination. Conversely, water quality criteria exceedances for chromium, nickel, and copper are not necessarily reflective of adverse impacts in the Estuary.

Mercury data generated by the RMP have been successfully used in identifying the major ongoing mercury pollutant source to the Estuary and placing the inactive New Almaden Mine in the Santa Clara Valley in perspective relative to inputs from other pollutant sources and pathways (Abu-Saba and Tang, 2000). A phased approach to mercury load reductions beginning with the New Almaden Mine as the most significant source is now being pursued based on a clear quantitative fine-grained sediment concentration target of 0.4 mg/kg dry weight total mercury.

In recognition of the high costs of TMDL implementation for legacy pollutants, such as the variety of chlorinated hydrocarbons that are still impairing designated uses of the Estuary despite use restrictions or outright bans decades ago, the parties involved in the RMP recognized that it is prudent to incorporate a pro-active surveillance component into the monitoring program. This includes an evaluation of the existing analyte list and possible inclusion of compounds that may be of environmental concern but which have not yet appeared on the regulatory "radar screen." Examples of these kinds of analytes are synthetic chemicals produced and approved for use since the development of the EPA Priority Pollutant List (last revised in 1977), such as polybrominated diphenyl ethers (see Darnerud *et al.*, 2001), alkylphenols, phthalates, and others. An effort is currently underway to identify and attempt to estimate concentrations of a select list of target compounds from archived chromatographic data for which some toxicological literature already exists.

5. Future Challenges

The RMP is not an ecosystem monitoring program but focuses fairly narrowly on contaminants. Therefore, the RMP partners increasingly recognize the need for close coordination with other, primarily state and federally funded, assessment programs that are more broadly addressing other parameters of "ecosystem health," such as hydromodification, biological invasions, and conversion of habitat used by a variety of species to more or less exclusive human uses. Most of the partners involved in the RMP recognize the challenge of integrating individual monitoring and assessment elements from within and outside the RMP and have made the resources available to meet this challenge. However, the institutional framework is not yet prepared to generate and act on a comprehensive picture of ecosystem integrity that could be used to focus the allocation of resources in areas where relief of pressures on the ecosystem could provide the greatest environmental benefit.

It is important to recognize that the information users with the greatest ability to adjust environmental management decisions and policy are not necessarily located within the traditional circle of scientific peers and agency staff. The case of unknown toxicity to aquatic test organisms is a prime example of a contaminant problem that either has not yet reached the appropriate audience (i.e., lawmakers) or has not managed to become persuasive enough to modify the approval and registration process of synthetic organic compounds. The monitoring system seems to be mired (regardless of how lavishly funded) in an endless reactive cycle of "providing sufficient scientific evidence" of adverse effects and their underlying causes, uncoupled from the registration and approval process of new synthetic compounds. Only serious revisions to the legislative framework may break the costly assessment-evaluation-remediation cycle for synthetic chemicals that may prove, after they have taken a foothold in the market, to have unexpected and unintended adverse environmental effects. The integrated coastal management approach, as outlined by the National Research Council (1993), has been adopted and largely implemented by all stakeholders in the RMP. As the monitoring results have shown, however, the next challenge of including legislative adjustments in the "adaptive management cycle" has yet to be met.

As a successful example of taking action in light of scientific uncertainty, the stakeholder process that led to a consensus-based development of site-specific water quality objectives for copper and nickel in the extreme South Bay (that will be higher than the national objectives) showed that it is possible to recognize that science and technology are not the answer to resolving environmental problems, although they can and should inform the solutions. As Ludwig *et al.* (1993) pointed out, "[resource] problems are not really environmental problems. They are human problems that we have created at many times in many places, under a variety of political, social, and economic systems."

Acknowledgements

The following people are gratefully acknowledged for assistance in field and laboratory, for data summaries, graphics, editorial expertise, hours of discussions, and manuscript preparation: David Bell, Jordan Gold, and Paul Salop, Captain Gordon Smith, Wally Jarman, François Rodigari, Rebecca Wood, Rob Mason, Mark Stephenson, Brian Anderson, John Hunt, Sarah Lowe, Patricia Chambers, Mike May, Kathryn Lucas, Brock Bernstein, and Bruce Thompson.

References

Abu-Saba, K.E. and Tang, L.W.: 2000, *Watershed Management of Mercury in the San Francisco Bay Estuary: Total Maximum Daily Load Report to U.S. EPA*. California Regional Water Quality Control Board, San Francisco Bay Region, Oakland, CA.

Bernstein, B.B., Thompson, B.E. and Smith, R.W.: 1993, 'A Combined Science and Management Framework for Developing Regional Monitoring Objectives'. *Coastal Management* **21**, 185–195.

Bernstein, B.B. and O'Connor, J.: 1997, Five-year program review of the Regional Monitoring Program for Trace Substances in the San Francsico Estuary, RMP Contribution #28, San Francisco Estuary Institute, Richmond, CA.

Bernstein, B.B., Hoenicke, R. and Tyrrell, C.: 1997, 'Planning Tools for Developing Comprehensive Regional Monitoring Programs', *Environmental Monitoring and Assessment* **48**, 297–306.

Darnerud, P.O., Eriksen, G.S., Johannesson, T., Larsen, P.B. and Viluksela, M.: 2001, 'Polybrominated Diphenyl Ethers: Occurrence, Dietary Exposure, and Toxicology', *Environmental Health Perspectives* **109**, 49–68.

Davis, J.A.: 2002, A PCB Budget for San Francisco Bay. San Francisco Estuary Institute, Richmond, CA.

Eichbaum, W.M. and Bernstein, B.B.: 1990, 'Current Issues in Environmental Management: A Case Study of Southern California's Marine Monitoring System', *Coastal Management* **18**, 433–445.

Healey, M.C. and Hennessy, T.M: 1994, 'The Utilization of Scientific Information in the Management of Estuarine Ecosystems', *Ocean and Coastal Management* **23**, 167–191

Johnson, G. W., Jarman, W. M., Bacon, C.E., Davis, J.A., Ehrlich, R. and Risebrough, R. W.: 2000, 'Resolving Polychlorinated Biphenyl Source Fingerprints in Suspended Particulate Matter of San Francisco Bay', *Environmental Science and Technology* **34**, 552–559.

Ludwig, D., Hilborn, R. and Walters, C.: 1993, 'Uncertainty, Resource Exploitation, and Conservation: Lessons from History', *Science* **260**, 17–36.

Mackay, D., Sang, S., Vlahos, P., Diamond, M., Gobas, F. and Dolan, D.: 1994, 'A Rate Constant Model of Chemical Dynamics in a Lake Ecosystem: PCBs in Lake Ontario', *Journal of Great Lakes Research* **20**, 625–642.

National Research Council (NRC): 1990a, Managing Troubled Waters, Washington, D.C.: National Academy Press.

National Research Council (NRC): 1990b, Monitoring Southern California's Coastal Waters: A Case Study. Washington, D.C.: National Academy Press.

National Research Council (NRC): 1993, Managing Wastewater in Coastal Urban Areas. Washington, D.C.: National Academy Press.

Ogle, S., Gunther, A., Hoenicke, R., Bell, D., Cotsifas, J, Gold, J. and Salop, P.: 2001, RMP Summary Report on Episodic Ambient Water Toxicity in the San Francisco Estuary. San Francisco Estuary Institute, Richmond, CA.

Olivieri, A., Hall, T., Bruinsma, D., Hoenicke, R., Munley, T., Grieb, T.:2000, 'TMDL as a Regulatory Process: Conducting the Copper and Nickel TMDL in South San Francisco Bay,' Proceedings of the Watershed 2000 Specialty Conference. Vancouver, B.C.

San Francisco Estuary Institute: 2001, The Pulse of the Estuary - Monitoring and Managing Contamination in the San Francisco Estuary 1993–1999. Richmond, CA

Thompson, B.E., Hoenicke, R., Davis, J.A. and Gunther, A.: 2000, 'An Overview of Contaminant-related Issues Identified by Monitoring in San Francisco Bay', *Environmental Monitoring and Assessment* **64**, 409–419.

U.S. Forest Service, 1999, The Strategy and Design of the Effectiveness Monitoring Program for the Northwest Forest Plan, General Technical Report PNW-GTR–437.

BI-NATIONAL ASSESSMENT OF THE GREAT LAKES: SOLEC PARTNERSHIPS

PAUL BERTRAM[1*], NANCY STADLER-SALT[2], PAUL HORVATIN[1], and HARVEY SHEAR[3]

[1]*U.S. Environmental Protection Agency, Great Lakes National Program Office, 77 W. Jackson Blvd., Chicago, IL 60604;* [2]*Environment Canada—Ontario Region, Office of the Regional Science Advisor, 867 Lakeshore Rd., Burlington, Ontario L7R 4A6;* [3]*Environment Canada—Ontario Region, Regional Science Advisor, 4905 Dufferin St., Downsview, Ontario M3H 5T4*

[*]*(author for correspondence, e-mail: bertram.paul@epa.gov)*

Abstract. Many administrative jurisdictions have authority over parts of the Great Lakes, sometimes with competing purposes as well as governance at differing scales of time and space. As demand increases for high quality information that is relevant to environmental managers, environmental and natural resource agencies with limited budgets must look to interdisciplinary, collaborative approaches for the collection, analysis and reporting of data. The State of the Lakes Ecosystem Conferences (SOLEC) were begun in 1994 in response to reporting requirements of the Great Lakes Water Quality Agreement between Canada and the U.S. The biennial conferences provide independent, science-based reporting on the state of health of the Great Lakes ecosystem components. A suite of indicators necessary and sufficient to assess Great Lakes ecosystem status was introduced in 1998, and assessments based on a subset of the indicators were presented in 2000. Because SOLEC is a multi-agency, multi-jurisdictional reporting venue, the SOLEC indicators require acceptance by a broad spectrum of stakeholders in the Great Lakes basin. The SOLEC indicators list is expected to provide the basis for government agencies and other organizations to collaborate more effectively and to allocate resources to data collection, evaluation and reporting on the state of the Great Lakes basin ecosystem.

Keywords: Great Lakes, indicators, SOLEC

1. Introduction

Governance in the Great Lakes basin is complex. Canada and the United States share a common border that runs through four of the five lakes. Both countries have several federal agencies with jurisdiction over some aspect of the Lakes. In the U.S., for example, there exists the U.S. Environmental Protection Agency (U.S. EPA), the U.S. Fish and Wildlife Service, U.S. Geological Survey, the National Oceanic and Atmospheric Administration, the National Park Service, and the U.S. Coast Guard, among others. In Canada, there are Environment Canada, Fisheries and Oceans Canada, Health Canada, Agriculture Canada, Natural Resources Canada, Transportation Canada, and others.

Eight U.S. states and two Canadian provinces contain part of the Great Lakes basin in their jurisdictions. Each of the states and provinces also has multiple agencies, including the state or provincial equivalent of an environmental quality agency and a natural resources management agency.

Within the Great Lakes basin are 83 U.S. counties, whose health departments, for example, monitor the swimming beaches. In Ontario, 41 Public Health Units monitor, among other things, swimming beaches and drinking water quality.

Environmental Monitoring and Assessment **81**: 27–33, 2003.

In the U.S., 29 Tribes have independent governance in the Great Lakes basin. In Canada, the relationship between the federal government and First Nations is more complex, but there are 64 recognized First Nations in the basin.

Also, the two countries have established special binational commissions, such as the International Joint Commission to assist the governments achieve their goals of restoring the integrity of the Great Lakes ecosystem and of preventing further pollution to the system, and the Great Lakes Fishery Commission to control infestation of the sea lamprey and to promote a multi-use fishery.

Despite governance in the Great Lakes basin being distributed among many jurisdictions with independent mandates, organizing frameworks do exist that facilitate cooperation and collaboration. One important element is the Great Lakes Water Quality Agreement (GLWQA) between Canada and the U.S. (United States and Canada, 1987). First signed in 1972, and subsequently revised in 1978 and 1987, the purpose of the GLWQA is to "restore and maintain the chemical, physical, and biological integrity of the waters of the Great Lakes Basin Ecosystem." The current Agreement calls for the development of Lakewide Management Plans for each of the Great Lakes and their connecting channels, Remedial Action Plans for each of the 42 designated Areas of Concern, ecosystem objectives for each of the Lakes, and indicators of progress toward those objectives. The GLWQA also assigns responsibility to the governments of Canada and the U.S. to report on progress toward the goals of the Agreement every two years. The process of assessing and reporting on the status and trends of Great Lakes ecosystem components provides a framework for effective partnerships to develop among Great Lakes stakeholders.

2. What is SOLEC?

As demand increases for high quality information that is relevant to environmental managers, environmental and natural resource agencies with limited budgets must look to interdisciplinary, collaborative approaches for the collection, analysis and reporting of data. Also, for the governments of Canada and the U.S. to meet their commitments to the GLWQA, the environmental protection and natural resource activities of federal, state, provincial, tribal and First Nations agencies must be coordinated. A Binational Executive Committee (BEC), co-chaired by U.S. EPA and Environment Canada, is comprised of senior-level executives from these agencies, and it fosters the binational coordination of environmental programs. The concept of a biennial State of the Lakes Ecosystem Conference (SOLEC) to report on the condition of the Great Lakes ecosystem components was created by the BEC to fulfill, in part, the GLWQA requirements for assessing and reporting progress toward the goals and objectives of the Agreement. The conferences are science-based, and they are a result of consultation and collaboration between the U.S. and Canada, and between federal, state, provincial and local government agencies, environmental groups, industry and the public. Following each conference, the information presented, the assessments provided, and the inferences that were explored, were compiled, and a State of the Great Lakes report published (Canada and the United States 1997; 1999; 2001; United States and Canada, 1995).

Four objectives were established for SOLEC: to assess the state of the Great Lakes ecosystem based on accepted indicators; to strengthen decision making and environmental management; to inform local decision makers of Great Lakes environmental issues; and to provide a forum for communication and networking among all stakeholders. The primary audience includes environmental managers and decision makers, but the information needed by senior administrators and the public is also considered.

SOLEC started in 1994 with a basic assessment of the state of the Great Lakes that included human health and socio-economics. In 1996, the conference looked more closely at the nearshore environment and at some land use issues. In both conferences, assessments were made of the health of the system using ad hoc indicators and expert opinion. For SOLEC 98, a comprehensive, basin-wide suite of indicators was developed. These indicators were to be easily understood and were to objectively reflect the condition of the Great Lakes basin, the stresses on the ecosystem, and the human responses to those stresses. These indicators would provide predictable signs of ecosystem health and the progress being made to remedy existing problems. At SOLEC 2000, the indicator approach was implemented, and reports were received for 33 of the 80 indicators in the suite.

3. SOLEC Indicators

The SOLEC indicators list is a compilation of indicators that were derived, for the most part, from pre-existing indicators proposed or in use by other Great Lakes programs. Two important features of the list are: 1) each of the indicators is necessary for the evaluation of the Great Lakes, i.e., redundant indicators are not included, and 2) the suite of indicators, taken as a whole, is sufficient to evaluate the ecosystem components for SOLEC purposes. The process for identifying the indicators on the list required about 2 years of work that involved over 130 people.

A Core Group and an Expert Panel were created for each of the following subject areas: open and nearshore waters, coastal wetlands, nearshore terrestrial, land use, human health, and stewardship. There was no requirement for particular agency representation on these groups other than to ensure that both U.S. and Canadian expertise were involved. Each of the Groups mined indicators and indicator ideas from existing sources. The groups then screened their long list and revised, combined or created new indicators as needed. The work of all the groups was combined into the proposed SOLEC Indicator List and presented for discussion at SOLEC 98 (Bertram and Stadler-Salt, 1999).

Before, during and after SOLEC 98, the process has been open. Four versions of the indicator list have gone through a cycle of stakeholder review and Core Group revisions (Bertram and Stadler-Salt, 2000), and a process has been established for biennial review and revisions to be incorporated. In order for the SOLEC indicators approach to be successful, however, there must be consensus among federal, state, provincial and local management organizations that the SOLEC information is useful, and there must be collaboration and cooperation to collect, analyze and report the data.

4. SOLEC Participants & Partners

The selection of SOLEC indicators was a study in collaboration. Most of the Great Lakes management jurisdictions were involved, as well as industry and environmental groups, academia, Aboriginal groups, and private citizens. The BEC sets the general direction for SOLEC and the themes for the conferences. A SOLEC Steering Committee is responsible for identifying strategies necessary for the development and delivery of the SOLEC themes, setting the agenda for each SOLEC, convening the conferences, and issuing a conference proceedings and a State of the Great Lakes report. Membership on the Steering Committee is open to the BEC agencies and to other sectors, including environmental non-government organizations (ENGOs), industry, binational commissions, and private citizens. An Executive Committee, chaired and staffed by U.S. EPA and Environment Canada, implements the day-to-day activities of SOLEC.

The SOLEC indicators do not represent the singular view of any one or two agencies, but rather they are a compilation of the best ideas of the experts who participated. Of the 133 people involved with the SOLEC indicator selection process, only 64 (48%) represented federal, state or provincial agencies (Figure 1 upper). The rest were affiliated with academia (30 people, 23%), ENGOs (15 people, 11%), and other sectors.

Within each sector, several agencies or organizations were represented. For example, there were 14 academic institutions, 10 ENGOs, nine federal agencies, seven state or provincial agencies, and seven private groups or citizens involved (Figure 1 upper). In addition, within some of the federal, state and provincial agencies, more than one Division, Office or Branch was involved, and other sectors such as ENGOs and industry share and compile information, thereby increasing participation beyond named representatives. Also uncounted are the staff people within agencies or organizations who reviewed and commented on drafts of the indicator descriptions and other documents.

SOLEC partnerships continued during the preparation of the indicator reports for SOLEC 2000. Each report included sections on the purpose for the indicator, ecosystem objectives, the state of the ecosystem component, future pressures on the ecosystem, future actions that could be taken, and further work necessary to

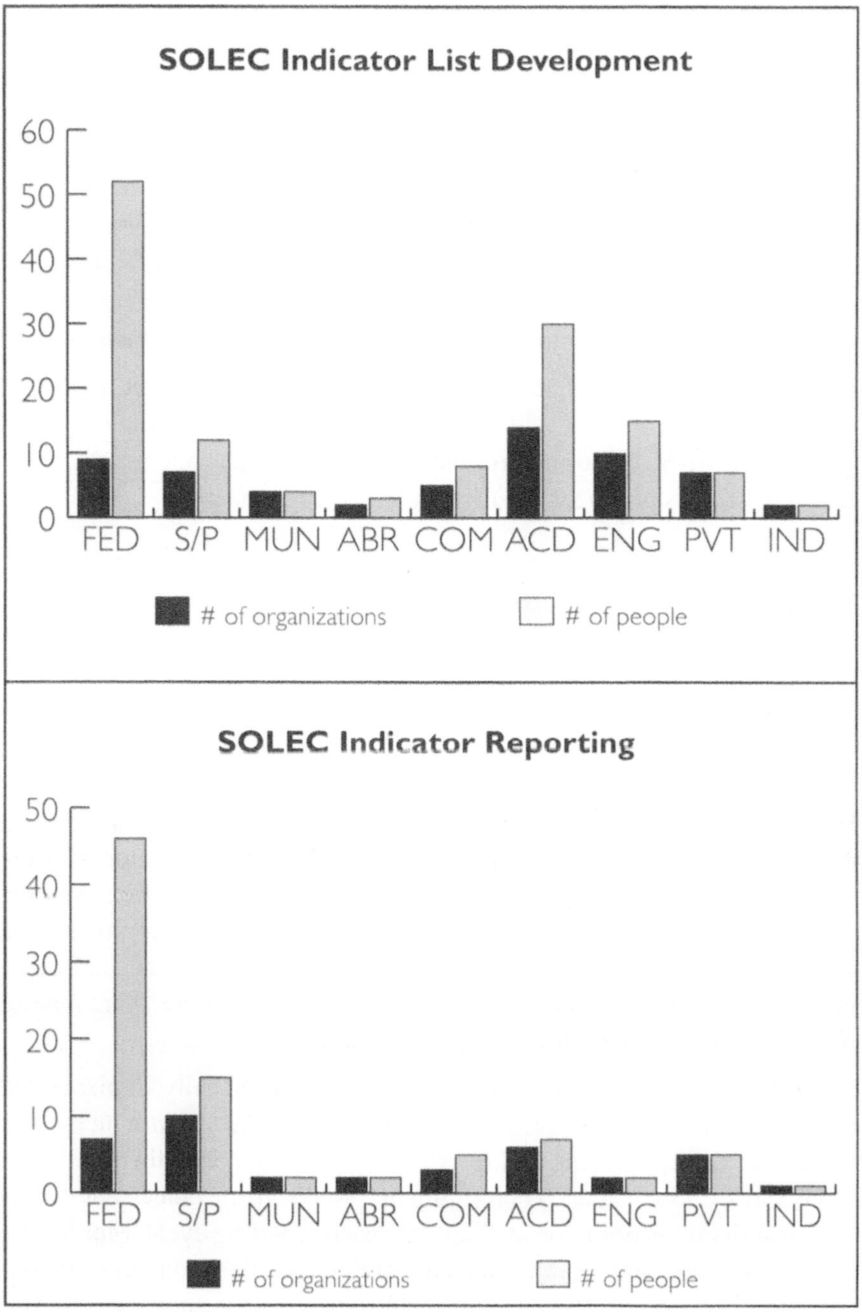

Figure 1. Number of organizations and number of people involved in the development and selection of SOLEC indicators (upper) and in the process of data collection, analysis and reporting (lower). FED = Federal, S/P = State or Provincial, MUN = Municipal, ABR = Aboriginal, COM = Commission, ACD = Academic, ENG = Environmental non-government, PVT = Private group or citizen, IND = Industry.

fully implement the indicator. The individual reports were then compiled and published electronically (Stadler-Salt and Bertram, 2000).

Similar to the indicator selection process, the reporting was shared by people affiliated with multiple agencies and organizations. Of the 85 people who were acknowledged in the indicator reports as authors, contributors, or sources for data, 46 (54%) represented federal agencies, 15 (18%) represented state and provincial agencies, 7 (8%) were affiliated with academia, and the rest were affiliated with other sectors (Figure 1 lower). The weighting toward federal authors was consistent with government sponsorship of many of the annual monitoring programs which provided data for the SOLEC indicators. Because government experts were usually the most familiar with the data, they were invited to prepare the reports.

Although only 7 federal agencies were identified to be involved in the indicator reporting, (Figure 1 lower), within those agencies at least 20 divisional or office units were identified. State and provincial agencies (10), academia (6), and private groups, contractors or citizens (5) were also well represented.

5. Sustained Partnerships

SOLEC is more than just a conference. It is a framework for unifying ecosystem objectives, for conducting monitoring programs, for information management, and for assessing and reporting on the integrity of the Great Lakes. The conference is part of the SOLEC process, and it provides participants the opportunity to discuss and comment on the data and interpretation of the indicators.

SOLEC itself is not a monitoring program. Those programs are delivered by government agencies or other partnering organizations to support GLWQA requirements (e.g. Remedial Action Plans for Areas of Concern, and the Lakewide Management Plans), fishery management agencies, or other environmental management purposes. The SOLEC approach is to identify the data needed and then to request appropriate agencies or organizations to collect and/or share that information through SOLEC and other reporting venues.

For SOLEC to be sustained, and the indicators to be fully implemented, each indicator must have a sponsor, i.e., an agency or organization which commits to gathering the data, analyzing and interpreting them, and reporting on the indicator through SOLEC. Such agency commitments would ensure consistency in the quality and quantity of information available for each reporting cycle. Monitoring agencies have their own mandates for data collection, and they do not necessarily have the resources or the motivation to quickly redirect their established programs to new indicators. However, following SOLEC 2000, some federal and state agencies have committed to sponsoring some of the indicators.

Because SOLEC has its origins in the GLWQA, those agencies and organizations whose mission includes activities to "restore and maintain the chemical, physical, and biological integrity of the waters of the Great Lakes Basin Ecosystem" have a common interest in periodically evaluating progress toward that goal. SOLEC provides a framework for coordinated, collaborative assessments of the state of the Great Lakes. The SOLEC Executive and Steering Committees are open to suggestions on the indicators, methodologies, interpretation of the indicator data, or on the conference itself. SOLEC is expected to continue to evolve as data needs and environmental issues change.

Acknowledgments

We thank the many SOLEC participants and reviewers who have helped shape the fabric of opportunities for Great Lakes organizations and citizens to create and sustain partnerships. We also thank Karen Rodriguez and Duane Heaton for their assistance and suggestions regarding this manuscript.

References

Bertram, P. and N. Stadler-Salt (eds): 1999, 'Selection of Indicators for Great Lakes Basin Ecosystem Health,' Version 3. Prepared for the State of the Lakes Ecosystem Conference 1998. U.S. Environmental Protection Agency, Great Lakes National Program Office, Chicago, and Environment Canada—Ontario Region, Office of the Regional Science Advisor, Burlington.

Bertram, P. and N. Stadler-Salt, (eds): 2000, 'Selection of Indicators for Great Lakes Basin Ecosystem Health,' Version 4. U.S. Environmental Protection Agency, Great Lakes National Program Office, Chicago, and Environment Canada—Ontario Region, Office of the Regional Science Advisor, Burlington. <www.on.ec.gc.ca/solec/indicators2000-e.html >.

Canada and the United States: 1997, 'State of the Great Lakes 1997,' Environment Canada, Toronto and U.S. Environmental Protection Agency, Chicago. <www.on.ec.gc.ca/solec/> or <www.epa.gov/glnpo/solec/>.

Canada and the United States: 1999, 'State of the Great Lakes 1999,' Environment Canada, Toronto and U.S. Environmental Protection Agency, Chicago. <www.on.ec.gc.ca/solec/> or <www.epa.gov/glnpo/solec/>.

Canada and the United States: 2001, 'State of the Great Lakes 2001,' Environment Canada, Toronto and U.S. Environmental Protection Agency, Chicago. <www.binational.net>.

Stadler-Salt, N., and P. Bertram (eds): 2000, 'Implementing Indicators,' Draft for Review November 2000. Environment Canada, Toronto, and U.S. Environmental Protection Agency, Chicago. <www.on.ec.gc.ca/solec/implementing2000-e.html>.

United States and Canada: 1987, 'Great Lakes Water Quality Agreement of 1978, as amended by Protocol signed November 18, 1987'.

United States and Canada: 1995, 'State of the Great Lakes 1995,' U.S. Environmental Protection Agency, Chicago and Environment Canada, Toronto. <www.cciw.ca/solec/> or <www.epa.gov/glnpo/solec/>.

THE MYSOUND PROJECT: BUILDING AN ESTUARY-WIDE MONITORING NETWORK FOR LONG ISLAND SOUND, U.S.A.

MARK TEDESCO[1*], W. FRANK BOHLEN[2], MARY M. HOWARD-STROBEL[2],
DAVID R. COHEN[2], AND PETER A. TEBEAU[3]

[1]*U.S. Environmental Protection Agency, Long Island Sound Office, Stamford, CT, U.S.A.;*
[2]*University of Connecticut, Department of Marine Sciences, Groton, CT, U.S.A.;*
[3]*Marine Research Associates, LLC, Groton, CT, U.S.A.*
[*]*(author for correspondence, e-mail: tedesco.mark@epa.gov)*

Abstract. A network of five water quality monitoring stations has been established in Long Island Sound, measuring temperature, salinity, and dissolved oxygen since 1999. The stations are located in areas of extreme water quality degradation (western Long Island Sound) as well as in pristine areas (eastern Long Island Sound). The data from these stations are collected every 15 minutes and posted to the project web site in real time as provisional data. After subsequent quality assurance procedures, the data are archived to the project File Transfer Protocol (FSP) site for downloading by the user community. The network of stations is in part supported logistically by a number of partners, including state and local agencies, schools, and non-governmental organizations. Data from the monitoring programs of some of these partners are also published to the project website providing a more comprehensive and complete picture of the status of the Sound than can be provided independently. This repository of information is used by marine educators, resource managers, scientists, and the general public, each with a different end purpose. We use the data from two of the stations to show that these high frequency time series measurements can be used to complement and enhance other monitoring programs within the Sound, documenting in greater detail the occurrence and duration of hypoxic events.

Keywords: automated, continuous water quality monitoring, estuarine monitoring, real-time data, MYSound, time-series data, Bridgeport Harbor, western Long Island Sound

1. Introduction

Long Island Sound is the largest and most heavily utilized estuary in the northeast United States. Its waters support large and diverse populations of fish, shellfish, waterfowl and marine mammals. With more than eight million people living in the Long Island Sound watershed, and thousands more regularly using the Sound for both commercial and recreational purposes, it is subject to a number of environmental stresses including nutrient enrichment and the resulting oxygen depletion; heavy metal contamination from prior industrial discharges; hydrocarbon pollution caused by oil spills and non-point source runoff; the introduction of non-indigenous nuisance species; and the disposal of marine debris (LISS, 1994). Periodic occurrences of elevated bacteria levels, harmful algal blooms, and outbreaks of fish and shellfish diseases within the Sound are often related to the pollution sources mentioned above, although the causes and dynamics of these outbreaks are poorly understood.

In September 1998, the U.S. Environmental Protection Agency (EPA) Long Island Sound Office and the University of Connecticut (UConn) initiated develop-

Environmental Monitoring and Assessment **81**: 35–42, 2003.

ment of a prototype marine environmental monitoring network for Long Island Sound. The project, known as the MYSound Project (for Monitoring Your Sound), is being funded under the EPA's Environmental Monitoring for Public Access and Community Tracking (EMPACT) Program. The goal of the project is to provide comprehensive and timely marine water quality monitoring data on Long Island Sound and its harbors and tributaries for use by managers and regulators, and to use these data as a means to increase public understanding of water quality, its importance and governing factors. The key component in the project is the establishment of an array of coastal water-quality monitoring stations throughout the Sound. These stations use in situ sensors and telemetry to capture and transmit data to shore where it is published to the project web site (www.mysound.uconn.edu) in real time. The project is in its third year.

This paper describes the current status of the project and discusses how the continuous, real-time monitoring data can be used with data from other established water quality monitoring efforts to investigate issues such as hypoxia in Long Island Sound. It also describes how partnerships have been formed to expand the scope of the network.

2. Methods

The MYSound network was established in collaboration with a number of agencies and organizations with a direct interest in the project. Partners in the project at present include the Connecticut Department of Environmental Protection (CT DEP), New York City DEP, Save the Sound, The Coalition to Save Hempstead Harbor, Bridgeport Regional Vocational Aquaculture School, The Maritime Aquarium at Norwalk, Mystic Aquarium, U.S. Coast Guard Academy, Connecticut Sea Grant, and New York Sea Grant. These partners assist by providing support logistics, contributing additional data for compilation on the project web site, and participating in communication and outreach activities.

Five stations form the current network. Three are located in the western half of Long Island Sound, historically the region of intense water quality degradation, and two are located at the eastern end (Figure 1), near the entrance where Sound waters mix with western Atlantic Ocean water. Three of the stations are moored buoys and two are fixed, one to a dock, the other to a navigational structure (see Table 1). The Hempstead Harbor station has been independently maintained by The Coalition to Save Hempstead Harbor as part of their own monitoring program for several years. UConn provided a receiving computer and radio frequency modem, enabling that station to be incorporated within the MYSound network.

All stations are outfitted with YSI 6600 and/or 6920 water quality sensors (measuring temperature, salinity and dissolved oxygen), a datalogger, and a radio frequency modem telemetering the data back to a receiving station every 15 minutes. The receiving stations are then polled every 15 minutes by a desktop computer, located within the University of Connecticut's Department of Marine Sciences campus at Groton, CT, that downloads, processes, archives and publishes the latest data to the project website. The initial processing involves a first order data quality check incorporating a wild point filter that converts data considered physically unrealistic for the region to a null value. The data are then published to the web with the caveat that they are provisional, i.e., the data are not considered final and may change after subsequent quality control and analysis procedures. After the final quality assurance procedures the data are uploaded to the project File Transfer Protocol (FTP) site, usually within 1–3 months of original data collection, where they are available for download by the user community.

3. Results

In this section, we analyze data from two of the stations to show how the MYSound monitoring network can be used to complement data from other agencies, providing a more comprehensive picture as to the extent and degree to which hypoxia impacts Long Island Sound. Figures 2 and 3 show the near bottom distribution of dissolved oxygen concentrations obtained by the CT DEP in the summers of 1999 and 2000. Data analyses indicate that for 1999 hypoxic conditions, defined as dissolved oxygen concentrations ≤ 3.5 mg/L (CT DEP, 1996), lasted 50 days in the western Sound, from July 2 to August 21, covering 314 sq. kilometers. For the summer of 2000, hypoxia lasted 36 days, from July 2 to August 6, covering 447 sq. kilometers of the western Sound. While these shipboard surveys are valuable in their ability to identify regions of water quality degradation, the time and space resolution is low and, therefore, unable to capture higher frequency perturbations. As an example of the nature of these higher variability events, the summer time series data from Bridgeport Harbor for the period July 1 through the end of August for both 1999 and 2000 are analyzed (see Figure 4). The data show a diurnal signal,

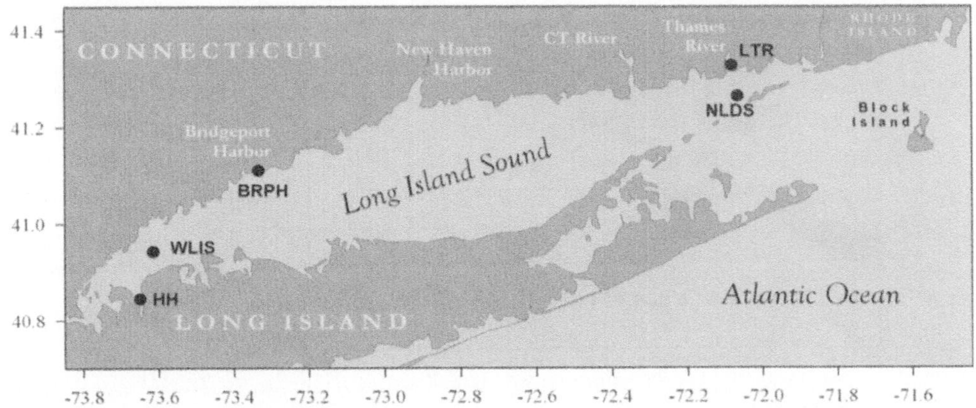

Figure 1. Location of the five stations which currently comprise the MYSound water quality monitoring network. HH: Hempstead Harbor, WLIS: Western Sound station, BRPH: Bridgeport Harbor, NLDS: Eastern Sound offshore (New London Disposal Site), LTR: Lower Thames River.

Table 1. Station Characteristics of the MYSound Monitoring Network.

Station Name	*Depth (m)*	*Location*	*Type*	*Sensor Status*
Hempstead Harbor	3.3	40° 15' N, 72° 05' W	fixed	mid-water only
Western Sound	15.2	40° 52' N, 73° 44' W	buoy	surface and bottom
Bridgeport Harbor	7.6	41° 10' N, 73° 10' W	fixed	surface and bottom
Eastern Sound	18.0	40° 16' N, 72° 04' W	buoy	surface and bottom
Lower Thames River	7.3	41° 22' N, 72° 05' W	buoy	surface and bottom

as well as a longer term fortnightly signal, at both the surface and bottom. This dissolved oxygen concentration data can be summarized as shown in Table 2, where all dissolved oxygen concentrations equal to or lower than 3.5 mg/L are compiled with dates and duration. The actual duration of these events, in Bridgeport Harbor, is on the order of hours and occurs at all stages of the tidal cycle.

Time series data from the western Sound station (WLIS), south of Greenwich, CT are shown in Figure 5. Although the station was not deployed until after the hypoxia surveys conducted by CT DEP, analysis of the time series data documents the evolution of destratification at this site. Figure 6 shows the time varying difference (delta) between the surface and bottom parameters of temperature, salinity and dissolved oxygen concentration (e.g., $\Delta T_t = T_{SFC_t} - T_{BTM_t}$, where T = Temperature in degrees Celsius at time, t, SFC is the surface sensor and BTM is the bottom sensor). From September 6th through the 25th, the magnitude of the delta for each parameter is much greater relative to the latter half (after September 26th) indicating a stronger vertical gradient and the existence of stratified conditions.

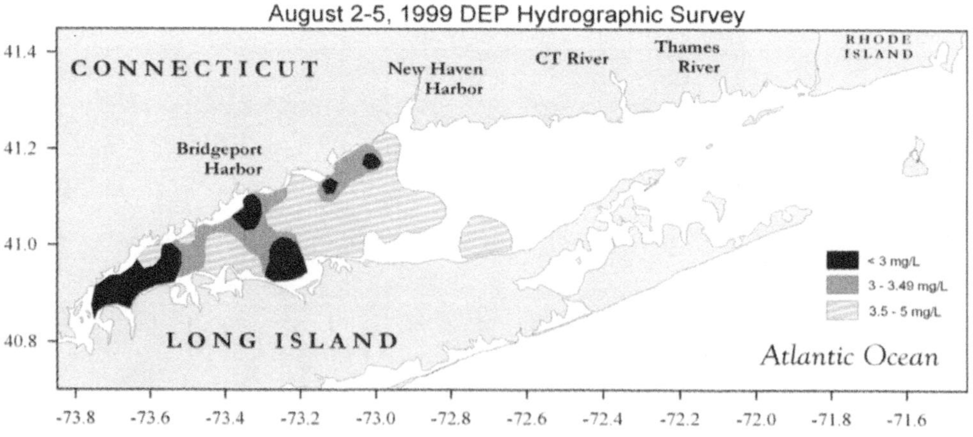

Figure 2. Summer bottom dissolved oxygen concentrations for Long Island Sound, collected over the period August 2–5, 1999. Hypoxia was estimated to have lasted 50 days, beginning on July 2 and ending on August 21, covering an area of 314 sq. kilometers. (Source: CT DEP, 2001)

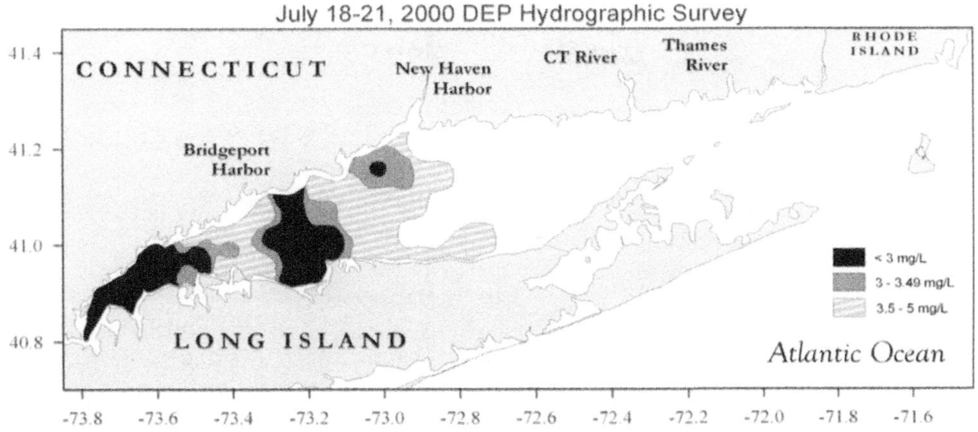

Figure 3. Summer bottom dissolved oxygen concentrations for Long Island Sound, collected over the period July 18–21, 2000. Hypoxia was estimated to have lasted 36 days, beginning on July 2 and ending on August 6, covering an area of 447 sq. kilometers. (Source: CT DEP, 2001)

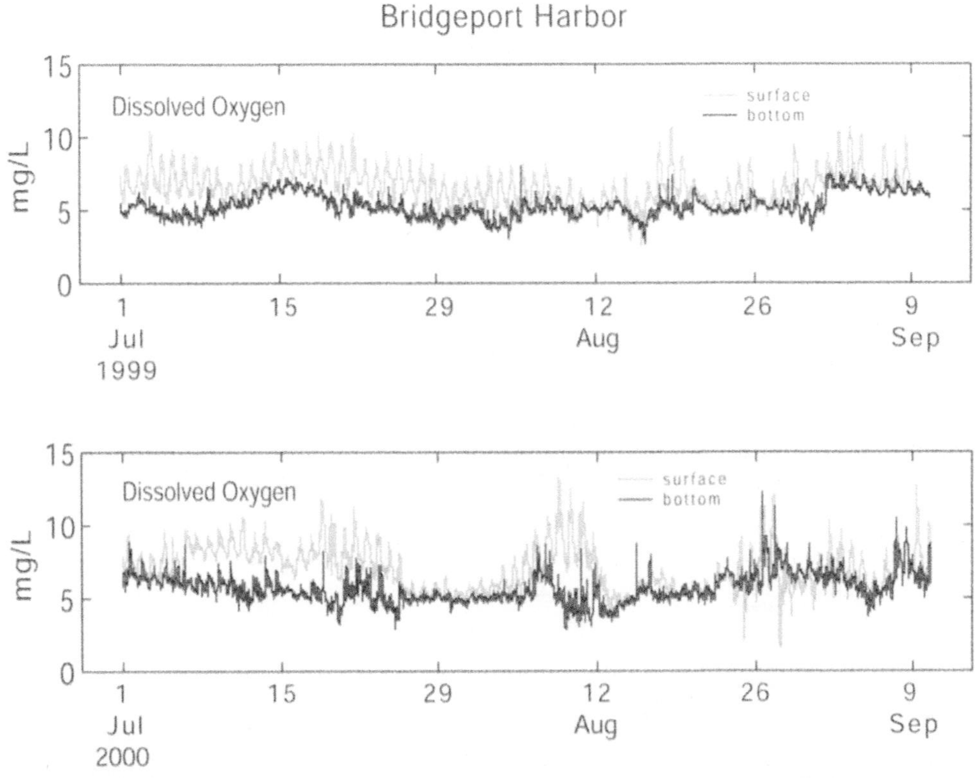

Figure 4. Dissolved oxygen time series from Bridgeport Harbor for the summer of 1999 (upper) and the summer of 2000 (lower).

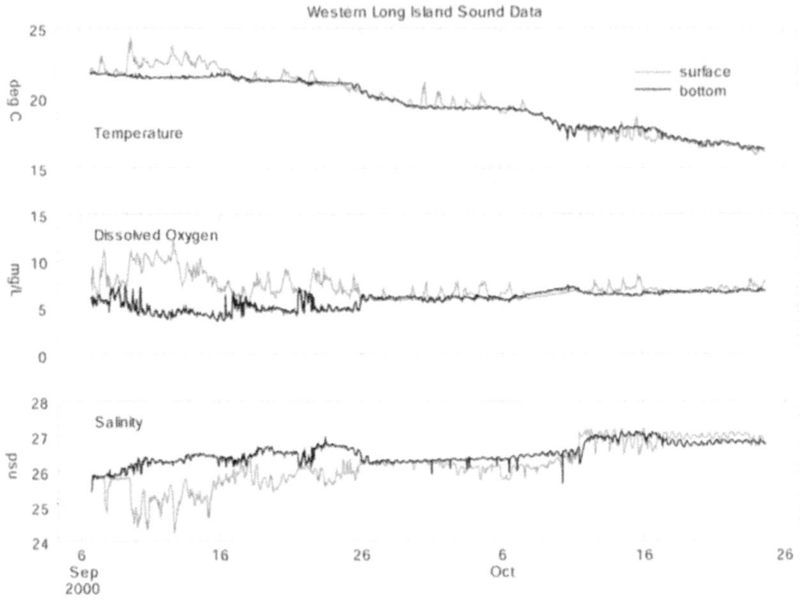

Figure 5. Time series data of temperature, dissolved oxygen, and salinity from the Western Sound station for the early fall, September 6th to October 26th, 2000.

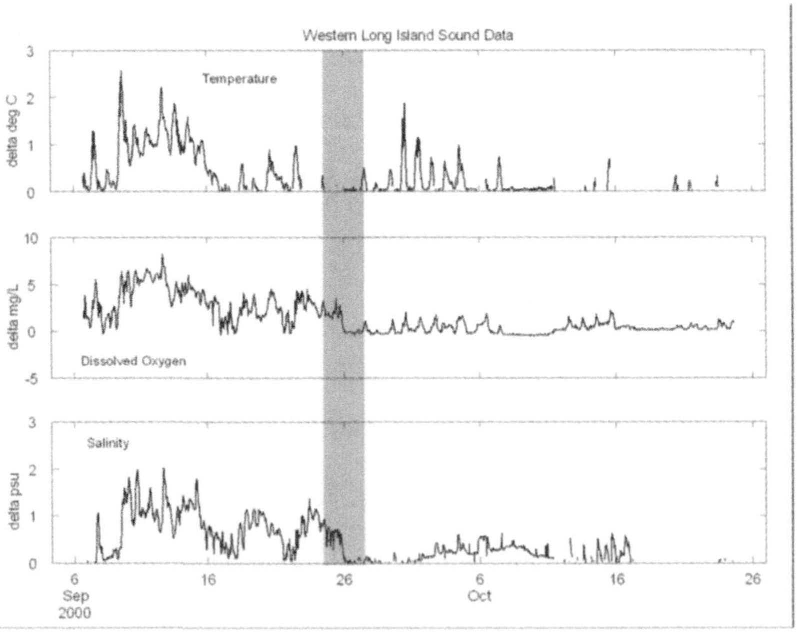

Figure 6. Time series of the data value for temperature, dissolved oxygen, and salinity from the Western Sound station; the time of destratification is highlighted by the gray bar.

Table 2. Occurrence and Duration of Hypoxic Events in Bridgeport Harbor (dissolved oxygen concentrations ≤ 3.5 mg/L)

	Day	*Start Time*	*Duration (in hours)*	*Tide*
1999	August 4	0700	1.00	mid ebb
	August 16	0515	2.00	mid ebb
2000	July 25	0330	0.75	mid flood
	August 9	0345	2.00	start flood
	August 10	0945	1.25	start ebb
	August 11	0845	3.50	end flood

Relatively high frequency daily fluctuations are superimposed on the record, which decrease in magnitude until about September 25th; at which point these fluctuations all but disappear. An examination of the meteorological data obtained from LaGuardia Airport (Long Island, NY) at that time, shows the onset of a high-pressure system with north-northwest winds of 15–20 knots over a 2-day period, conditions that effectively destratified the western end of the Sound.

4. Conclusions

High frequency time series provide accurate estimates of the time variability and duration of water quality degradation at the monitoring locations and allow a more accurate assessment of the stress experienced by the bottom dwelling biota than can be deduced from the shipboard surveys alone. The shipboard surveys, while documenting the presence and spatial extent of stressful conditions, imply large-scale spatial and temporal homogeneity. In contrast, the time series data show high frequency variability. The integration of both techniques can and should be used to complement one another—the surveys identifying adversely impacted areas, and the high frequency time series detailing the hydrography and dynamics of those areas.

The successful implementation of a prototype monitoring network capable of transmitting a high frequency data stream from within Long Island Sound has broad implications that directly complement ongoing resource management efforts. The combination of field observations and numerical modeling in an inverse modeling scheme (e.g., Bogden and O'Donnell, 1998) can significantly improve the prediction capabilities of hydrodynamic and water quality models currently in use within Long Island Sound (HydroQual, 1996; Blumberg *et al.*, 1999). The results have the potential to increase our understanding of the Long Island Sound ecosystem and to improve resource management decisions. The real-time posting of the time series data to the MYSound project website offers resource managers, scientists and other users a more complete picture of the environmental status of Long Island Sound and the sensitivity of this system to a variety of natural and societal stresses.

5. References

Blumberg, A.F., Khan, L.A., and J. P. St. John: 1999, 'Three-dimensional hydrodynamic model of New York Harbor Region,' *Journal of Hydraulic Engineering* **125**(8), pp. 799–816.

Bogden, P.S. and J. O'Donnell: 1998, 'Generalized inverse with shipboard current measurements: Tidal and nontidal flows in Long Island Sound,' *Journal of Marine Research* **56**(5), pp. 995–1027.

CT DEP (Connecticut Department of Environmental Protection): 2001, http://www.dep.state.ct.us/wtr/lis/monitoring/lis_page.htm

CT DEP (Connecticut Department of Environmental Protection): 1996, Connecticut Water Quality Standards. CT DEP Planning and Standards Division, Hartford, CT. 50 pp.

HydroQual: 1996, Water quality modeling and analysis of hypoxia in Long Island Sound using LIS 3.0. Job No. NENG0035, Management Committee Long Island Sound Estuary Study and New England Water Pollution Control Commission. Mahwah, NJ.

LISS (Long Island Sound Study): 1994, Comprehensive Conservation and Management Plan. U.S. EPA Long Island Sound Office, Stamford, CT, 168 pp.

CONSERVATION AND MANAGEMENT APPLICATIONS OF THE REEF VOLUNTEER FISH MONITORING PROGRAM

CHRISTY V. PATTENGILL-SEMMENS[1*] AND BRICE X. SEMMENS[2]

[1]*REEF Environmental Education Foundation;* [2]*University of Washington, Zoology Department*
*(author for correspondence, e-mail: christy@reef.org)

Abstract. The REEF Fish Survey Project is a volunteer fish monitoring program developed by the Reef Environmental Education Foundation (REEF). REEF volunteers collect fish distribution and abundance data using a standardized visual method during regular diving and snorkeling activities. Survey data are recorded on preprinted data sheets that are returned to REEF and optically digitized. Data are housed in a publicly accessible database on REEF's Web site (http://www.reef.org). Since the project's inception in 1993, over 40,000 surveys have been conducted in the coastal waters of North America, tropical western Atlantic, Gulf of California and Hawaii. The Fish Survey Project has been incorporated into existing monitoring programs through partnerships with government agencies, scientists, conservation organizations, and private institutions. REEF's partners benefit from the educational value and increased stewardship resulting from volunteer data collection. Applications of the data include an evaluation of fish/habitat interactions in the Florida Keys National Marine Sanctuary, the development of a multi-species trend analysis method to identify sites of management concern, assessment of the current distribution of species, status reports on fish assemblages of marine parks, and the evaluation of no-take zones in the Florida Keys. REEF's collaboration with a variety of partners, combined with the Fish Survey Project's standardized census method and database management system, has resulted in a successful citizen science monitoring program.

Keywords: marine fish, volunteer monitoring, conservation, citizen science, coral reef, reef fish

1. Introduction

Effective conservation and management of marine resources requires a comprehensive understanding of ecosystem structure and function. Through concerted and persistent data collection, researchers and resource managers can gain an understanding of these ecosystem components. The monumental task of surveying, recording and cataloging an immense liquid wilderness can be daunting for managers. Establishing monitoring programs at a scale large enough to appropriately monitor marine communities is frequently cited as a stumbling block to effective management (e.g. Baird *et al.*, 2000). One solution is to use volunteers to help collect information. Volunteer data collection, or "citizen science", has become a widespread alternative for scientists and resource agencies needing information but lacking sufficient resources to gather it. With citizen science, a portion of monitoring costs is born by volunteers and sampling effort and geographic coverage are much larger than would otherwise be feasible. Additionally, volunteer involvement leads to increased resource stewardship.

The Reef Environmental Education Foundation (REEF) enlists thousands of recreational divers and snorkelers visiting coastal areas each year to provide meaningful information while learning about the environment. REEF was founded as a non-profit organization in 1990 out of growing concern for the health of the marine

Environmental Monitoring and Assessment **81**: 43–50, 2003.

environment and the desire to provide the SCUBA diving community a way to contribute to the understanding and protection of marine environments. REEF achieves this goal through its volunteer fish-monitoring program, the REEF Fish Survey Project. The Project was developed with support from The Nature Conservancy and guidance by the Southeast Fisheries Science Center of the National Marine Fisheries Service and the University of Miami.

The Fish Survey Project started in Florida in 1993, and since then has expanded to include the entire tropical western Atlantic (Florida, Caribbean, Bahamas, and Gulf of Mexico), southern Atlantic states (Georgia through North Carolina), the Northeast (Virginia through Newfoundland), the West Coast of the United States and Canada (California through British Columbia), the tropical eastern Pacific (Gulf of California to the Galapagos Islands), and most recently the Hawaiian Islands. By the end of 2001, over 40,000 surveys had been conducted by REEF members.

Volunteers conduct surveys on their own or during organized field surveys using the Roving Diver Technique (RDT; Schmitt and Sullivan, 1996). During RDT surveys, divers swim freely throughout a dive site and record every observed fish species. At the conclusion of each survey, divers assign each recorded species one of four log10 abundance categories [single (1); few (2–10); many (11–100); and abundant (>100)]. The species data along with survey time, depth, temperature, and other environmental information are then transferred to a REEF scansheet. Complete scansheets are returned to REEF headquarters and are optically scanned. After being processed through a set of error checking/quality control steps, the data are loaded into REEF's publicly accessible database. Summary reports of the data can be generated through the Internet (www.reef.org) and raw datafiles are provided when requested by researchers and resource managers.

Volunteers are categorized as expert or novice according to survey experience and performance on a series of fish identification exams. The RDT survey data provide species lists, frequency of occurrence, and relative abundance data. Percent sighting frequency (%SF) for each species is the percentage of all dives in which the species was recorded. An estimate of abundance is calculated as:

$$\text{abundance score} = D \times \%SF \qquad (1)$$

where the density score (D) for each species is a weighted average index based on the frequency of observations in different abundance categories. Density score is calculated as:

$$D = [(n_S x1)+(n_F x2)+(n_M x3)+(n_A x4)] \,/\, (n_S + n_F + n_M + n_A) \qquad (2)$$

where n_S, n_F, n_M, and n_A represent the number of times each abundance category was assigned for a given species.

In this paper we describe the application of Fish Survey Project data to a variety of conservation and management projects. We first illustrate how REEF's program has been incorporated into monitoring and baseline data gathering efforts by management agencies through contracts and training programs. Next we summarize two regional analyses of the data, and discuss the contribution REEF data have made to the general knowledge of reef fish natural history. Finally, REEF's role in educating the public about the issues and threats facing marine resources is discussed. The intent is to outline the benefits that managers, scientists and the public derive from the REEF Fish Survey Project, to highlight the role partnerships have had in the Project's success, and to promote citizen science programs as a viable source of monitoring data and stewardship development.

2. REEF In Monitoring Programs

Useful monitoring programs should provide the means to compare the characteristics of different sites under the care of a managing agency through space and time. Ideally, monitoring includes sites outside the management area so that local trends can be distinguished from regional trends. For most local managing agencies, developing a monitoring program with such traits is difficult given limitations to budget, person-hours, and expertise. In such instances, REEF provides a solution.

REEF provides a large and willing pool of surveyors, a standard methodology, data management and reporting mechanisms. Several management agencies have taken advantage of REEF's services by incorporating the Fish Survey Project into monitoring programs.

2.1 Training and Local Applications

REEF provides government agencies and local conservation organizations fish identification and survey training. By incorporating the Fish Survey Project methodology into a monitoring program, an agency gains access to a data management system at no expense. Additionally, the data they collect are consistent with data collected by others over a broad geographic area. The following are examples of partnerships that REEF has formed with both international and local management agencies.

2.1.1 Marine Parks

The Negril Coral Reef Preservation Society and the Negril Environmental Protection Trust in Jamaica worked with REEF to provide training for marine park rangers, biologists and local dive leaders. The week-long program, funded by the United Nations Environmental Program, involved seminars in identification, survey methodology and applications of survey data. Park staff and volunteers now conduct regular survey dives as part of an ongoing monitoring program. In addition, local dive leaders are able to train visiting tourists to take part in the monitoring effort.

The National Marine Park in Cozumel became interested in incorporating the Fish Survey Project program into its regular monitoring efforts after learning of the large quantity of data available in the area. REEF members regularly visit this area and conduct fish surveys during their dives. REEF worked with the Marine Park to train staff and biologists in field identification and survey methodology. Following a one-week staff training program, a separate program was conducted to train and involve key marine park volunteers. Local volunteers now conduct training on a regular basis for visiting divers as well as take part in regular monitoring dives scheduled by the Marine Park.

2.1.2 State Agency

Since 1980, the Florida Fish and Wildlife Conservation Commission (FWCC) has funded an artificial reef program using manmade materials to attract and replenish fish stocks throughout Florida's coastal waters. In 1999, the FWCC approached REEF to help meet the program's need for standardized methods and data reporting. The agency had contracted with county dive teams to conduct periodic fish surveys on the artificial reefs. However, the survey methods and data formats of the county dive teams were inconsistent, and comparing information among counties was difficult. Once involved, REEF conducted five regional training workshops with both classroom and field sessions. The FWCC now requires all county teams to conduct RDT fish surveys as a minimum level of data collection. REEF provides standardized format datafiles back to the FWCC on a quarterly basis.

2.2 MONITORING AND ASSESSMENT CONTRACTS

Resource agencies that do not have the expertise or person-power to monitor all the taxa and areas important to conservation and management can contract REEF to help meet their monitoring needs. REEF has established an Advanced Assessment Team (AAT) to fulfill monitoring and assessment contracts it is awarded. The AAT is made up of REEF's most experienced and knowledgeable surveyors. To be a part of the Team, members have conducted a minimum of 35 surveys, have passed the Advanced Exam, and have kept their survey experience up to date. In 1997, the Florida Keys National Marine Sanctuary (FKNMS) contracted REEF to begin monitoring fishes at select sites as part of the Sanctuary's zone monitoring program. The objective is to collect data from no-take areas and from reference sites in order to evaluate the effect of harvest restrictions. REEF formed the AAT in direct response to this contract, inviting highly trained and active members to become a part of the project. The annual FKNMS contract employs eight AAT to survey 37 sites during a two-week time frame each year. The contract also ensures that a minimum number of REEF surveys are conducted each year within the FKNMS. In addition to the data collection, the contract includes funding for data summaries and an annual report. Other National Marine Sanctuary sites have also initiated AAT monitoring contracts in order to fill data gaps, while engaging their constituents.

Through a recent partnership between REEF, The Ocean Conservancy (formally known as the Center for Marine Conservation) and NOAA's Coastal Zone Management Program, a series of training workshops have been organized on US Island Territories (Puerto Rico and the U.S. Virgin Islands) to enable local stakeholders to collect data on their island's coral reefs. In addition to the workshops, the contracts include the gathering of baseline data for each area using REEF's AAT. AAT participation in the workshops provides additional training personnel, while increasing the amount of data collected from the area.

3. Data Analysis

As the REEF program grows, the data collected by volunteers increases accordingly. In program areas with particularly high survey effort, it is possible to "mine" the data to explore trends or anomalies among species, or sites. Since the inception of the REEF program, large amounts of data have been collected in both the Bonaire Marine Park and the Florida Keys National Marine Sanctuary. Here we describe analyses conducted on the data from these two areas.

3.1 Bonaire Marine Park (BMP)

The BMP is the oldest marine park in the Caribbean. However, the fishes of Bonaire have not been systematically studied. Since 1993, the BMP has been the focus of eight REEF Field Surveys (week-long survey trips by a team of volunteer surveyors). Additionally, individual survey effort there has been intense. The resulting data provide a thorough reef fish species inventory, as well as a baseline of information that the BMP can use to detect future changes in the resource. Site-specific data provide a tool to evaluate relationships between sites with different physical characteristics and levels of diver use.

The REEF dataset was used to provide a description of the fish assemblage of the BMP (Pattengill-Semmens, 1998). The paper evaluated REEF surveys conducted between December 1993 and September 1999. A total of 1,557 novice and 457 expert RDT surveys had been conducted by REEF volunteers on the reefs of Bonaire and Klein Bonaire, representing 1,937 survey hours at 77 sites. A total of 362 species was reported. At one of the sites, Bari Reef, volunteers reported 270 species, the highest species richness of all sites in the REEF database. In addition to the taxonomic inventory, relationships among sites on Bonaire and the neighboring island of Klein Bonaire were examined. Despite the close proximity of these two areas, results suggested that the fish assemblages of Bonaire sites were relatively distinct from those on Klein.

The use of REEF Fish Survey Project data to describe the fish assemblages of the BMP is a step toward better understanding the Park's resources and the results highlight the statistical power of a large survey effort. The paper represents the most comprehensive taxonomic inventory of fishes within the Park, and it will serve as a baseline for comparison after another 20 years of management.

3.2 FLORIDA KEYS NATIONAL MARINE SACTUARY

REEF has a strong foundation in the Florida Keys. As of January 2001, REEF members had completed 6,630 fish surveys in the Florida Keys. These surveys are the result of several organized projects, individual survey activities, and the implementation of the annual AAT zone monitoring in 1997. Two projects that used the REEF data from the Florida Keys are described below.

3.2.1 Biogeography

Correlating benthic habitat variables with the distribution, abundance, and size of reef fishes can provide the basis of a model for estimating habitat affinities for individual species. The spatial distribution of benthic habitats often mediates the effects of ecological processes (e.g., predation, competition, dispersal) that determine the distribution and abundance of tropical fishes. The relationship between fishes and habitat has recently received a good deal of attention, especially with regard to identifying Essential Fish Habitat (EFH), as mandated by the Magnusen-Stevens Act (H.R. 2061).

In 1999 NOAA's Biogeography Program completed a benthic habitat GIS and wanted to include spatially referenced living marine resources data. To accomplish this, the Program initiated a joint project with REEF and NOAA's Marine Sanctuaries Division. The project's goal was to describe the distribution and abundance of reef fishes in the FKNMS, and to use that information to evaluate fish-habitat interactions and the performance of management zones. The first step has been completed (Jeffrey *et al.*, 2001). Current efforts focus on correlating reef fish distribution and abundance with benthic habitat parameters and the development of predictive habitat affinity models for selected fish species (Jeffrey *et al.*, in press). These efforts will provide a diagnostic tool that can be used in a variety of management and conservation applications, including marine reserve siting and locating essential fish habitat.

3.2.2 Trend Analysis Tool

Using the REEF FKNMS dataset, researchers from the University of Washington took an approach that addressed both the need for data-synthesis tools and the need for analysis techniques that are robust to large, volunteer generated datasets (Sauer *et al.*, 1994). Semmens *et al.* (in press) developed a method for calculating multi-species trends in reef fish populations. The analysis method was adapted from techniques used to analyze data from breeding bird surveys in North America (James *et al.*, 1996). This across-site comparison generated an objective categorization of sites, and provided coral reef managers with a powerful management tool to identify sites of management concern. This work also highlighted the increasing role that the REEF dataset will play in providing a clear picture of which FKNMS sites are changing rapidly, versus sites that sustain relatively consistent reef fish assemblages.

4. Conservation Contributions

The data collected by volunteers and the REEF field survey expeditions have resulted in several significant contributions to the general knowledge of reef fish natural history. In collaboration with taxonomic experts, the bluestriped goby (*Ptereleotris randalli*; L. Rocha, pers. comm.), red banner blenny (*Emblemariopsis tayrona*; P. Hastings, pers. comm.), giraffe snake eel (*Heteroconger camelopardalis*), and yellow garden eel (*Heteroconger luteolus*) have all been described as new species in the Caribbean basin as a result of REEF surveyors. The REEF database is also a powerful tool for documenting species distribution patterns. As part of the partnership with the NOAA Biogeography Program, a distribution atlas is currently being produced.

The threat of exotic fish species has gained increased attention among scientists and resource agencies, and REEF surveyors have been enlisted to help track exotic sightings. This has been particularly useful in tropical waters where amateur aquarists release fish too big or undesirable for a fish tank. In 2000, REEF partnered with the New England Aquarium and the FKNMS to remove three Indo-Pacific orbicular batfish (*Platax orbicularis*) from a reef in the Florida Keys. Two were successfully captured and are now in an exhibit at the Aquarium that educates the public about exotic species.

5. Conclusions

REEF's standardized census method and database management system, combined with partnerships, have resulted in a successful citizen science monitoring program. Collaborations with international, federal, and state agencies, scientists, conservation organizations, and private institutions have led to a variety of conservation and monitoring applications. The Fish Survey Project has been incorporated into existing monitoring programs, its data have been used in analyses, and the database has provided an increased understanding of the distribution and natural history of fishes in the tropical western Atlantic.

In addition to the direct applications of the Fish Survey Project and its data, REEF has made significant contributions to public education and marine resource stewardship. Participation in the Fish Survey Project enhances a diver's ability to discern details about the marine environment. The excitement of finding a rare fish can only be appreciated if one knows it's rare. By learning identification techniques and recording observations, REEF surveyors become keen observers and naturalists. The benefits, however, extend beyond enhancing an individual diver's underwater experience. The sense of stewardship that arises from involvement in citizen science programs such as REEF's Fish Survey program raises the public's awareness of and involvement in marine resource issues. Ultimately, REEF's efforts empower volunteers to take an active role in support of effective marine resource management. Volunteer or "citizen" science allows all those who are interested in the resource to contribute to its understanding. Beyond providing valuable data, the increased stewardship that comes from participation in the Fish Survey Project is vital to the protection of coastal marine resources.

References

Baird, B. E., Miller-Henson, M., and Semmens, B. X.: 2000, Improving California's System of Marine Managed Areas: Final Report of the State Interagency Marine Managed Areas Workgroup, Resource Agency of California, Ocean Resources Management Program, Sacramento, CA. http://ceres.ca.gov/cra/ocean/.

James, F. C., McCulloch, C. E., and Wiedenfeld, D. A.: 1996, 'New Approaches to the Analysis of Population Trends in Land Birds', *Ecology* **77**(1), 13–27.

Jeffrey, C. F. G., Pattengill-Semmens, C. V., Buja, K., Christensen, J. D., Coyne, M., Monaco, M. E., and Gittings, S.: in press, 'Benthic Habitat Associations of Reef Fishes in the Florida Keys: Coupling of Benthic Habitats and Fish Distributions via GIS Technology', Proceedings International Coral Reef Symposium, October 2000.

Jeffrey, C. F. G., Pattengill-Semmens, C. V., Gittings, S., and Monaco, M. E.: 2001, Distribution and Sighting Frequency of Reef Fishes in the Florida Keys National Marine Sanctuary, Marine Sanctuaries Conservation Series MSD-01-1, US Dept. of Commerce, NOAA, Silver Spring, MD.

Pattengill-Semmens, C. V: 1998, 'The Reef Fish Assemblage of Bonaire Marine Park: an Analysis of REEF Fish Survey Project Data', Proceedings 52nd Gulf Caribbean Fisheries Institute Meeting.

Sauer, J. R., Peterjohn, B. G., Link, W. A.: 1994, 'Observer Differences in the North American Breeding Bird Survey', *Auk* **111**(1), 50–62.

Schmitt, E. F. and Sullivan, K. M.: 1996, 'Analysis of a Volunteer Method for Collecting Fish Presence and Abundance Data in the Florida Keys', *Bulletin of Marine Science* **59**(2), 404–416.

Semmens, B. X., Ruesink, J. L., and Pattengill-Semmens, C. V.: In Press, 'Multi-site Multi-species Trends: a New Tool for Coral Reef Managers', Proceedings International Coral Reef Symposium, October 2000.

THE COASTAL COMPONENT OF THE U.S. INTEGRATED OCEAN OBSERVING SYSTEM

THOMAS C. MALONE[1*]

[1]Horn Point Laboratory, University of Maryland Center for Environmental Science, P.O. Box 775, Cambridge, MD 21613

[*](author for correspondence, e-mail: malone@hpl.umces.edu)

Abstract. The combined effects of human activities and natural variability present significant challenges to the goals of protecting, restoring, and sustaining coastal ecosystems. Meeting these challenges and resolving conflicts in an informed fashion will require: (1) more timely detection and prediction of environmental changes and their consequences; and (2) more timely access to relevant environmental information. The achievement of these goals depends on the development of a sustained and integrated coastal ocean observing system (ICOOS) that insures timely access to the data and information required to improve: (1) climate predictions and the effects of changes in the weather on coastal populations; (2) efforts to sustain and restore healthy coastal marine ecosystems and living marine resources; and (3) compliance monitoring and evaluations of the efficacy of environmental policies. Although the responsible federal and state agencies all require similar environmental information, many separate programs have evolved for collecting, managing, and analyzing data for various purposes. Consequently, there is too much redundancy; access to diverse data from disparate sources is limited and time consuming; and individual programs are inevitably underfunded and too limited in scope. A system is needed that coordinates and integrates many of the elements of these programs to minimize redundancy, be more comprehensive, provide more timely access to data and information, and satisfy the information needs of a greater number of user groups in a more cost-effective fashion. This is the purpose of the ICOOS.

Keywords: integrated coastal ocean observing system, detecting and predicting environmental change, anthropogenic effects, sustained observations, coastal monitoring, ecosystem assessments

1. Introduction

Living resources are concentrated in coastal marine and estuarine systems and the number of people living in coastal drainage basins is increasing rapidly (Costanza *et al.*, 1997; Daily *et al.*, 2000; Hinrichsen, 1998). Consequently, the demands on these systems to provide goods and services will continue to grow (e.g., shipping and recreation; provide resources and living space; receive, process, and dilute the effluents of human society). At the same time, coastal ecosystems are experiencing unprecedented changes (Table 1). The phenomena of interest include a broad spectrum of variability from changes in sea state, coastal flooding, and harmful algal events to habitat loss (e.g., coral reefs, sea grass beds and tidal wetlands) and sea level rise associated with climate change. Together, these changes are making the coastal zone more susceptible to natural hazards, more costly to live in, and of less value to people and the national economy.

The purpose of this contribution is to describe a major effort that is underway to organize an integrated coastal observing system by coordinating, enhancing and supplementing existing programs that meet the design requirements of the

Environmental Monitoring and Assessment **81**: 51–62, 2003.

Table 1. Natural and anthropogenic forcings and associated phenomena of interest in coastal marine ecosystems where forcings refer to inputs of matter and energy from sources external to the system in question.

Forcings

Natural & Anthropogenic

- Storms & other extreme weather events
- Ocean currents and waves
- Climate change & sea level rise
- River & ground water discharges
- Physical restructuring of the environment
- Construction of impervious surfaces & dams
- Harvesting living & nonliving resources
- Nutrient enrichment
- Sediment inputs
- Chemical contamination
- Introductions of non-native species

Phenomena of Interest

Marine Services & Coastal Hazards

- Changes in sea state & sea ice
- Coastal currents
- Coastal flooding & erosion
- Changes in shallow water bathymetry
- Susceptibility to natural hazards

Public Health and Status of Ecosystems & Living Resources

- Human health risks, beach and shellfish bed closures
- Loss of biodiversity
- Declines in living marine resources
- Harmful algal events
- Invasive species
- Loss of essential fish habitat
- Accumulations of organic matter
- Oxygen depletion
- Disease & mass mortalities of fish, birds & mammals
- Chemical contamination of sediments and organisms

system (see section 3). The goal is to develop an integrated, coastal ocean observing system (ICOOS) that will enable: (1) more cost-effective use of infrastructure, knowledge and expertise; (2) more rapid access to environmental data and information; (3) more timely detection and prediction of changes and events; (4) broader applications of data and information; and (5) increases in the value of and funding for coastal marine and estuarine research and monitoring.

2. Rationale for an Integrated Approach

The nation's efforts to understand, detect and predict the changes occurring in coastal ecosystems suffers from the paradox of too much (redundancy), too little (not comprehensive in terms of both the measurements that are made and the data that reside in accessible data bases) and too late (the time required to acquire and analyze data is too long). A large number of state and federal agencies are currently engaged in and responsible for different but overlapping aspects of environmental change in coastal ecosystems. At least 8 federal agencies are responsible for collecting environmental data from estuarine and coastal marine systems (Commerce, Navy, Interior, Transportation, Energy, National Science Foundation, National Aeronautics and Space Administration, and the Environmental Protection Agency), and agency budgets and programs are reviewed and approved by 47 different Congressional Committees (Weisberg *et al.*, 2000). Agency missions range from forecasting the weather, enabling safer and more efficient marine operations, and predicting the effects of climate change to managing the health of coastal ecosystems and the resources they support, protecting public health, and mitigating the effects of natural disasters and human activities. In response to these needs and in a case-by-case fashion, individual agencies and departments within agencies (both federal and state) have developed separate systems for collecting, managing, analyzing and applying environmental data, e.g., compliance monitoring, resource management, coastal zone management, coastal engineering. This has led to much redundancy (from measurements to data management) and to inefficiencies in the detection and prediction of environmental changes and their consequences. At the same time, individual programs, by themselves, are inevitably underfunded, limited in scope, and target a limited number of user groups. Consequently, major gaps exist in our understanding of variability in coastal environments; our ability to assess the status of coastal ecosystems and predict future conditions is inadequate; and our capacity to mitigate the effects of environmental changes is rudimentary at best.

It is likely that, in the absence of an organized and integrated system for improved detection and prediction of environmental changes and their effects on the environment and people, conflicts between commerce, recreation, development, environmental protection, and the management of living resources will become increasingly contentious and politically charged, and the social and economic costs of uninformed decisions will become more expensive. Although, the challenges of developing such a system are significant, we are witnessing a convergence of societal needs and technical capabilities that suggest the time is right to develop an ICOOS. It is time to design and implement a coastal system that is: (1) based on sound science and modern technologies; (2) provides more timely access to data and information on environmental change; (3) makes more effective use of existing resources, knowledge and expertise; and (4) develops in concert with the integrated ocean observing system for climate (Nowlin *et al.*, 1996; Malone and Cole, 2000).

3. The Design of An End-to-End, Integrated and Sustained Observing System

3.1 ELEMENTS OF AN END-TO-END OBSERVING SYSTEM

Both detection and prediction depend on the development of an integrated (multidisciplinay, end-to-end) and sustained observing system that effectively links measurements and data analysis to the needs of government agencies for more timely access to data and environmental information. Linking user needs to measurements to form an end-to-end, user-driven system requires a managed, two-way flow of data and information among three essential subsystems:

- The monitoring subsystem (networks of platforms, sensors, sampling devices, and measurement techniques) to measure the required variables on the required time and space scales to detect and predict changes in coastal indicators;
- The communications network (data dissemination and access) and data management subsystem (telemetry, protocols and standards for quality assurance and control, data dissemination and exchange, archival, user access); and
- The modeling and applications subsystem (data assimilation, synthesis and analysis; procedures for translating data into products).

The monitoring subsystem is the measurement end of the system. It consists of the national infrastructure required to measure the required variables and transmit data to the communications network and data management subsystem. The infrastructure consists of the mix of platforms, samplers, and sensors required to measure the common variables with sufficient spatial and temporal resolution to capture important scales of variability in 4 dimensions. This will require the synthesis of data from remote sensing and in situ measurements involving six interrelated categories of monitoring elements: (1) coastal observing networks for the near shore; (2) fixed platforms, moorings and drifters; (3) research and survey vessels, ships of opportunity (SOOP) and voluntary observing ships (VOS); (4) remote sensing from satellites and aircraft; and (5) remote sensing from land-based platforms (e.g., high frequency radar).

Data communications and management link measurements to applications. The objective is to develop a system for both real-time and delayed mode data that allows users to exploit multiple data sets from disparate sources in a timely fashion. A hierarchical system of local, regional and national organizations is envisioned to provide data, information and access to users at each level. Regional Information Centers (RICs) are needed that will provide processed products (e.g., assimilating data from remote and in situ sources for numerical model predictions and GIS applications that require substantial computing power). The development of this component of the system should be of the highest priority.

Data assimilation and modeling are critical components of the observing system. Real-time data from remote and in situ sensors will be particularly valuable in that data telemetered from these sources can be assimilated to: (1) produce more accurate estimates of the distributions of state variables, (2) develop, test and validate models, and (3) initialize and update models for improved forecasts of coastal environmental conditions and, ultimately, changes in ecosystem health and living resources. A variety of modeling approaches (statistical, empirical, theoretical) will be required. The challenge of developing a cost-effective observing system underscores the importance of the interaction between measurements and modeling. Due to the complexity of coastal ecosystems and the cost of observing them, Observation System Simulation Experiments (OSSEs) will become increasing valuable as tool for assessing the efficacy of different sampling schemes and the usefulness of measuring different variables (Walstad and McGillicuddy, 2000).

3.2 SCIENTIFIC FOUNDATIONS

Given the reality that the phenomena of interest are diverse and exhibit a broad spectrum of variability from hours to decades (cf. Powell, 1995; Steele, 1995), is it realistic to believe that an ICOOS can be developed that will provide the data and information required to detect and predict the causes and consequences of changes in the status of coastal ecosystems? Although this may seem to be an insurmountable goal, there is reason to believe that a cost-effective, sustained and integrated system is achievable. Many of the changes occurring in coastal ecosystems are local expressions of regional to global scale changes that are related through ecosystem dynamics. Marine ecosystems are structured by physical processes and changes in biological, chemical and geological properties are related through a hierarchy of physical-ecological interactions that can be represented by robust models of ecosystem dynamics (e.g., numerical models of physical processes and coupled physical-ecological models) (cf. Brink and Robinson, 1998; Robinson and Brink, 1998). This suggests that there is a relatively small set of common variables that, if measured with sufficient resolution for extended periods over sufficiently large areas, will serve many needs from forecasting the effects of tropical storms and harmful algal events on short time scales (hours to days) to predicting the environmental consequences of human activities and climate change on longer time scales (years to decades).

The phenomena to be addressed by the ICOOS fall into three general categories (Table 1): (1) coastal marine services and natural hazards, (2) ecosystem and public health, and (3) living marine resources (LMRs). The data requirements of coastal marine services are, for the most part, common to all three categories (Figure 1). Safe and efficient coastal marine operations require accurate nowcasts and timely forecasts of storms, coastal flooding and precipitation; of coastal wind-, current-, wave-, and ice-fields; and of water depth, temperature and visibility. The set of variables that must be measured and assimilated in near real time include barometric pressure, winds, air and water temperature, sea level, and surface currents and waves. In addition to these variables, protecting, sustaining and restoring "ecosystem health" require timely information on measures of ecosystem status that are related to nutrient enrichment and habitat modification (including coastal erosion, oxygen depletion, loss of critical fish habitats such as submerged attached vegetation and coral reefs); harmful algal events and mass mortalities of marine organisms; chemical and bacterial contamination of habi-

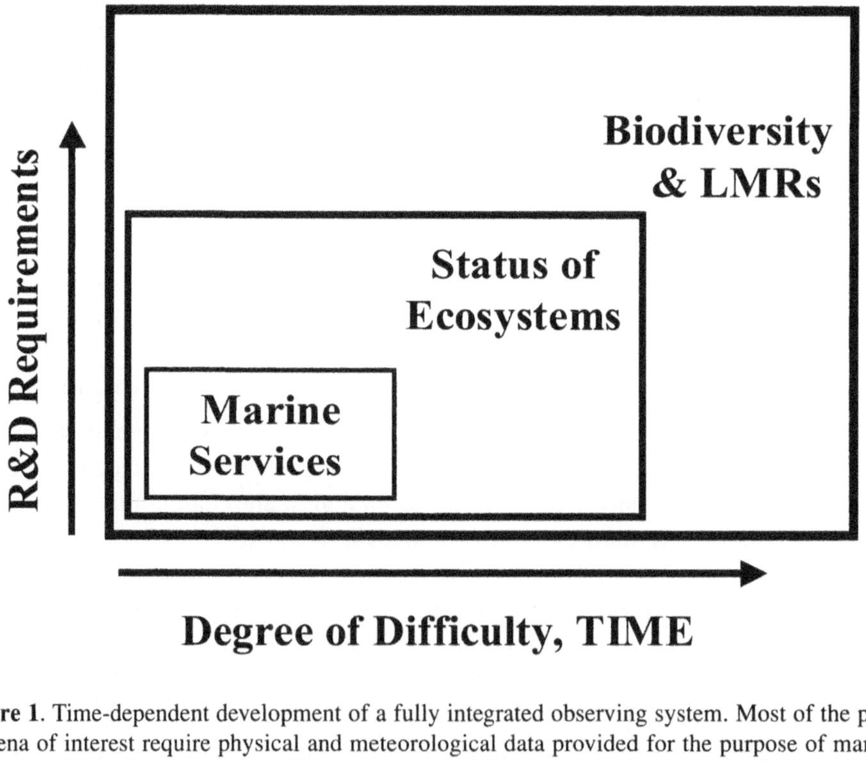

Figure 1. Time-dependent development of a fully integrated observing system. Most of the phenomena of interest require physical and meteorological data provided for the purpose of marine services and weather forecasting. Given current capabilities and the importance of physical processes to issues of ecosystem health and living marine resources, the initial backbone will emphasize those variables that are most useful to detecting and predicting the largest number of the phenomena of interest. As technologies for rapid measurements of biological and chemical variables are developed they will be incorporated into the system. Note that biological and chemical variables can and should be incorporated now as regional enhancements.

tats and organisms; and changes in the abundance and distribution of native and non-native species of benthic and pelagic organisms. The demands of protecting "living marine resources" and managing harvests (of wild and farmed stocks) in an ecosystem context require data on all of the above as well as timely information on population (stock) abundance, distribution, age (size)-structure, reproduction, year-class strength, migratory patterns, and mortality (including catch statistics). While all three categories must be addressed from the beginning, current capabilities dictate an emphasis on marine operations and natural hazards. The achievement of a comprehensive and integrated system that provides the full spectrum of data and information requires a system that is designed to adapt as new capabilities become available and a major investment in research and development (Figure 1).

3.3 DESIGN FRAMEWORK FOR AN OPERATIONS SYSTEM

Given the interdependence of the physical and ecological processes that govern the status of estuarine and marine ecosystems and the reality that priorities vary regionally and among states, the coastal ocean observing system is conceived as a national network (the backbone) for the measurement of common variables that is regionally and locally enhanced (e.g., more variables, greater resolution, additional products) to detect and predict state changes that are of greatest concern to participating entities (State agencies, industry, academic institutions). ICOOS will initially develop through the selected incorporation of existing programs and the establishment of regional pilot projects (proof of concept) that incorporate existing programs and new initiatives into a coordinated and integrated end-to-end system that is consistent with the ICOOS design principles (see below).

An operational observing system requires that measurements be routine, long-term (sustained into the foreseeable future), and systematic (stable delivery of data and information with sufficient precision and accuracy on appropriate time and space scales). As discussed above, many of the measurements and models required for a comprehensive, fully integrated, multi-disciplinary observing system are not operational. In addition, much work is needed to identify and develop those products that are most useful and to build capacity at the state level. To these ends, the design will be guided by the following principles:

1. The observing system will provide the data and information on changes in the status of coastal ecosystems that require regional to national approaches (e.g., local expressions of larger scale changes). It will produce data-products that address a broad spectrum of user needs.

2. The development of ICOOS must involve a more cost-effective use of existing data, expertise and infrastructure than is currently realized, i.e., the entire process from measurements to products will be cost-effective. ICOOS will build a national backbone of common observations that will be regionally enhanced to

address regional and local priorities as determined by the participating users in the region. The system will develop by incorporating, enhancing and supplementing existing programs as appropriate. It will become a comprehensive system of observations through shared use of infrastructure from measurement systems and platforms to communication networks, data management systems, assimilation techniques and modeling.

3. ICOOS will enable a constructive and timely synergy between hypothesis-driven research, the detection of patterns of variability and the generation of information in response to user needs.

4. The observing system must be both integrated and sustained (Figure 2). Observations will be sustained in perpetuity to capture episodic events and long-term trends (document both high and low frequency variability), enhance scientific analysis and support model predictions. The observing system will be integrated from measure-

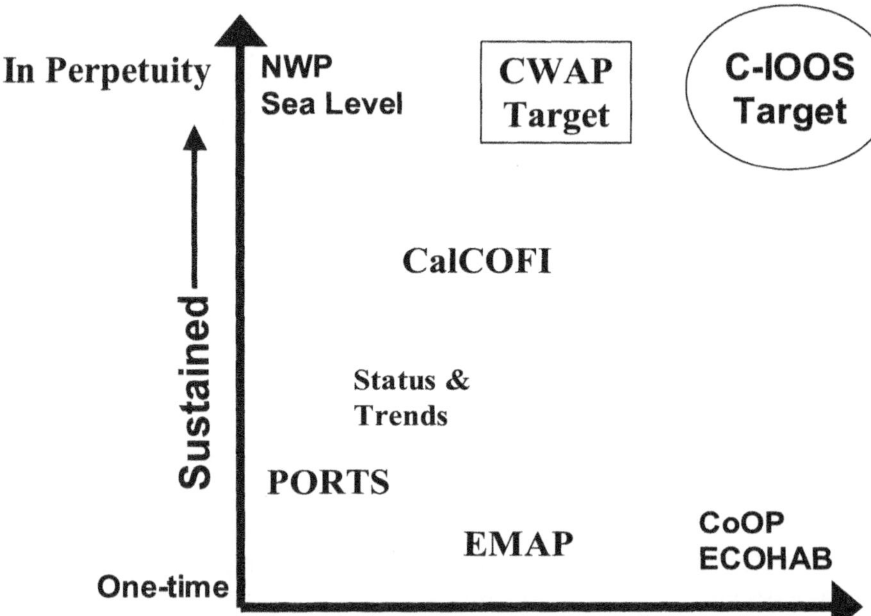

Figure 2. Some programs are sustained [Numerical Weather Predictions (NWP) and Sea Level networks] and some are integrated [research programs such as the Coastal Ocean Program (COOP) and the Ecology and Oceanography of Harmful Algal Blooms (ECOHAB)]. Other programs such as NOAA's Status and Trends, EPA's EMAP, and the NOAA's PORTS (Physical Oceanography Real-Time System) are intermediate along one of the axes. The monitoring program proposed as part of the Clean Water Action Plan (CWAP; www.cleanwater.gov) is intended to be sustained but it does not achieve the level of integration that is the target of the ICOOS. C-IOOS is the Coastal Component of the U.S. Integrated Ocean Observing System; and CalCOFI is the California Cooperative Fisheries Investigation.

ments (synoptic measurements of physical, biological and chemical properties over a broad range of time and space scales) to data management (multiple data types from disparate sources) and analyses that are responsive to the needs of multiple end-users, i.e., it will be a user-driven, end-to-end system.

To date, few, if any, programs are both integrated and sustained. A sustained commitment will be required of ICOOS partners to establish, maintain, validate, make accessible, and distribute high quality data that meet nationally agreed upon standards. By building on existing activities, capabilities, and infrastructure, and by using a phased implementation approach, work can start immediately to achieve the vision. New technologies, past investments, evolving scientific understanding, advances in communications and data processing, and pressing societal needs combine to provide the opportunity to initiate an integrated observing system for coastal waters immediately. The major pieces missing are a nationally accepted design; national and state commitments of assets and funds; and partnerships among states, institutions, data providers and users.

4. Implementation

Detecting, assessing, predicting and mitigating the effects of natural variability and human activities on coastal ecosystems requires a regional perspective that transcends political boundaries and provides the means to evaluate local changes in marine ecosystems in terms of larger scale changes in climate, ocean circulation, fishing pressure and land-use practices. Regional approaches also provide a practical framework to coordinate the efforts of local, state, and federal programs, involve stakeholders, and enable the timely analysis of data. A nation-wide network can thus be built by coordinating the development and linking of regional observing systems that measure a common set of core variables and are regionally and locally enhanced (additional variables, greater resolution in time and space, data management and synthesis centers) to address those issues that are of greatest interest to user groups in the region.

Many of the elements of an integrated coastal ocean observing system (ICOOS) are already in place or in development, and the NOAA Coastal Services Center has established a web site that identifies and provides links to them (http://www.csc.noaa.gov/cts/coos/). In this regard, it is clear that a more cost-effective approach must be found that coordinates and integrates many of the elements of these systems. As ICOOS develops, it must become more than the sum of its parts. By formulating and implementing a plan for regional (often interstate) observing systems that are nationally coordinated and locally relevant, a wider array of users will be more effectively served with relatively modest increases in costs relative to the additional benefits. If this is to be achieved, the system will not be an opportunistic assembly of whatever might be available. It will develop by selectively incorporating, enhancing and supplementing existing programs consistent with the ICOOS design principles described above.

4.1 Governance Considerations

The ICOOS must be nationally coordinated and locally relevant, and it must enable government agencies and other user groups to fulfill their respective missions and goals more effectively. At present, there is no coherent governance structure that provides an efficient mechanism to achieve these goals. National leadership and coordination is required to enable and promote:

- Cooperation and collaboration among federal and state agencies to enable the nation-wide development and implementation of economically and ecologically sound environmental policies;
- Efficiencies in the design and implementation of regional programs and the timely incorporation of new technologies, models and products;
- Capacity building through training programs and infrastructure development;
- The measurement of core variables by all regional observing systems using nationally accepted methods and QAQC standards;
- The dissemination and management of data for the benefit of all;
- The development of ICOOS in the international framework of the Global Ocean Observing System (GOOS); and
- Sustained, predictable and performance based funding to insure uninterrupted data streams and the routine provision of data-products.

Given these goals, and the reality that the boundaries of most environmental issues of local importance are not confined to legal jurisdictions or to the missions of any one government agency, new mechanisms are needed that enable federal and multi-state collaboration in the allocation and management of funds and the periodic assessment of each regional system. To be successful, the governance of regional programs must harmonize "bottom-up" programmatic development through regional organizations of stakeholders (data providers and users) with "top-down" coordination by federal agencies and national organizations. The success of this approach will depend on the development of programs that are comprehensive in design and enjoy continuity of support that is not susceptible to short term political decisions and the annual funding cycles of state and federal governments.

4.2 A National Federation of Regional Systems

The goal is to develop a national federation of regional coastal ocean observing systems, each of which meets national performance standards for data quality and dissemination, is responsive to user needs, enjoys strong public support, and improves as new technologies become available and new needs emerge. Clearly, no single government agency is capable of planning and supporting such systems.

Thus, in addition to providing the mechanism for building and managing observing systems that ensure ongoing, constructive feedback between data providers and users, the governance structure must clearly define and harmonize the roles of federal, state, and local agencies.

It has been proposed that the National Ocean Research Leadership Council (NORLC) should oversee the coordinated implementation of an ICOOS that conforms to the ICOOS design principles described above. In this capacity the NORLC would approve standards and protocols for the administration of the system including: (1) the establishment of a common set of observations that would form the basis of the national backbone and be nationally distributed according to accepted protocols; (2) quality control and assurance; (3) the development and application of models for nowcasting and forecasting the status of coastal marine and estuarine ecosystems; and (4) data management for timely access to and archival.

The federation of regional observing systems would include, for example, seven regions: northeast, southeast, Gulf of Mexico, West Coast, Hawaii, Alaska and the Great Lakes. Regional governing bodies would be established to provide leadership, management and accountability. The governing body would consist of representatives from user groups in the regions (e.g., the shipping, fishing, tourist, and construction industries; state and federal agencies; environmental research institutions; indigenous populations; and Non-Government Organizations and would be responsible for:

- Developing sustained state, regional, and federal financial support;
- Overseeing the design, implementation and management of the observing system;
- Ensuring that the system is driven by local priorities and begins by incorporating existing programs, expertise and infrastructure, and
- Engaging stakeholders at all stages, from the establishment of priorities and planning to implementation, program evaluation, and product development.

Mechanisms for the management of funds, technical operations, research, data (including QAQC), product development, system review and evaluation, and public outreach will have to be established.

Acknowledgments

This paper reflects the thinking of many people including members of the U.S. Global Ocean Observing System (GOOS) Steering Committee (especially Worth Nowlin, Evan Richert, and Steve Weisberg) and the Intergovernmental Oceanographic Commission Coastal GOOS Panel (especially Keith Thompson and Julie Hall). Funding for this and related efforts was provided by NOAA, NSF, NASA and EPA.

References

Brink, K.H. and Robinson, A.R. (eds.): 1998, The Global Coastal Ocean I. Processes and Methods, John Wiley & Sons, Inc., NY, 606 pp.

Costanza, R. and others: 1997, 'The value of the world's ecosystem services and natural capital,' *Nature* **387**, 253–260.

Daily, G.C. and others: 2000, 'The value of nature and the nature of value,' *Science* **289**, 395–396.

Hinrichsen, D.: 1998, 'Coastal Waters of the World: Trends, Threats, and Strategies,' Island Press, Washington, D.C., 275 pp.

Malone, T.C. and Cole, M.: 2000, 'Toward a global scale coastal ocean observing system,' *Oceanography* **13**, 7–11.

Nowlin, W.D. and others: 1996, 'An ocean observing system for climate,' *Bull. Am. Meteor. Soc.*, **77**, 2243–2273.

Powell, T.M.: 1995, Physical and biological scales of variability in lakes, estuaries, and the coastal ocean, In Powell, T.M. and Steele, J.H. (eds.), Chapman & Hall, NY, pp. 119–138

Robinson, A.R. and Brink, K.H. (eds.): 1998, The Global Coastal Ocean. II. *Regional Studies and Synthesis*, John Wiley & Sons, Inc., NY, 1062 pp.

Steele, J.H.: 1995, Can ecological concepts span the land and the ocean domains?, In Powell, T.M. and Steele, J.H. (eds.), Chapman & Hall, NY, pp. 5–19.

Walstad, L.J. and McGillicuddy, D.J.: 2000, 'Data assimilation for coastal observing systems,' *Oceanography* **13**: 47–53.

Weisberg, S.B., Hayward, T.L., and Cole, M.: 2000, 'Towards a US GOOS: A synthesis of lessons learned from previous coastal monitoring efforts,' *Oceanography* **13**: 54–61.

GREAT LAKES MONITORING RESULTS—COMPARISON OF PROBABILITY BASED AND DETERMINISTIC SAMPLING GRIDS

GLENN J. WARREN* AND PAUL J. HORVATIN

U.S. EPA, Great Lakes National Program Office, 77 W. Jackson Blvd., Chicago, IL 60604, U.S.A.
**(author for correspondence, e-mail: warren.glenn@epa.gov)*

Abstract. The Great Lakes may be viewed as a coastal environment, affected by the same meteorological and physical forces as the coastal ocean. The U.S. EPA, Great Lakes National Program Office (GLNPO) has monitored the open waters of the lakes, annually, since 1983. As part of the U.S. EPA Environmental Monitoring and Assessment Program (EMAP), a pilot study was performed in Lake Michigan to compare the existing GLNPO deterministic sampling grid with the EMAP probabilistic grid. Results of chemical analyses of trophic status indicators (total phosphorus and chlorophyll a) as well as nutrients and conventional limnological measurements, from spring and summer surveys in 1992 indicate little difference between the grids in the offshore region of the lake. The few statistically significant differences may be due to station distribution throughout the lake, or simple chance. This might be expected due to the well mixed nature of the open waters of Lake Michigan. The detection of a long-term trend for total phosphorus in Lake Michigan benefits from an annual program: viewing cumulative frequency distributions based on a four year EMAP interval does not convey information on the decrease in phosphorus in the lake. If the EMAP sampling grid were to be used in the Great Lakes, pilots in each of the lakes would be necessary for utilization of the existing long-term record as a basis for trend detection.

Keywords: Great Lakes, monitoring, EMAP, total phosphorus, Lake Michigan, trophic status indicators

1. Introduction

The Environmental Monitoring and Assessment Program-Great Lakes (EMAP-GL) was initiated in 1990 as part of the Environmental Monitoring and Assessment Program nationwide monitoring initiative. Four resource classes were defined for the program: offshore, nearshore, harbors and embayments, and wetlands (Hedtke *et al.*, 1992). Pilot programs were planned for all resource classes. In 1992 EMAP-GL began a series of pilot studies to determine the suitability of a group of indicators which had been selected for the offshore and nearshore areas of the Great Lakes, and to determine the spatial sampling intensity appropriate to the offshore area. The largest pilot effort went into verification of offshore indicators and comparison of sampling techniques.

The history of monitoring in the Great Lakes includes a long term sampling program on the lower four Great Lakes through the U.S. Environmental Protection Agency, Great Lakes National Program Office (GLNPO). Evaluation of the data from GLNPO was viewed as a key to piloting the offshore portion of the EMAP-GL implementation plan. These data are utilized on a regular basis to determine whether the lakes are changing in relation to trophic status guidelines recommended by the International Joint Commission (1980). Compatibility of these data with those to be generated by EMAP was viewed as important to the EMAP-GL program. The GLNPO sampling record begins in the mid-1970s, and although not

Environmental Monitoring and Assessment **81**: 63–71, 2003.

continuous, is extremely informative for looking at historical trends. Although selected by very different means, the GLNPO and EMAP designs for Lake Michigan coincide in that all of the stations are in at least 85 meters of water and were selected to represent the homogeneous offshore waters.

This paper reports on the comparison of sampling results from the offshore of Lake Michigan from a GLNPO sampling grid selected from a large number of offshore stations based on statistical and logistic considerations (deterministic) and an EMAP grid which is probability-based (Figure 1). Two surveys were conducted in 1992 during which the GLNPO and EMAP sampling grids were sampled for measures of trophic status as well as other nutrients and parameters of limnological interest. Long-term data are also presented as cumulative distribution functions and as trend lines to compare these two methods.

2. Materials and Methods

The comparison study began in the spring of 1992, using the eleven GLNPO stations in the offshore (defined as deeper than 90 meters) and the 12 base grid EMAP stations in the offshore (defined as deeper than 85 meters). Sampling was conducted from the R/V Lake Guardian. Samples were taken from several depths at

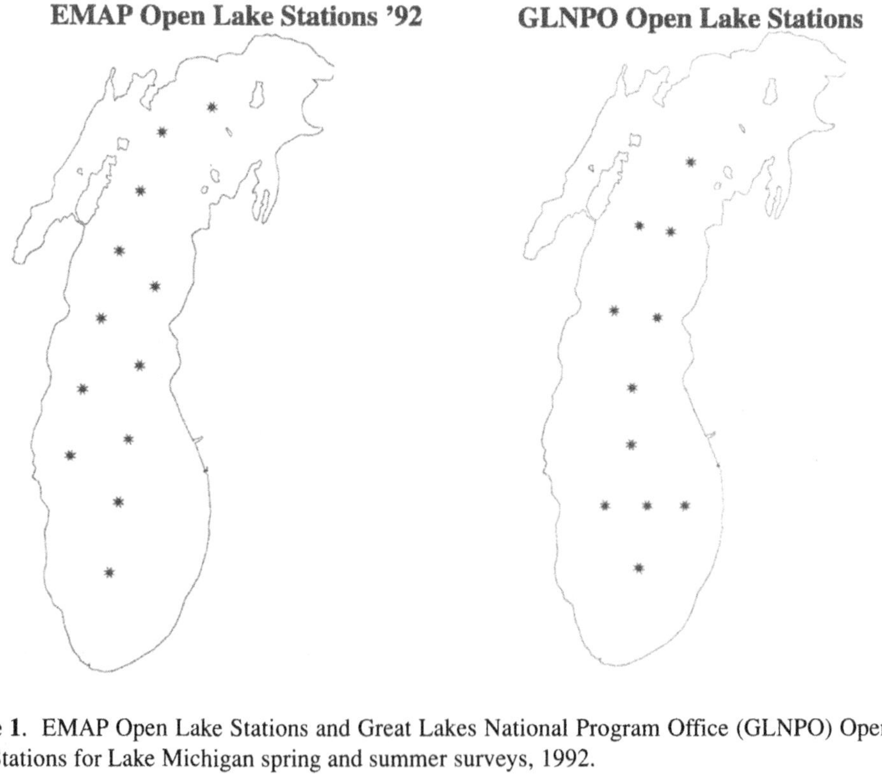

Figure 1. EMAP Open Lake Stations and Great Lakes National Program Office (GLNPO) Open Lake Stations for Lake Michigan spring and summer surveys, 1992.

each station, using a remotely triggered "rosette" sampler with 8 liter Niskin bottles. Routine sampling depths during spring were: 1 meter, mid-depth, 10 meters above bottom and 2 meters above bottom. Water samples were processed aboard the vessel, and some analyses were performed immediately in shipboard laboratory. Trophic status indicators, chlorophyll a and total phosphorus were measured at each site. The nutrients NO_3–NO_2 and Si were also measured. For purpose of comparing GLNPO and EMAP stations, several other chemical and physical measurements, part of the GLNPO sampling plan, were included for each sample.

During the summer survey, a similar set of parameters was measured. Samples were taken in the epi-, meta- and hypolimnion; however, only data from the epilimnion samples were included in the analysis. These were viewed as representative of the biologically active photic zone.

The chemical analytical methods are described in Lesht and Rockwell (1987). Total phosphorus, total dissolved phosphorus, silica, nitrate-nitrite nitrogen, chloride, sulfate, and total Kjeldahl nitrogen were analyzed using a Technicon Autoanalyzer II. Magnesium, sodium and calcium were analyzed using inductively coupled argon plasma emission spectroscopy. Potassium was analyzed by flame atomic absorption. Chlorophyll a was analyzed using a Turner fluorometer on acetone-extracted samples.

In order to provide uniform weighting to the data associated with each station, the station average concentrations are reported. For spring, the station average was calculated using results from all sampling depth. During summer, only samples from the epilimnion were averaged and reported for each station. Comparisons between sampling networks were determined with parametric t-tests on untransformed data. Tests of bias in variability between networks and for long-term trend in total phosphorus employed a non-parametric sign test and Sen regression (Sen, 1968), respectively.

3. Results

3.1 SPRING SURVEY

Chloride and specific conductance, generally viewed as conservative tracers of water mass, show small differences in absolute values (Table 1). However, there is a nearly statistically significant difference ($p < 0.059$) between specific conductance at EMAP and GLNPO stations. One possible explanation for this difference is the western-half bias (9 of 12 stations west of the center line) in the 1992 EMAP grid stations. This difference, however is not evident in the chloride comparison. Of the nutrients sampled, only nitrite-nitrate nitrogen displays a significant difference between sampling networks, albeit small in absolute value. For trophic status evaluation, total phosphorus and chlorophyll a, exhibit no significant difference between networks. A sign test of the standard deviations of the parameters shows no significant bias in variability in either network.

Table 1. Comparison of sixteen measurements from spring 1992, Lake Michigan GLNPO (n = 11); and EMAP (n = 12) stations.
**indicates statistical significance*

Parameter	*GLNPO (mean)*	*S.D.*	*EMAP (mean)*	*S.D.*	*t-probability*
Alkalinity (mg/L)	107.227	7.5987	106.208	9.7432	0.782
Specific Conductance mmho/cm	288.002	1.4683	286.708	1.6336	0.059
Turbidity (FTU)	0.454	0.2621	0.441	0.1132	0.874
Chlorophyll a (mg/L)	1.654	0.7957	1.971	0.9880	0.405
Dissolved Organic Carbon (mg/L)	1.379	0.0880	1.426	0.0961	0.230
Si (mg/L)	0.623	0.0617	0.591	0.0849	0.308
Nitrate + Nitrite (mg/L)	0.289	0.0091	0.276	0.0192	0.042*
Total Dissolved Phosphorus (mg/L)	0.0023	0.0006	0.0022	0.0007	0.607
Total Phosphorus (mg/L)	0.0037	0.0005	0.0039	0.0005	0.408
Chloride (mg/L)	8.895	0.1524	8.896	0.1252	0.983
Sulfate (mg/L)	22.202	0.5447	22.013	0.7680	0.501
Total Kjeldahl Nitrogen (mg/L)	0.103	0.0497	0.111	0.0449	0.681
Calcium (mg/L)	36.354	0.6314	36.509	0.6842	0.578
Magnesium (mg/L)	11.394	0.1568	11.432	0.1345	0.542
Sodium (mg/L)	5.278	0.0947	5.283	0.0611	0.869
Potassium (mg/L)	1.213	0.0129	1.220	0.0329	0.542

3.2. SUMMER SURVEY

Specific conductivity and chloride exhibit no significant difference between networks. Calcium was the only parameter to display a significant ($p < 0.024$) difference between networks (Table 2). The absolute difference was small, and may be due to the "whiting" phenomenon of precipitation of calcium carbonate during summer. The only nutrient to demonstrate a large relative difference between sampling networks was silicon, and this may be explained by a group of three stations with elevated concentrations. The parameters used for trophic status evaluations were comparable between networks. Again, neither sampling network results in a consistently higher or lower variability of measures for the parameters.

3.3 LONG-TERM TRENDS

The GLNPO monitoring program assesses trophic status of the Great Lakes on an annual basis. Annual data, shown as box plots of station means (Figure 2), indicates the gradual, but significant (Spearman rank: $p < 0.001$) decrease in total phosphorus concentration in the offshore region of the lake. This decrease, estimated using the non-parametric Sen's slope estimator is approximately 0.05 μg/L per year. This value is much less, on an annual basis, than the 20 percent inter-annual change that would be detectable with the GLNPO sampling grid. Long-term sampling, on an annual basis, at the same stations, provides us with sufficient data to detect the trend.

Table 2. Comparison of fifteen measurements from summer epilimnion, 1992, Lake Michigan GLNPO (n = 11) and EMAP (n = 12) stations.
**indicates statistical significance*

Parameter	*GLNPO (mean)*	*S.D.*	*EMAP (mean)*	*S.D.*	*t-probability*
Alkalinity (mg/L)	107.3485	1.7424	104.6250	13.2517	0.507
Specific Conductance mmho/cm	280.6970	2.8243	281.9583	2.0721	0.233
Turbidity (FTU)	0.9392	0.4422	0.7133	0.3098	0.168
Chlorophyll a (mg/L)	1.8282	1.4458	1.3108	0.6850	0.279
Dissolved Organic Carbon (mg/L)	1.4762	0.0675	1.5488	0.1542	0.165
Si (mg/L)	0.1004	0.0505	0.0841	0.0298	0.351
NO3 (mg/L)	0.1486	0.0349	0.1378	0.0349	0.468
Total Dissolved Phosphorus (mg/L)	0.0020	0.0004	0.0020	0.0007	0.944
Total Phosphorus (mg/L)	0.0035	0.0011	0.0035	0.0009	0.909
Chloride (mg/L)	9.7621	0.2738	9.5624	0.3694	0.159
Total Kjeldahl Nitrogen (mg/L)	0.1842	0.0301	0.1834	0.0552	0.964
Calcium (mg/L)	33.9000	0.4893	34.5617	0.7711	0.024*
Magnesium (mg/L)	10.7436	0.0890	10.8275	0.1433	0.110
Sodium (mg/L)	5.0982	0.0890	5.1258	0.1259	0.553
Potassium (mg/L)	3.4364	0.1167	3.4400	0.1382	0.947

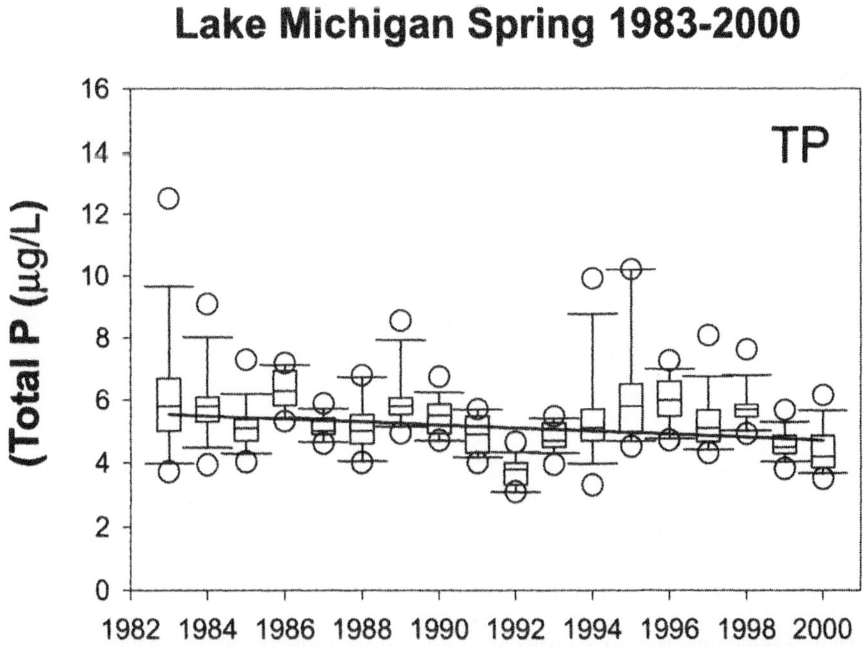

Figure 2. Lake Michigan total phosphorus (TP) concentrations (µg P/L), 1983–2000. Sen (1968) slope indicates a significant long-term decrease

If one assumes that the GLNPO data are representative of what would be gathered by sampling the EMAP-GL grid, an evaluation of the EMAP graphical reporting format, for long term trends, can be made. For trend determination in EMAP, only samples taken at the same locations (every fifth EMAP year) are compared (Figure 3). Cumulative frequency distributions (cfd) from 1983 through 1999 do not indicate a long-term trend in total phosphorus in Lake Michigan. For example, 1983 and 1995 cfd's have two or more total phosphorus values above the 7 µg/L target set for the lake. Viewing these data, separated by four years, does not inform environmental manager of what is occurring in the lake.

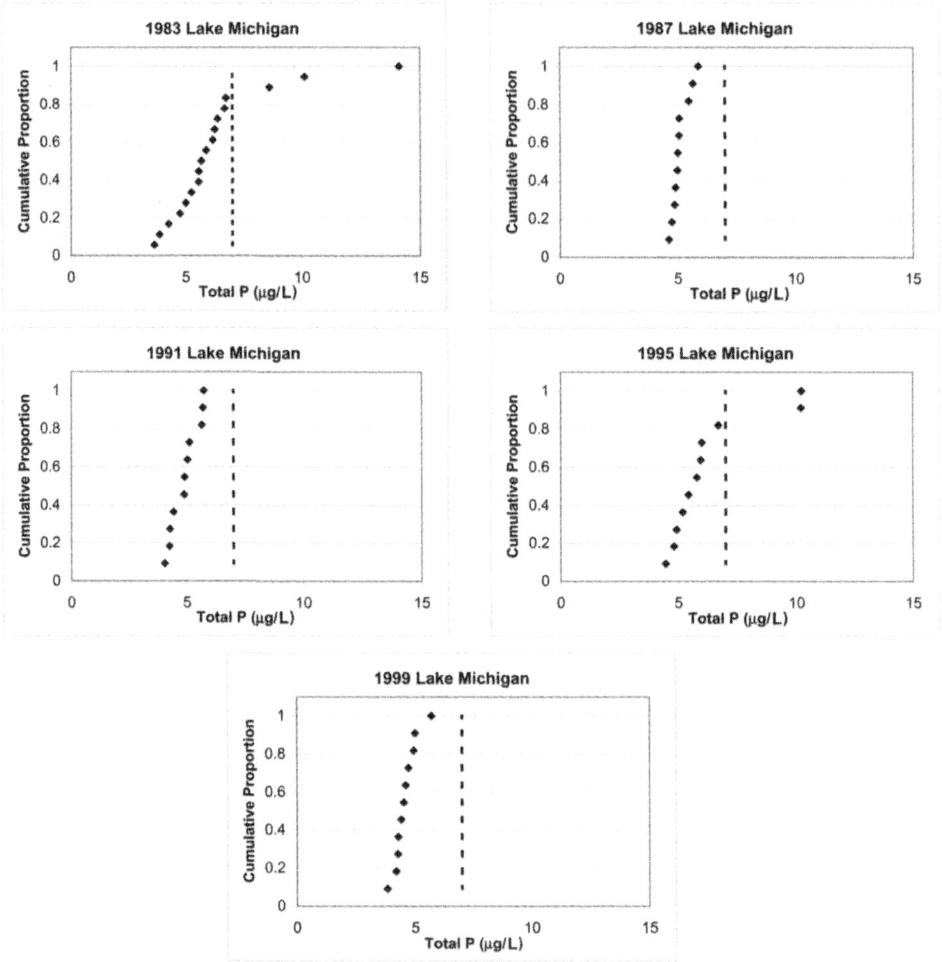

Figure 3. Cumulative frequency distributions of total phosphorus in the offshore region of Lake Michigan. Comparison of four year intervals as if reporting on EMAP stations revisited during the normal base grid rotation. Dashed vertical lines indicate the 7 µg/L total phosphorus target for Lake Michigan. Diamond symbols are station average total phosphorus concentrations

3.4 ADEQUACY OF GRID DENSITY—OFFSHORE

The offshore EMAP grid (base grid) has been compared qualitatively to the GLNPO network and it was noted that the density is roughly the same. The means and standard deviations from Table 2 were used to check the required sample size and minimum detectable difference for all parameters using the GLNPO data quality objective of detecting an inter-annual difference of 20% of the mean for total phosphorus with type one error alpha = 0.1 and beta = 0.2. These tests were repeated for beta = 0.1. The results are presented in Table 3. The results provide support for the adequacy of the EMAP grid for total phosphorus and nitrate plus nitrite. The minimum detectable differences for two levels of beta show the magnitude of the difference from the mean that the grid for 1992 can detect. For chlorophyll a and dissolved reactive silica, this difference is greater than 20%. This is due, in part to the variability of these parameters as they relate to primary production, even in the spring.

Table 3. Required n and Minimum Difference for EMAP Grid

c = 0.1	*CHA (og/L)*	*TN (mg/L)*	*TP (mg/L)*	*SI (mg/L)*
Variance	0.9661	3.935e-1	2.997e-7	9.113e-3
VAR/DSQR	5.842	0.05171	0.5523	0.6644
DIFF, fl = .2	0.635	0.0128	0.000353	0.0617
DIFF, fl = .1	0.774	0.0156	0.000430	0.0751
n	25	<12	<12	14

4. Discussion

A comparison of data from EMAP and GLNPO stations serves as more than a simple test of data comparability over time. It is a test of the comparability of a deterministic sampling design versus a probability-based design. The basis of the current GLNPO sampling network for Lake Michigan is intensive monitoring work preformed during 1976 and 1977. Several (12 in the south basin and 5 in the north basin) surveys were performed on a seasonal basis to determine the overall spatial and temporal variability of the chemistry and biology of the lake (Rockwell, *et al.*, 1980). Approximately 104 sites were sampled on each survey. Subsequent analysis (Rockwell et.al., 1980; Lesht 1984) of the survey data indicated a relatively homogeneous offshore area which could, for operational purposes, be defined as the 90 meter depth contour. Additional analysis of the data (Lesht, 1984) led to a sampling network somewhat larger than that currently used. The current GLNPO network was developed by consideration of the inter-station sample variability and a data quality objective of an 80% ability to detect an interannual 20% change in trophic status at the 90% confidence level. Our current network gives us that ability.

The intensive surveys in 1976–77, and development of the subsequent sampling network provide information crucial to the definition of resource classes as required by EMAP. The considerable effort that has gone into the current GLNPO sampling network, and the level of knowledge of the offshore chemistry and biology derived from it, provide an ideal resource in which to compare the model based and probability based sampling designs.

The offshore area of Lake Michigan, deeper than 85 to 90 meters, is in a well-mixed condition for much of the winter and spring (Mortimer, 1971). While there are two or three subbasins in the lake, based on general circulation patterns, the differences among these basins, in nutrients and biological productivity are rather small (Rockwell et. al., 1989). For modelling purposes, during this time, the lake is often treated as a one to three segment mixed reactor. During conditions of thermal stratification, the lake is more readily divided into nearshore and offshore areas, with boundaries that can be defined by depth as well as other physical characteristics having to do with circulation and internal wave patterns similar to oceanic coastal phenomenon.

Given the well mixed nature of the offshore area, especially during the spring, the comparison of EMAP and GLNPO station networks presents a nearly optimum test of a statistically based and a model based sampling design. There are few situations where one could sample from a population as homogeneous as this one. The results from the study indicate that the EMAP and GLNPO sampling networks yield similar data on trophic status indicators, in most cases parameters are not statistically different. Given the amount of effort that went into the development of the current GLNPO, deterministic, network this agreement is exceptional. However, the EMAP population definition for the offshore region benefited from the historical data available from GLNPO and associated sampling efforts. The fact that there are differences between networks for some paramaters, and that these differences may be related to a bias in sampling one part of the lake, indicates that the EMAP grid may be less than perfect in the Lake Michigan case. Additional comparison work could reduce this problem.

5. Conclusions and Recommendations

Based on similar physical criteria, both the GLNPO and EMAP networks appear to give a reasonable assessment of the offshore of Lake Michigan. The differences in average values of chemical parameters between the networks is small in absolute value, although sometimes statistically significant. For expansion of the EMAP program into the Great Lakes, additional comparison studies between the GLNPO and EMAP networks should be conducted, and for the full four year EMAP rotation. If we are to continue to use chlorophyll a as an indicator of trophic state, the spatial intensity of sampling should be further evaluated.

6. References

Hedtke, S., Pilli, A., Dolan, D.., McRae, G., Goodno, B., Kreis, R., Warren, G., Swackhamer, D., and Henry, M. 1992. Great Lakes Monitoring and Research Strategy: Environmental Monitoring and Assessment Program. USEPA, Office of Research and Development, ERL-Duluth, Duluth, Minnesota. EPA/602/R-92/001. 204 pp.

International Joint Commission. 1980. Phosphorus Management for the Great Lakes. Final Report of the Phosphorus Management Strategies Task Force. 125 pp.

Lesht, B.M. 1984. Lake Michigan Database Analysis and Model Development. In: Radiological and Environmental Research Division Annual Report, January–December, 1980, Report ANL 80–115, Part III, Argonne National Laboratory, IL, pp. 53–57.

Lesht, B.M., and Rockwell, D.C. 1985. The State of the Middle Great Lakes: Results of 1983 Water Quality Survey of Lakes Erie, Huron, and Michigan. Report ANL/ER-85-2, Environmental Research Division, Argonne National Laboratory, Argonne, IL, 170 pp.

Lesht, B.M., and Rockwell, D.C. 1987. The State of the Middle Great Lakes: Results of 1984 Water Quality Survey of Lakes Erie, Huron, and Michigan. Report ANL/ER-87-1, Environmental Research Division, Argonne National Laboratory, Argonne, IL, 141 pp.

Mortimer, C.H. 1971. Large scale oscillatory motions and seasonal temperature change in Lake Michigan and Lake Ontario. Special Report No. 12, Center for Great Lake Studies, University of Wisconsin, Milwaukee, WI, Part 1. 111pp.

Rockwell, D.C., DeVault, D.S., Palmer, M.F. Marion, C.V., and R.J. Bowden. 1980. Lake Michigan Intensive Survey 1976–1977. Report EPA-905/4-80-003-A, Great Lakes National Program Office, U.S. Environmental Protectio Agency, Chicago, IL, 155 pp.

Rockwell, D.C., Salisbury, D.K., and Lesht, B.M.. 1989. Water Quality in the Middle Great Lakes: Results of the 1985 USEPA Survey of Lakes Erie, Huron and Michigan. Report EPA-905/6/89-001, Great Lakes National Program Office, U.S. Environmental Protection Agency, Chicago, IL, 207 pp.

Sen, P.K. 1968. 'Estimates of the regression coefficient based on Kendall's tau', *Journal of the American Statistical Association* **63**: 1379–1389.

A HYDROLOGIC NETWORK SUPPORTING SPATIALLY REFERENCED REGRESSION MODELING IN THE CHESAPEAKE BAY WATERSHED

JOHN W. BRAKEBILL[1*] AND STEPHEN D. PRESTON[2]

[1]U.S. Geological Survey, Baltimore, MD, USA;
[2]U.S. Geological Survey, Chesapeake Bay Program, Annapolis, MD, USA
[*](author for correspondence, email: jwbrakeb@usgs.gov)

Abstract. The U.S. Geological Survey has developed a methodology for statistically relating nutrient sources and land-surface characteristics to nutrient loads of streams. The methodology is referred to as SPAtially Referenced Regressions On Watershed attributes (SPARROW), and relates measured stream nutrient loads to nutrient sources using nonlinear statistical regression models. A spatially detailed digital hydrologic network of stream reaches, stream-reach characteristics such as mean streamflow, water velocity, reach length, and travel time, and their associated watersheds supports the regression models. This network serves as the primary framework for spatially referencing potential nutrient source information such as atmospheric deposition, septic systems, point-sources, land use, land cover, and agricultural sources and land-surface characteristics such as land use, land cover, average-annual precipitation and temperature, slope, and soil permeability. In the Chesapeake Bay watershed that covers parts of Delaware, Maryland, Pennsylvania, New York, Virginia, West Virginia, and Washington D.C., SPARROW was used to generate models estimating loads of total nitrogen and total phosphorus representing 1987 and 1992 land-surface conditions. The 1987 models used a hydrologic network derived from an enhanced version of the U.S. Environmental Protection Agency's digital River Reach File, and course resolution Digital Elevation Models (DEMs). A new hydrologic network was created to support the 1992 models by generating stream reaches representing surface-water pathways defined by flow direction and flow accumulation algorithms from higher resolution DEMs. On a reach-by-reach basis, stream reach characteristics essential to the modeling were transferred to the newly generated pathways or reaches from the enhanced River Reach File used to support the 1987 models. To complete the new network, watersheds for each reach were generated using the direction of surface-water flow derived from the DEMs. This network improves upon existing digital stream data by increasing the level of spatial detail and providing consistency between the reach locations and topography. The hydrologic network also aids in illustrating the spatial patterns of predicted nutrient loads and sources contributed locally to each stream, and the percentages of nutrient load that reach Chesapeake Bay.

Keywords: SPARROW, Chesapeake Bay, nutrients, loads, regression modeling, reach, stream, RF1, DEM

1. Introduction

SPAtially Referenced Regressions On Watershed attributes (SPARROW) use a nonlinear statistical method that defines relations among upstream nutrient-sources, downstream nutrient loads, and the land-surface characteristics that potentially affect nutrient delivery to streams. The SPARROW methodology provides a statistical basis for estimating stream-nutrient loads (predictions) as well as additional spatial detail on environmental factors and transport processes included in the regression models. This method, developed by the U.S. Geological Survey (USGS) at a national scale, addresses data-interpretation limitations of regional water-quality assessments caused by sparse sampling, sampling bias, and basin

Environmental Monitoring and Assessment **81**: 73–84, 2003.

heterogeneity (Smith *et al.*, 1997). In the Chesapeake Bay watershed (Figure 1), a 165,000 km^2 (square-kilometer) watershed that covers parts of Delaware, Maryland, New York, Pennsylvania, Virginia, West Virginia, and Washington, D.C., the regression models are used to explore the application at regional scales, while providing resource managers with useful information about nutrients entering Chesapeake Bay and its tributaries (Preston and Brakebill, 1999).

SPARROW models estimating nutrient loads of total nitrogen and total phosphorus were developed in the Chesapeake Bay watershed representing 1987 and 1992 land-surface conditions. These models estimate the amount of total nitrogen and total phosphorus generated locally in tributary watersheds, as well as how much of the nitrogen and phosphorus generated locally reaches the Chesapeake Bay, taking into account the amount of in-stream decay that may occur during transit (Preston and Brakebill, 1999).

Supporting the models is a geographically referenced hydrologic network of connected streams and associated watersheds. The stream network is based on an enhancement of the U.S. Environmental Protection Agency (USEPA) digital River Reach File (RF1) (U.S. Environmental Protection Agency; 1996; Alexander *et al.*, 1999). RF1, a vector database based on 1:500,000-scale mapping containing locations of stream reaches and associated characteristics, defines a stream reach with a single line representing surface-water pathways that extends either from headwater to stream junction, or from one stream junction to another stream junction. Enhancements to the USEPA RF1 database (ERF1) improved stream-reach characteristics such as mean water velocity, streamflow, time of travel, and reservoir information that are necessary to calculate stream-channel transport parameters for the SPARROW models (Smith *et al.*, 1997; Alexander *et al.*, 1999).

Watershed boundaries for each stream-reach provide the basis for spatially referencing potential nutrient sources, land-surface characteristics, and nutrient predictions. Atmospheric deposition, septic systems, point-sources, land use, land cover, and agricultural sources such as applied commercial fertilizer and manure are all potential nutrient sources examined by SPARROW. Land use, land cover, average-annual precipitation and temperature, slope, and soil permeability are land-surface characteristics that potentially affect nutrient transport and delivery. By retaining spatial referencing, the geographical distribution and relative contribution of nutrient sources and the factors that affect nutrient transport can be examined at various scales (Preston and Brakebill, 1999; Brakebill and Preston, 1999).

The 1:500,000-scale ERF1 data and watersheds generated from derivative products of a 1-km (kilometer) cell Digital Elevation Model (DEM) (U.S. Geological Survey, 1997) provided the basis for the network supporting the 1987 models. These data presented limitations in spatial location and detail, but possessed stream-reach characteristics necessary for the models. The limitations in the characterized stream-reach data and higher resolution elevation data on a regional scale

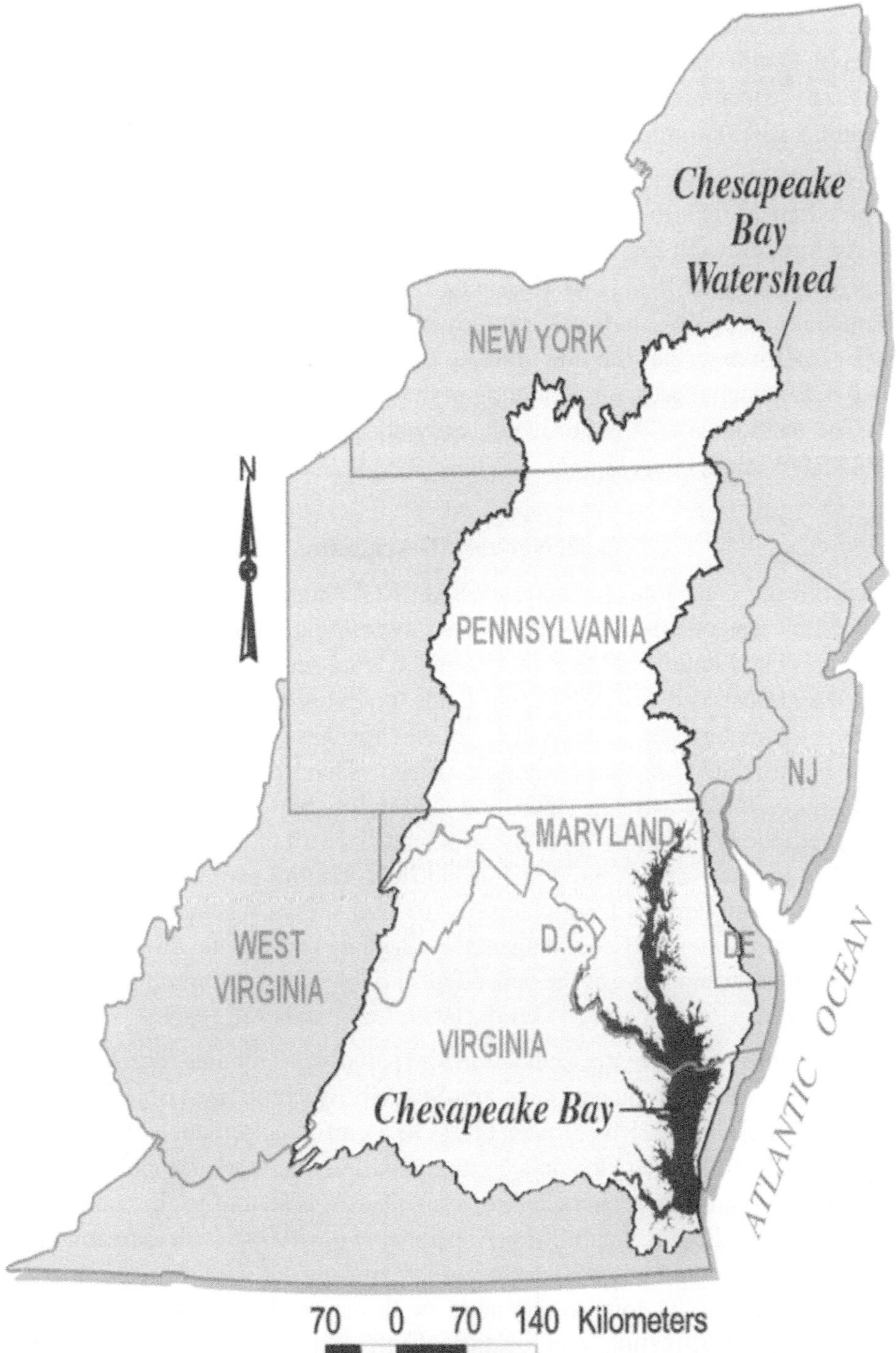

Figure 1. Location of the Chesapeake Bay watershed.

limited the accuracy and precision of the 1987 network. Improvements to stream locations and their associated watersheds were warranted in order to increase the level of spatial detail in areas where stream reaches were not present in the RF1 data set (coastal areas) and to improve the continuity between the stream-reach locations and topographic features such as elevation and slope.

The 1992 models introduced improvements to the network by applying more detailed DEM data to recreate a stream-reach network and its associated watersheds. Stream-reach characteristics (mean streamflow, velocity, and time of travel) from the enhanced RF1 (ERF1) data and seamless 30-m (meter) DEMs are key elements of the 1992 network's improved utility (Brakebill *et al.*, 2001). This paper briefly describes the functionality of a hydrologic network supporting spatially referenced regression modeling in the Chesapeake Bay watershed (Section 2). The methods used to improve the network and examine the implications for SPARROW model development are described in Section 3.

2. Network Description

Total nitrogen and total phosphorus SPARROW models applied in the Chesapeake Bay watershed utilize a connected hydrologic network of stream reaches and associated characteristics and watersheds to geographically reference predicted and observed stream-nutrient loads, potential nutrient sources, and land-surface characteristics to individual stream reaches. Nitrogen and phosphorus sources associated with an individual stream reach, the relative contribution of each source to downstream loads, and the relation between the model's nutrient load predictions can all be spatially evaluated since all of the information is referenced to the network. Annual stream nutrient-loading estimates, referred to as "observed", are derived from water-quality and stream-discharge data collected by numerous State and Federal agencies. The data provide the downstream loading information and serve as the dependent variable in the calibration of the SPARROW models (Preston and Brakebill, 1999; Langland *et al.*, 1995).

Stream-reach data are the foundation for spatially referencing nutrient load (observed), source, and transport data, and SPARROW predictions of nutrient loads. Based on the Enhanced River Reach File (ERF1), an enhancement of the 1:500,000-scale USEPA RF1, the reach data contains spatially networked topological properties and stream-reach characteristics (attributes) essential to the model's calibration and prediction capabilities (Alexander *et al.*, 1999). Every stream reach is consistently oriented in the direction of streamflow and is connected to at least one other reach at its downstream node. Nodes are endpoints of lines that maintain the identity, direction, and location of intersected linear features. This topological information is used to define reach-to-reach connectivity and allows for the identification of each reach upstream or downstream of any location along the stream network [Environmental Systems Research Institute (ESRI), 1992a]. Linear connection is important because it allows for systematic climbing upstream from each monitoring station, aggregating the observed load to every reach upstream to the next monitoring station. This aggregation ensures that each reach upstream of a monitoring station receives representation of the observed load from the downstream location. In the model prediction phase, the design of the network also permits the accumulation of nutrients from each reach as they move downstream toward Chesapeake Bay.

Important stream-reach characteristics include estimates of mean streamflow and water velocity, reach length and travel time (calculated as a ratio of reach length over mean stream-water velocity), and a unique reach identification number. Because SPARROW statistically relates upstream nutrient sources to observed downstream loads, the methodology takes into account nutrient losses due to natural in-stream processes such as sedimentation and denitrification. In-stream nutrient-loss rates, which can vary by stream size, are estimated statistically for stream classes using the mean annual streamflow and travel time values from the stream-reach data (Smith *et al.*, 1997). The unique identification number is used to associate all load, source, transport, and prediction information to each reach. A unique number identifies each reach as a single unit, and allows the aggregation of nutrient source, transport, and prediction data to these units.

Associated watershed boundaries generated from coarse resolution DEMs for each stream reach provides the spatial detail of the 1987 SPARROW models (Brakebill and Preston, 1999). Whereas the stream-reach network provides the linear connection of surface-water pathways, watersheds attributed with a unique identification number of their associated reach provide the ability to spatially reference nutrient source, load, transport, and load prediction data for each reach on an area basis. Nutrient-source data merged with the watersheds produce load values for each reach watershed and are used as input values for the models. Land-surface characteristics also are merged with the watersheds to produce average watershed characteristics for each stream reach. Nutrient sources, and predicted nutrient loads displayed by individual watersheds as yields, are evaluated for spatial distribution, relative importance, and potential for delivery to Chesapeake Bay.

3. Network Improvements

Limitations in the ERF1 stream-reach data and the coarse resolution of the DEMs used to generate associated watersheds from the 1987 network necessitated further evaluation. The ERF1 stream reaches poorly represent the actual locations of stream channels when compared to other topographic features such as elevation and slope. Stream-reach densities varied throughout the Chesapeake Bay watershed due to different mapping procedures used to compile the data. Coastal area streams along major estuaries are not represented; in the Potomac River estuary, only major tributaries draining directly into the estuary are represented in the

ERF1 data set. This representation could exclude potential nutrient point sources such as sewage treatment plants within the current network. Single lines in the center of major estuaries are present in the original RF1 and were not used because they represent tidal processes that were too complex to incorporate into the calibration phase of the models. Errors in the stream-network connectivity and the accuracy in stream locations at confluences also were a concern.

When the 1987 models were developed, the 1-kilometer cell DEM acquired to generate watershed boundaries for each reach was the only consistent source of topographic information available on a regional scale (U.S. Geological Survey, 1997). Other elevation data at various resolutions were considered, but required extensive processing time or cost for compilation.

In order to improve the network for the 1992 Chesapeake Bay SPARROW models, a method was developed that increased the level of spatial detail in areas where stream reaches were not present while improving the continuity among the stream-reach locations, watershed boundaries, and topographic features. This method required the construction of synthetic stream reaches and watershed areas generated using seamless DEMs from the USGS National Elevation Data (NED) (U.S. Geological Survey, 1999a), and the transfer of existing stream-reach characteristics from the ERF1 file (Brakebill *et al.*, 2001).

3.1 STREAM REACHES

Improvements to the stream-reach network utilized seamless 30-meter cell DEMs to generate the water pathways that would make up the stream-reach network (U.S. Geological Survey, 1999a). Components key to constructing water pathways are the direction of surface-water flow from each elevation cell (flow direction), and the accumulation of cells flowing into any given cell (flow accumulation). Flow direction represents the steepest downslope direction water on a surface will flow. Once this direction is known, the identification and the number of cells flowing into a given cell can be calculated, and used to generate stream networks and watershed boundaries (Environmental Systems Research Institute, 1992b).

Flow direction from each 30-meter elevation cell within the Chesapeake Bay watershed was calculated to one of its eight adjacent or diagonal neighboring cells. Flow accumulation was calculated for each 30-meter cell using the flow-direction information. Based on the flow-accumulated cells, a streamwater pathway was generated by applying a threshold of 5,000 cells that will flow into any single cell. Any cell with more than 5,000 cells flowing into it represents a water pathway, or stream reach. The number 5,000 was chosen as a threshold because it yielded desirable pathways that were comparable in scale to the ERF1 stream reaches. This threshold was necessary so that attributes from the ERF1 data could be used with the newly created stream reaches, simplifying the transfer of attributes from ERF1.

The newly generated stream reaches were converted from a raster form to a vector form. Because the synthetic reaches were generated using elevation information, this process maintained the directional topology of the reaches, such that the direction of the topology was consistent with the direction of streamflow, allowing for the movement upstream and downstream of any given reach. Positional accuracy and density of the reach data were evaluated by comparing them to the USGS National Hydrography Dataset (NHD) 1:100,000-scale stream data (U.S. Geological Survey, 1999b). Where the DEMs failed to yield satisfactory stream reaches (typically in flat coastal areas or near wide rivers, lakes, and reservoirs), or to improve the distribution of reaches where ERF1 data did not exist, the NHD vector data were inserted.

The synthetic reach generation process created more water pathways than were necessary to build the improved network. Stream reaches corresponding to the scale of the ERF1 stream-reach data were identified and selected out of the data set to produce a subset of generated synthetic stream reaches. Each new reach was attributed with the same unique reach identifier that corresponds to the ERF1 reach, creating a one-to-one relation between the two data sets. This relation provided the means to transfer stream-reach characteristics from one data set to another. A comparison of location among ERF1, topography, and the newly generated stream reaches using DEMs is shown in Figure 2.

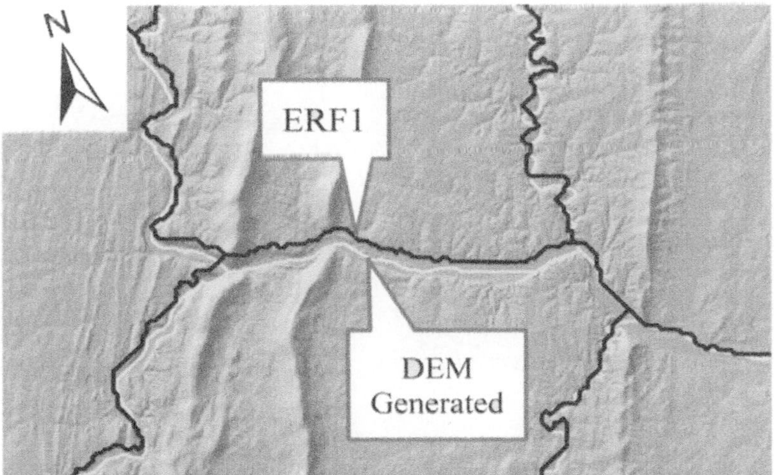

Figure 2. Spatial relation among newly generated stream-reaches using 30-meter DEM (Digital Elevation Models), ERFI reaches, and topographic features.

To ensure that load estimates used for model calibration were referenced to the downstream end of a reach, a node was placed at each streamflow sampling location. By adding a node at the sampling location, the reach was segmented into two separate stream reaches, creating a new reach upstream of the sampling site and maintaining the reach connectivity. This segmentation ensured that a watershed from each sampling site location along its associated reach would be generated, allowing for the distribution of observed loads upstream of the sampling sites.

3.2 WATERSHEDS

Watersheds are important to the SPARROW models because they provide the ability to reference nutrient sources and predicted results in spatial detail. To improve the resolution in coastal areas for spatial referencing, nodes were placed at arbitrary locations along the estuary shoreline boundaries, creating new reach segments. This improved resolution ensures that during the watershed generation process, areas draining directly to the coastal estuaries would be included in the network. These new segments represent the end of the transport processes to the estuaries, and only are used to spatially reference and display potential nutrient sources and predicted results (Figure 3).

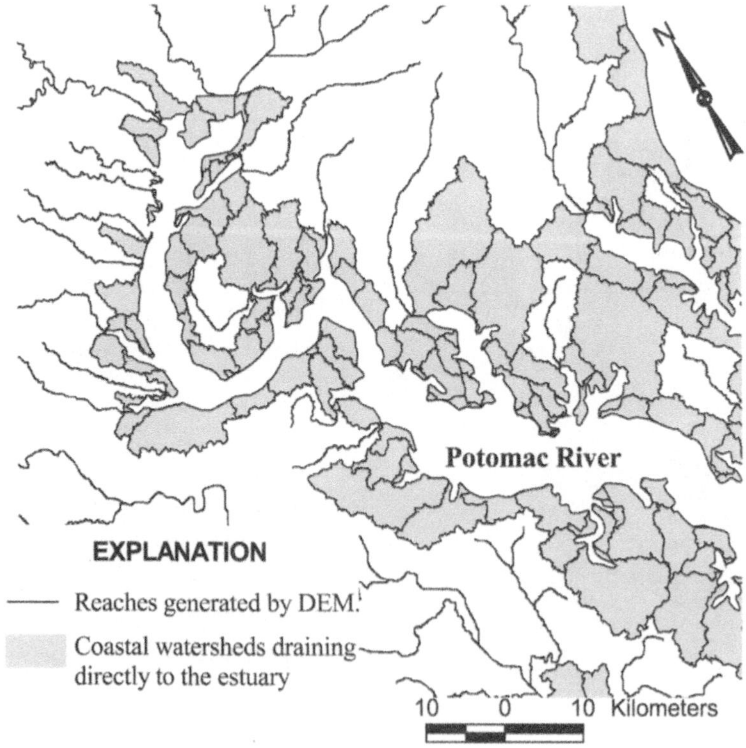

Figure 3. Coastal watersheds that drain directly into the Potomac River estuary.

The 30-meter flow-direction data and the new attributed synthetic-reach network were used to generate 2,249 watershed drainage areas for each stream reach and segmented shoreline boundary. This result was accomplished by converting the reach network back into a 30-meter grid using the unique reach identifier number to populate multiple cells that represent a stream reach with the same corresponding unique value. Watershed areas for each reach were generated using all reach cells, which represent the water pathway, or the lowest points within the watershed (Environmental Systems Research Institute, 1992b). In this method, all cells that represent a single reach are used as locations for the watershed generation process to begin, instead of a single cell representing the absolute lowest point on the downstream end of a reach. This method also maintains the unique reach value from the associated reach network in the watershed data set, serving as an identification tool as well as a common field to related data sets.

The watersheds generated for the 1992 SPARROW models represent a vast improvement from the 1987 network. Finer resolution elevation data improved the locations of the boundaries with respect to the topography, and eliminated noticeable boundary generalizations (Figure 4).

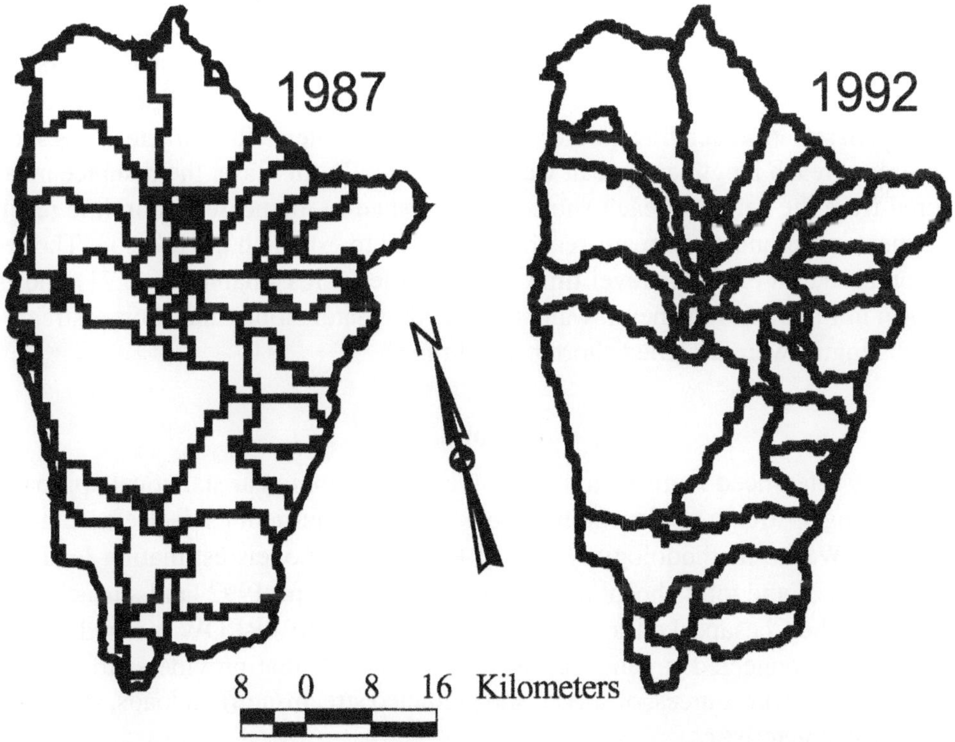

Figure 4. Comparison watersheds of the 1987 network based on a 1-km DEM (Digital Elevation Model) and the 1992 network based on a 30-m DEM.

3.3 ATTRIBUTED NETWORK

ERF1 contains stream-reach characteristics essential to the SPARROW modeling applications. This information is necessary to calculate stream-channel transport parameters used in by models. ERF1 is the main component of the hydrologic network supporting SPARROW, and will continue to be a valuable asset to modeling networks until better methods for estimating necessary stream characteristics are developed (Smith et al, 1997; Alexander *et al.*, 1999).

Each newly generated stream reach was populated with ERF1 stream characteristics, creating an attributed stream network. This process utilized the unique reach identification number in both data sets as a common field. Characteristics necessary to the model's functionality then were transferred from ERF1 to the new stream-reach data set. These characteristics include mean water velocity, mean streamflow, and travel time for reaches within reservoirs. Travel time for stream reaches not within reservoirs was calculated separately for each reach, as was reach length. This calculation included reaches with nodes placed at sampling site locations. Once the attributes were transferred and travel times calculated, the new reaches upstream of the sampling sites were issued a new unique identification number and attributed with the sampling site identification number of their immediate downstream site.

Travel time values for shoreline watersheds that drain directly to major estuaries were estimated by establishing a relation between travel time and watershed characteristics (Figure 5). This relation was based on data from coastal drainages for which stream-reach travel time was defined and watershed characteristics such as stream-reach length and mean watershed slope. In this case, the distance measured from the center of each shoreline watershed to the nearest shoreline reach segment represents the stream-reach length used to establish the relation. The relation based on existing travel time data provided a reasonable R^2 (0.71) and is shown in Figure 5. In general, watersheds with greater slopes and shorter stream-reach lengths were assigned shorter travel times.

4. Summary

Spatially referenced regression modeling utilizes a nonlinear statistical approach for relating upstream nutrient sources to downstream nutrient loads. Referred to as SPARROW, the methodology was used to generate models estimating loads of total nitrogen and total phosphorus representing 1987 and 1992 land-surface conditions in the Chesapeake Bay watershed. Supporting SPARROW is a hydrologic network of connected stream reaches and watersheds that provide spatial referencing of nutrient sources, observed and predicted stream-nutrient loads, and land-surface characteristics.

A network supporting the 1987 models was based on stream locations and characteristics from an enhanced version of the USEPA's River Reach File (ERF1) and watersheds generated using a 1-kilometer cell DEM. Although successful, this process produced a network containing general reach locations and watershed boundaries. A new method was developed to generate a hydrologic network supporting the 1992 models that used seamless 30-meter DEMs to generate water pathways representing stream-reach locations and their associated watersheds. This result produced spatial improvements from the 1987 network including water pathways and their associated watershed boundary locations matching topographic features. Stream-reach characteristics necessary for SPARROW's functionality such as mean streamflow and travel time were transferred from the ERF1 data set used in the 1987 network to the newly generated stream reaches. Coastal areas of the network were improved spatially by adding reaches to the data set, generating watersheds for areas that drain directly to coastal estuaries, and estimating necessary travel time information for these areas based on stream reaches with similar watershed characteristics.

$$\text{Travel Time} = \beta_0 + \beta_1 \text{Centroid} + \beta_2 \text{Slope}$$

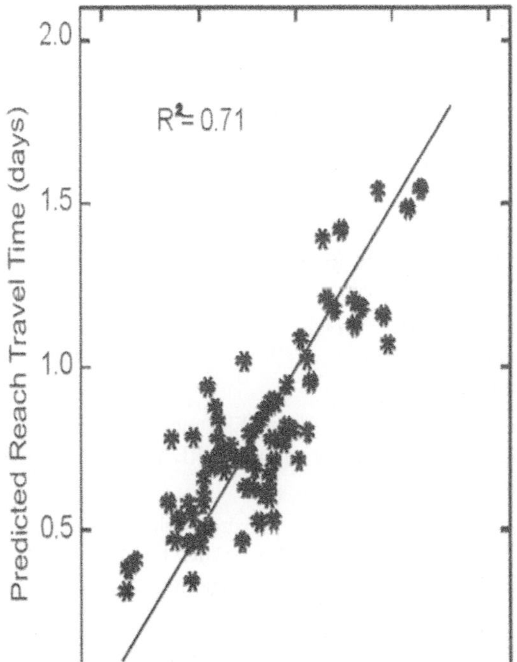

Figure 5. Relation of predicted and observed stream-reach travel time.

5. References

Alexander, R.B., Brakebill, J.W., Brew, R.E., and Smith, R.A.: 1999, Enhanced River Reach File 1.2 (ERF1), U.S. Geological Survey Open-File Report 99-457 (scale 1:500,000) accessed January 1999 at URL http://water.usgs.gov/GIS/metadata/usgswrd/erf1.html

Brakebill, J.W., and Preston, S.D.: 1999, Digital Data Used to Relate Nutrient Inputs to Water Quality in the Chesapeake Bay Watershed, Version 1.0: U.S. Geological Survey Open-File Report 99-60, accessed May 2000 at URL http://md.usgs.gov/publications/ofr-99-60/report.html

Brakebill, J.W., Preston, S.D., and Martucci, S.K.: 2001, Digital Data Used to Relate Nutrient Inputs to Water Quality in the Chesapeake Bay Watershed, Version 2.0: U.S. Geological Survey Open-File Report 01-251, accessed July 2001 at URL http://md.usgs.gov/publications/ofr-01-251/report.html

ESRI (Environmental Systems Research Institute, Inc): 1992a, Arc/Info Users Guide, ARC/INFO Data Model, Concepts and Key Terms, 2d ed.: Redlands, California, 201 pp.

ESRI (Environmental Systems Research Institute, Inc): 1992b, Arc/Info Users Guide, Cell-based modeling with GRID, 2d ed.: Redlands, California, 267 pp.

Langland, M.J., Lietman, P.L., and Hoffman, S.A.: 1995, Synthesis of Nutrient and Sediment Data for Watersheds Within the Chesapeake Bay Drainage Basin: U.S. Geological Survey Water-Resources Investigations Report 95-4233, 121 pp.

Preston, S.D. and Brakebill, J.W.: 1999, Application of Spatially Referenced Regression Modeling for the Evaluation of Total Nitrogen Loading in the Chesapeake Bay Watershed: U.S. Geological Survey Water-Resources Investigations Report 99-4054, 12 p. Available online at URL http://md.usgs.gov/publications/wrir-99-4054/report.html

Smith, R.A., Schwarz, G.E., and Alexander, R.B.: 1997, 'Regional Interpretation of Water-Quality Monitoring Data': *Water Resources Research* **33** (12) 2781–2798, available online at URL http://water.usgs.gov/nawqa/sparrow/wrr97/results.html

USEPA (U.S. Environmental Protection Agency): 1996, Reach File Version 1.0 (RF1) for the conterminous United States: U.S. Environmental Protection Agency accessed January 1998 at http://www.epa.gov/ost/basins/metadata/rf1.htm

USGS (U.S. Geological Survey): 1997, HYDRO1K Elevation Derivative Data Base: U.S. Geological Survey accessed January 1998 at URL http://edcwww.cr.usgs.gov/landdaac/gtopo30/hydro/namerica.html

USGS (U.S. Geological Survey): 1999a, National Elevation Data Base (NED): U.S. Geological Survey Fact Sheet 148–99, available online at URL http://mapping.usgs.gov/mac/isb/pubs/factsheets/fs14899.html

USGS (U.S. Geological Survey): 1999b, National Hydrography Data Base (NHD): U.S. Geological Survey Fact Sheet 106–99, available online at URL http://mapping.usgs.gov/mac/isb/pubs/factsheets/fs14899.html http://mapping.usgs.gov/mac/isb/pubs/factsheets/fs10699.html

THE IMPORTANCE OF CONSIDERING SPATIAL ATTRIBUTES IN EVALUATING ESTUARINE HABITAT CONDITION: THE SOUTH CAROLINA EXPERIENCE

ROBERT F. VAN DOLAH[1*], DAVID E. CHESTNUT[2], JOHN D. JONES[1], PAMELA C. JUTTE[1], GEORGE RIEKERK[1], MARTIN LEVISEN[1], AND WILLIAM MCDERMOTT[2]

[1]*South Carolina Department of Natural Resources, Marine Resources Research Institute, P.O. Box 12559, Charleston SC 29412;* [2]*South Carolina Department of Health and Environmental Control, Bureau of Water, 2600 Bull St. Columbia SC 29201*

[*](author for correspondence, email: *vandolahr@mrd.dnr.state.sc.us*)

Abstract. The South Carolina Estuarine and Coastal Assessment Program (SCECAP) was initiated in 1999 to assess the condition of the state's coastal habitats using multiple measures of water quality, sediment quality, and biological condition. Sampling was subsequently expanded to include components required for the National Coastal Assessment (Coastal 2000) Program. Habitats are classified as either "tidal creeks" (< 100 meters in width) or larger "open water" bodies. Approximately 30 sites are sampled within each habitat during the summer months using a probability-based random sampling design. Results obtained from the first two years of sampling documented significant differences in several water quality parameters (DO, salinity, pH, turbidity, fecal coliform bacteria, total nitrogen, TKN, total phosphorus) and biological measures (chlorophyll-a, finfish and crustacean abundance and biomass and a number of benthic species) between the tidal creek and open water habitats. These differences highlight the value of partitioning shallow water habitats separately from the larger open water bodies traditionally sampled in estuarine monitoring programs, especially since tidal creeks serve as critical nursery areas for many species. Based on the differences observed, there is a clear need to identify different physical and biological thresholds for evaluating the condition of each habitat type.

Keywords: water quality, South Carolina, estuarine monitoring, tidal creek, finfish, benthos, sampling design, watershed scale

1. Introduction

Historical estuarine monitoring programs conducted in South Carolina and other coastal states, as well as those conducted over larger regional areas by NOAA and the USEPA, have generally focused on evaluating conditions in relatively large-scale water bodies such as tidal rivers, bays, and sounds (e.g., SCDHEC, 1997a; Carlton *et al.*, 1998; Summers *et al.*, 1993; Strobel *et al.*, 1995; Hyland *et al.*, 1998). Although some of these programs have included sites in smaller watersheds, such as tidal creeks, the number of stations sampled has been very limited.

These small creek watersheds serve as critical nursery habitats for many finfish and crustacean species as well as wading birds (Hackney *et al.*, 1976; Weinstein, 1979; Shenker and Dean, 1979; Ogburn *et al.*, 1988; Dodd and Murphy, 1996). Additionally, tidal creeks are often the first point of entry of non-point source runoff from upland areas. Therefore, these drainage systems can provide an early indication of anthropogenic stress (Holland *et al.*, 1997; Sanger *et al.*, 1999a,b; Van Dolah *et al.*, 2000).

Environmental Monitoring and Assessment **81**: 85–95, 2003.

In South Carolina, a comprehensive new monitoring program was initiated in 1999 as a cooperative effort between the Department of Natural Resources (DNR) and the South Carolina Department of Health and Environmental Control (SCDHEC). The South Carolina Estuarine and Coastal Assessment Program (SCECAP) significantly expands historical monitoring activities conducted by both agencies in the state's coastal zone and is designed to integrate multiple measures of water quality, sediment quality, and biological condition at a large number of sites sampled each year throughout the state. Due to the importance of tidal creeks as critical estuarine habitats, the program includes an equal level of effort in monitoring conditions in creeks versus the larger open water bodies, even though our analysis of the coastal hydrography indicates that tidal creeks represent only about 17% of the state's estuarine waters.

This paper compares several measures of water quality and biological condition observed in tidal creeks versus larger water bodies during the first two years of SCECAP. The differences we observed highlight the value of including these smaller watersheds in estuarine monitoring programs, and partitioning habitats by spatial scale when interpreting the data.

2. Sampling Design and Methods

Approximately 60 stations were sampled each year throughout the coastal regions of the state (30 in creeks, 30 in larger water bodies) using a probability-based, random tessellation, stratified sampling design for each habitat type (Stevens, 1997; Stevens and Olsen, 1999). New stations were selected each year. Tidal creeks are defined as those estuarine water bodies less than 100 meters in width from marsh bank to marsh bank. Open water habitats include all other larger water bodies. Station locations were recorded using a differentially corrected Global Positioning System (GPS). Sites were sampled within ± 3 hours of mean low water and had a minimum depth of 1 m at low tide to qualify for sampling. All data reported here were obtained during the summer months of 1999 and 2000 (primarily July and August).

Water quality measurements and samples were collected at all stations prior to deployment of other sampling gear. Both instantaneous (surface, mid, bottom) and 25-hour time-series measurements (bottom only) of dissolved oxygen, salinity, and temperature were obtained at each site. Only the bottom time-series data are presented here and were obtained using either Yellow Springs Instruments, Inc. Model 6920 multiprobes or Hydrolab DS-3 and DS-4 datasondes. Measurements were logged at 15-minunte intervals to obtain a record over two complete tidal cycles.

Water quality samples included measures of near-surface concentrations of total nitrate/nitrite nitrogen, total Kjeldahl nitrogen, ammonia, total phosphorus, total organic carbon (TOC), turbidity, biological oxygen demand (BOD_5), and fecal coliform bacteria. All samples were collected at a depth of 0.3 meter and stored on ice until they were preserved (nutrient and TOC samples only) for later processing. Laboratory processing of total nutrient samples, total organic carbon, total alkalinity, turbidity, BOD_5 and fecal coliform bacteria were completed at SCDHEC laboratories using standardized procedures described by SCDHEC (1997b, 1997c).

Near-surface (0.3 meter) chlorophyll-a samples were also collected during the water quality sampling effort. Samples collected in 1999 were filtered in the field (three replicate 50 mL aliquots) and the filters were placed in centrifuge tubes containing 25 mL of acetone and $MgCO_3$ and stored on ice in the dark. In the laboratory, the samples were centrifuged and the supernatant analyzed on a Turner Model 10-AU fluorometer. In 2000, the chlorophyll-a samples were measured using processing methods described by SCDHEC (2000). Comparison of the two protocols on 1999 samples showed similar results.

Benthic grab samples (8–10) were collected using a stainless steel 0.04 m^2 Young grab sampler to evaluate sediment composition, contaminant levels, toxicity, and benthic community composition. Three of the grab samples were sieved separately through a 0.5 mm sieve and preserved in a 10% buffered formalin-seawater solution with rose bengal stain. Fauna from two of those samples were identified to the lowest possible taxonomic level and the third sample was archived. Surficial sediments (0–5 cm) of the remaining grab samples were composited to obtain samples for analysis of composition, contaminants, and sediment toxicity. Results of those analyses are not presented here.

Fish and large crustaceans (primarily penaeid shrimp and blue crabs) were collected at each site following the benthic sampling to evaluate community composition. Two replicate tows were made at each site using a 4-seam trawl (18 ft foot rope, 15 ft head rope and 3/4 in bar mesh throughout). Trawl tow lengths were standardized to 0.5 kilometer for open water sites and 0.25 kilometer for creek sites. Tows were made only during daylight hours with the current and speeds standardized as much as possible. Tows were limited to periods when the marsh was not flooded (approx. 3 hrs ± mean low water). Catches were sorted to lowest practical taxonomic level, counted, and checked for gross pathologies, deformities or external parasites (not reported here). All organisms were measured to the nearest 1.0 cm and weighed to the nearest 0.01 kg. When more than 30 individuals of a species were collected, the species was sub-sampled.

Statistical comparisons reported here were generally completed using t-tests on non-transformed or transformed (where necessary) data. A Mann-Whitney U test was used if data transformations did not meet t-test criteria. An analysis of the cumulative distribution function (CDF) of the values observed for each parameter was also performed using procedures described by Diaz-Ramos *et al.* (1996). The CDF analyses provide an estimate of the percent of the state's overall creek and open water habitat that fall within ranges of values selected based either on state water quality criteria, or historical measurements collected by SCDHEC from 1993–1997 in the state's larger open water bodies (SCDHEC, 1998a) when state water quality criteria were not available.

3. Results and Discussion

Sampling was successfully completed at 57 tidal creek sites and 59 open water sites over the two-year period (Figure 1). Average depth of the tidal creek sites was 2.2 meters versus 4.8 meters at the open water sites.

Mean values of the various water quality measures documented several statistically significant differences between the two habitats sampled (Table 1). The average bottom dissolved oxygen concentration in tidal creeks was 4.1 mg/L versus 4.9 mg/L at the open water sites. Both averages were below the state water quality standard (based on a daily average not less than 5.0 mg/L and no values below 4.0 mg/L; SCDHEC, 1998b). About 55% of the open water sites and 98% of the tidal creek sites had minimum DO values less than 4.0 mg/L. Based on the CDF analyses, approximately 46% of the state's tidal creek habitat had mean DO concentrations less than 4.0 mg/L compared to only 9% of the open water habitat (Table 2).

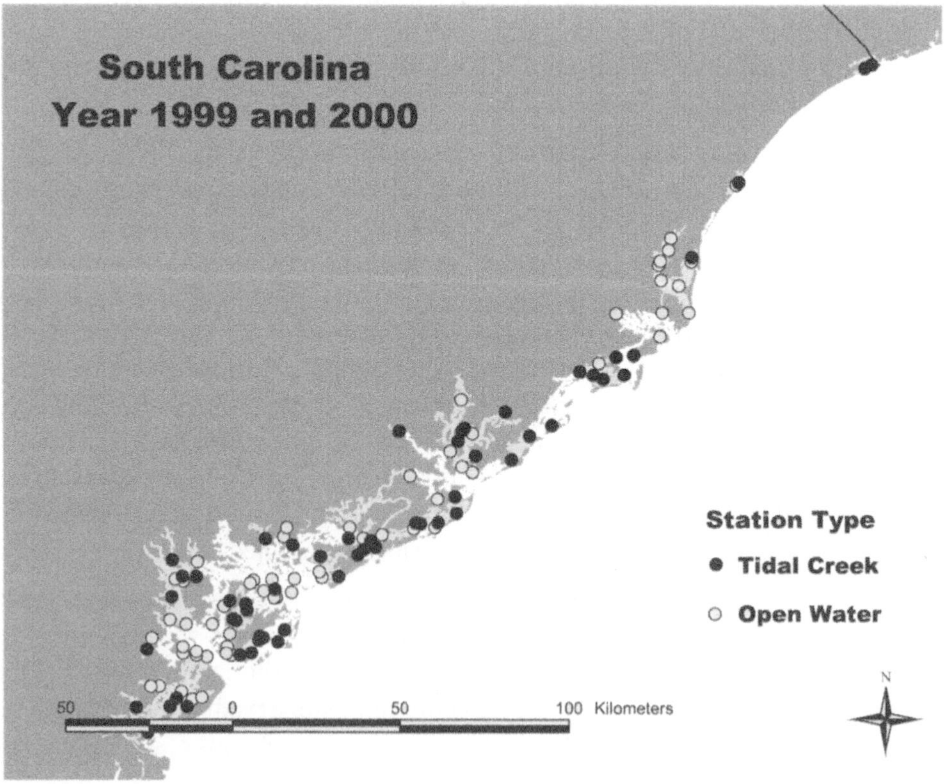

Figure 1. Distribution of SCECAP stations sampled in 1999 and 2000.

Table 1. Summary of mean water quality values collected in tidal creeks and larger open water bodies during 1999 and 2000

Parameter	*Creeks*	*Open*	*Significance*
Dissolved Oxygen (mg/L)	4.1	4.9	$p < .001$
Salinity (ppt)	31.3	27.2	$p = .002$
pH	7.5	7.6	$p = .013$
Temperature (°C)	29.9	29.8	NS
BOD_5 (mg/L)	1.8	1.6	NS
TOC (mg/L)	3.6	4.0	NS
Turbidity (NTU)	21.0	14.2	$p < .001$
Fecal Coliform Bacteria	43.0	27.1	$p = .053$
Total Nitrogen (mg/L)	0.65	0.53	$p = .009$
NO_2/NO_3 (mg/L)	0.02	0.04	NS
Ammonia (mg/L)	0.24	0.15	NS
TKN (mg/L)	0.64	0.48	$p < .001$
Total Phosphorus (mg/L)	0.10	0.07	$p < .001$
Chlorophyll-a (µg/L)	12.78	9.68	$p = .022$

The state DO standard is generally based on monthly surface water measures collected year-round. Summer-time values would be expected to approach or slightly exceed the lower limit. Collection of bottom water data does not significantly affect our conclusion since we observed very little difference between surface and bottom instantaneous DO measures (mean difference = 0.17 mg/L at open water sites and 0.14 mg/L at tidal creek sites).

Since the majority of creek and open water sites sampled in both years were located in relatively pristine environments, there are clearly natural differences in DO concentrations in the smaller- versus larger-scale watersheds and the existing state standards are not very applicable for the purposes of the SCECAP program. New thresholds of potential DO stress during the summer months need to be developed. Ideally, these should be based on biological response data, but until sufficient data are available to identify DO effects levels, we consider a mean < 4.0 mg/L to be indicative of marginal DO conditions (i.e., average is below state lower limit), and a mean < 3.0 mg/L as evidence of poor DO conditions. Sites with mean DO values < 3.0 mg/L had an average of 26% of the time-series records with values < 2.0 mg/L, which is considered to be stressful for most organisms (USEPA, 1999).

Average measures of biological oxygen demand (BOD_5) in creek and open water habitats were very similar (Table 1) and only slightly above the 50th percentile value (1.4 mg/L) of all BOD_5 measures collected state-wide in 1993–1997 by SCDHEC (1998b). Approximately 23% of the state's tidal creek waters and 17% of the open water habitat had BOD_5 values that exceeded the 90th percentile of values observed in the earlier study (Table 2).

Six of the other 11 water quality measures reported here showed significantly higher average values in tidal creeks versus open water habitats and one (pH) showed a significantly lower value (Table 1). The higher turbidity values observed in creeks probably reflects greater tidal mixing in these shallow, well-flushed environments. The higher mean salinity observed in tidal creeks was surprising since these habitats would be expected to have greater freshwater input, lower average

Table 2. Summary of the percent of South Carolina's estuarine habitat that represents various water quality conditions based on cumulative distribution function analyses. See text for selection of criteria used to define value ranges.

		Percent of Habitat	
Parameter	*Criteria*	*Creeks*	*Open*
Dissolved Oxygen (mg/L)	≥ 0 < 3	7	0
	≥ 3 < 4	39	9
	≥ 4 < 5	45	45
	≥ 5	9	46
BOD_5 (mg/L)	≤ 1.8	56	66
	> 1.8 ≤ 2.6	21	17
	> 2.6	23	17
Fecal Coliform Bacteria	> 0 ≤ 4	82	94
	> 43 ≤ 400	16	4
	> 400	2	2
Total Nitrogen (mg/L)	≤ 0.95	88	96
	> 0.95 ≤ 1.29	12	4
	> 1.29	0	0
Total Phosphorus (mg/L)	≤ 0.09	45	80
	> 0.09 ≤ 0.17	47	20
	> 0.17	8	0
Chlorophyll-a (µg/L)	> 0 ≤ 5	6	7
	> 5 ≤ 20	81	90
	> 20 ≤ 60	13	3
	> 60	0	0

salinities, and greater salinity ranges. South Carolina experienced drought conditions during 1999 and 2000 and we would anticipate salinities to be lower in creeks versus the larger water bodies during normal rainfall conditions.

Mean fecal coliform concentrations were significantly higher in tidal creeks compared to open water areas (Table 1). State fecal coliform standards are based on repetitive sampling, and since only one sample was collected at each site, a strict interpretation of the standards is not possible. However, the maximum criteria portion of the standards (43 colonies/100 mL for shellfish harvesting and 400 colonies/100 mL for primary contact recreation), can be used as thresholds for evaluation purposes. Approximately 18% of the state's creek habitat exceeded 43 colonies/100 mL compared to only 6% of the open water habitat (Table 2). Only 2% of either habitat type had fecal concentrations exceeding 400 colonies/100 mL.

The mean total nitrogen (N) value observed in creek habitats was also significantly higher than the average open water concentration (Table 1), but both concentrations were well below the total dissolved N concentration considered to be high (1 mg/L) by Bricker *et al.* (1999). Based on the 5-yr database compiled for South Carolina estuarine water bodies (SCDHEC, 1998a), total N > 0.95 mg/L represents moderately enriched conditions (> 75th percentile) and concentrations > 1.29 mg/L represent highly enriched conditions (> 90th percentile). Approximately 12% of the state's tidal creek habitat was moderately enriched using these criteria compared to only 4% of the open water habitat (Table 2). No sites were highly enriched.

The average total phosphorous (P) concentrations measured in tidal creeks was slightly above dissolved phosphorous concentration reported by Bricker *et al.* (1999) as "high" concentrations (0.1 mg/L), whereas average P concentrations were lower at the open water sites (Table 1). Based on the five-year database available for South Carolina waters, 47% of the state's creek habitat was moderately enriched (> 75th percentile) and 8% was very enriched (> 90th percentile). In comparison, only 20% of the state's open water habitat was moderately enriched (Table 2).

The higher turbidity, fecal coliform, and nutrient values observed in tidal creeks would be expected due to the closer proximity of these habitats to sediment, nutrient and fecal inputs resulting from land runoff, combined with the lower dilution capacities in tidal creek versus larger-scale water bodies. Additionally, the shallower depths and strong tidal flushing typically observed in most creeks would tend to re-mix bottom sediments. This bottom re-mixing may have further influenced turbidity and the nutrient measures since most of the creek samples were collected within 2 m of the bottom. The significant differences observed between tidal creeks and open water habitats suggest that different thresholds of enrichment should be developed for each habitat type.

Biological measures also reflected significant differences between the two types of water bodies sampled. Average chlorophyll-a concentrations were significantly higher in creeks versus open water habitats (Table 1). Additionally, our CDF analysis indicated that approximately 13% of the state's tidal creek habitat had > 20 μg/L of chlorophyll-a (Table 2), which is considered to be high by Bricker *et al.* (1999). This higher chlorophyll concentration may be reflective of the higher nutrient concentrations observed in the creeks. It may also reflect possible re-suspension of benthic algal mats from the creek bottoms and marsh surfaces that would not be as likely to be observed in the surface waters of deeper, larger water bodies.

Tidal creeks also supported a significantly higher abundance and biomass of finfish and crustaceans ($p < 0.001$), and a significantly higher number of species per trawl ($p < 0.05$) than observed at the open water sites (Figure 2). In contrast, the mean abundance and mean number of species of benthic infauna was lower at the creek sites compared to the open water sites ($p < 0.2$, abundance; $p < 0.1$, no. spp; Figure 2). The lower benthic measures may reflect the effects of higher predation pressure since the abundance and biomass of demersal biota was 2-4 times higher than observed at the open water sites. Lower DO conditions and other water quality conditions observed in the creeks may have also influenced the benthic fauna, but this is less likely since most creek sites did not experience extended periods (> 6-8 hours) of hypoxic conditions.

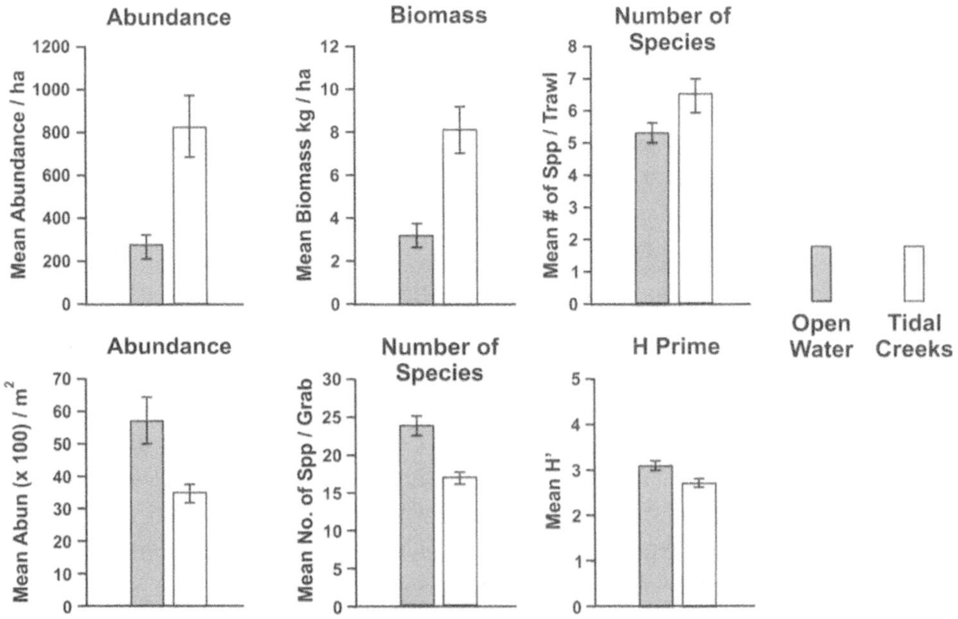

Figure 2. Summary of selected biological measures collected for (top) demersal fish and crustacean assemblages sampled by trawl, and (bottom) benthic assemblages sampled by grab (1999–2000).

As noted for the water quality variables measured, our analysis of several biological measures indicates very different conditions in creeks versus the larger open water bodies. Development of both water quality and biological measures that would be indicative of stress must account for the natural differences associated with large and small watersheds in order to effectively evaluate habitat quality. Based on our findings in South Carolina's coastal waters: (1) monitoring programs that don't incorporate small watersheds, such as tidal creeks, may miss an important habitat that should be monitored due to the value of these habitats as nursery areas, and (2) programs that do include small watersheds need to recognize the natural differences present between large and small water bodies, and consider partitioning these habitats or risk the strong probability of establishing inappropriate criteria for tidal creeks that could inaccurately identify water body impairment.

Acknowledgements

We wish to thank several individuals who assisted in the field and laboratory efforts reported here: P. Maier, L. Forbes, S. Burns, J. Jefferies, G. Hay, T. Snoots, D. Graves, M. Giffin, B. Rabon, J. Bickley, and staff of the SCDHEC Bureau of Environmental Services, Analytical Services Division. We also thank T. Olsen and staff at the USEPA NHEERL, Corvallis, OR for their assistance in developing the sampling design and CDF routines. This study was funded in part by the USEPA National Coastal Assessment Program under Cooperative Agreement R-82847201-0. This manuscript is contribution number 470 from the South Carolina Marine Resources Center.

References

Bricker, S.B., Clement, C.G., Pirhalla, D.E., Orlando, S.P., and Farrow, D.R.G.: 1999, National estuarine eutrophication assessment: Effects of nutrient enrichment in the nation's estuaries. NOAA, National Ocean Service, Special Projects Office and the National Centers for Coastal Ocean Science. Silver Spring, MD. 71 pp.

Carlton, J., Brown, J.S., Summers, J.K., Engle, V.D., and Bourgeois, P.E.: 1998, A report on the condition of the estuaries of Alabama in 1993–1995: A program in progress. Alabama Department of Environmental Management, Mobile, AL. 19 pp.

Diaz-Ramos, S., Stevens, D.L., Jr. and Olsen, A.R.: 1996, EMAP statistical methods manual, EPA/620/R-96/002. U.S. Environmental Protection Agency, Office of Research and Development, Office of Research and Development, NHEERL-Western Ecology Division, Corvallis, Oregon.

Dodd, M.G. and Murphy, T.M.: 1996, 'The status and distribution of wading birds in South Carolina, 1988–1996', Final report prepared by S.C. Department of Natural Resources, Columbia SC. Submitted to the Charleston Harbor Project, Charleston, SC. 66 pp.

Hackney C.T., Burbanck, W.D., and Hackney, O.P.: 1976, 'Biological and physical dynamics of a Georgia tidal creek', *Chesapeake Science* **17**, 271–280.

Holland, A.F., Riekerk, G.H.M., Lerberg, S.B., Zimmerman, L.E., and Sanger, D.M.: 1997, 'Assessment of the impact of watershed development on the nursery functions of tidal creek habitats', pp. 110–115 in: *Management of Atlantic Coastal Marine Fish Habitat: Proceedings of a Workshop for Habitat Managers*, ASMFC Habitat Management Series #2, April 1997. 223 pp.

Hyland, J.L., Balthis, L., Hackney, C.T., McRae, G., Ringwood, A.H., Snoots, T.R., Van Dolah, R.F. and Wade, T.L.: 1998, Environmental quality of estuaries of the Carolinian Province: 1995. Annual statistical summary for the 1995 EMAP Estuaries Demonstration Project in the Carolinian Province. NOAA Technical Memorandum NOS ORCA 123 NOAA/NOS, Office of Ocean Resources Conservation and Assessment, Silver Spring, M.D. 143 pp.

Ogburn, M.V., Allen, D.M. and Michener, W.K.: 1988, Fishes, shrimps, and crabs of the North Inlet Estuary, SC: A four-year seine and trawl survey. *Baruch Institute Technical Report* No. 88-1. 299 pp. University of South Carolina, Columbia.

Sanger, D.M., Holland, A.F. and Scott, G.I.: 1999a, 'Tidal creek and salt marsh sediments in South Carolina coastal estuaries. I. Distribution of trace metals', *Environmental Contamination and Toxicology* **37**, 445–457.

Sanger, D.M., Holland, A.F. and Scott, G.I.: 1999b, 'Tidal creek and salt marsh sediments in South Carolina coastal estuaries. II. Distribution of organic contaminants', *Environmental Contamination and Toxicology* **37**, 458–471.

Shenker, J.M. and Dean, J.M.: 1979, 'The utilization of an intertidal salt marsh creek by larval and juvenile fishes: abundance, diversity and temporal variation', *Estuaries* **2**, 154–163.

South Carolina Department of Health and Environmental Control.: 1997a, State of South Carolina monitoring strategy for fiscal year 1998. Technical Report No. 002-97, Bureau of Water, Columbia, SC. 77 pp. plus appendices.

South Carolina Department of Health and Environmental Control.: 1997b, Environmental investigations standard operating procedures and quality assurance manual. Office of Environmental Quality Control. Columbia, SC.

South Carolina Department of Health and Environmental Control.: 1997c, Procedures and quality control manual for chemistry laboratories. Analytical Services Division, Bureau of Environmental Services, Columbia, S.C.

South Carolina Department of Health and Environmental Control.: 1998a, Summary of selected water quality parameter concentrations in South Carolina waters and sediments. January 1, 1993–December 31, 1997. Technical Report 004-98. Bureau of Water. Columbia, SC. 122 p.

South Carolina Department of Health and Environmental Control.: 1998b, Water classifications and standards (Regulation 61–68) and classified waters (Regulation 61–69). Office of Environmental Quality Control. 36 pp.

South Carolina Department of Health and Environmental Control.: 2000, Standard operating and quality control procedures for phytoplankton and chlorophyll a. Technical Report No 014-00. 20 pp. plus appendices.

Stevens, D.L., Jr.: 1997, 'Variable density grid-based sampling designs for continuous spatial populations', *Environmetrics* **8**, 167–195.

Stevens, D.L., Jr and Olsen, A.R.: 1999, 'Spatially restricted surveys over time for aquatic resources', *Journal of Agricultural, Biological, and Environmental Statistics* **4**, 415–428.

Strobel, C.J., Buffum, H.W., Benyi, S.J., Petrocelli, E.A., Reifsteck, D.R. and Keith, D.J.: 1995, Statistical summary EMAP-Estuaries Virginian Province—1990 to 1993, EPA/620/R-94/026. U.S. Environmental Protection Agency, National Health and Environmental Effects Research Laboratory, Atlantic Ecology Division, Narragansett, RI, 72 pp.

Summers, J.K., Macauley, J.M., Heitmuller, P.T, Engle, V.D., Adams, A.M. and Brooks, G.T.: 1993, Annual statistical summary: EMAP—Estuaries Louisianian Province—1991. EPA/600/R-93/001. U.S. Environmental Protection Agency, Office of Research and Development, Environmental Research Laboratory, Gulf Breeze, FL, 101 pp. plus appendices A–C.

USEPA (U.S. Environmental Protection Agency): 1999, Draft ambient water quality criteria for dissolved oxygen (saltwater): Cape Cod to Cape Hatteras, EPA 822-D-99-002, U.S. Environmental Protection Agency, Office of Water, 55 pp. plus appendices.

Van Dolah, R.F., Chestnut, D.E. and Scott, G.I.: 2000, A baseline assessment of environmental and biological conditions in Broad Creek and the Okatee River, Beaufort County, South Carolina. Final Report to the Beaufort County Council, Beaufort County, South Carolina. 281 pp.

Weinstein, M.P.: 1979, 'Shallow marsh habitats as primary nurseries for fishes and shellfish, Cape Fear River, North Carolina', *Fisheries Bulletin* **77**, 339–357.

LIVING WITH A LARGE REDUCTION IN PERMITED LOADING BY USING A HYDROGRAPH-CONTROLLED RELEASE SCHEME

PAUL A. CONRADS[1*], WILLIAM P. MARTELLO[2], AND NANCY R. SULLINS[3]

[1]*Hydrologist, U.S. Geological Survey, 720 Gracern Road, Columbia, SC, 29210;*
[2]*Vice President, Jordan, Jones, and Goulding, 745 S. Milledge Ave., Athens, GA 30605;*
[3]*Water Resource Engineer, Tetra Tech., Inc., 10306 Eaton Place, Fairfax, VA, 22030*
**(author for correspondence, email: pconrads@usgs.gov)*

Abstract. The Total Maximum Daily Load (TMDL) for ammonia and biochemical oxygen demand for the Pee Dee, Waccamaw, and Atlantic Intracoastal Waterway system near Myrtle Beach, South Carolina, mandated a 60-percent reduction in point-source loading. For waters with a naturally low background dissolved-oxygen concentrations, South Carolina anti-degradation rules in the water-quality regulations allows a permitted discharger a reduction of dissolved oxygen of 0.1 milligrams per liter (mg/L). This is known as the "0.1 rule." Permitted dischargers within this region of the State operate under the "0.1 rule" and cannot cause a cumulative impact greater than 0.1 mg/L on dissolved-oxygen concentrations. For municipal water-reclamation facilities to serve the rapidly growing resort and retirement community near Myrtle Beach, a variable loading scheme was developed to allow dischargers to utilize increased assimilative capacity during higher streamflow conditions while still meeting the requirements of a recently established TMDL.

As part of the TMDL development, an extensive real-time data-collection network was established in the lower Waccamaw and Pee Dee River watershed where continuous measurements of streamflow, water level, dissolved oxygen, temperature, and specific conductance are collected. In addition, the dynamic BRANCH/BLTM models were calibrated and validated to simulate the water quality and tidal dynamics of the system. The assimilative capacities for various streamflows were also analyzed.

The variable-loading scheme established total loadings for three streamflow levels. Model simulations show the results from the additional loading to be less than a 0.1mg/L reduction in dissolved oxygen. As part of the loading scheme, the real-time network was redesigned to monitor streamflow entering the study area and water quality conditions in the location of dissolved-oxygen "sags." The study reveals how one group of permit holders used a variable-loading scheme to implement restrictive permit limits without experiencing prohibitive capital expenditures or initiating a lengthy appeals process.

Keywords: Total Maximum Daily Load, hydrograph-controlled release, water-quality model, real-time data monitoring, variable loading, Pee Dee River, Waccamaw River, Atlantic Intracoastal Waterway

1. Introduction

The Grand Strand is a rapidly growing resort area on the northeastern coast of South Carolina (Figure 1). The municipalities of Myrtle Beach and North Myrtle Beach have experienced tremendous growth in the 1990's and have become the second largest summer tourist destination on the East Coast. As a result, demands on the water resources continue to increase.

The Total Maximum Daily Load (TMDL) for ammonia and biochemical oxygen demand determined by the South Carolina Department of Health and Environmental Control for the Pee Dee, Waccamaw, and Atlantic Intracoastal Waterway system mandated a 60-percent reduction in point-source loading (South Carolina

Environmental Monitoring and Assessment **81**: 97–106, 2003.

Department of Health and Environmental Control, 1999). Antidegradation rules in the South Carolina Department of Health and Environmental Control regulations allow a maximum deficit of 0.1 milligrams per liter (mg/L) where waters do not meet the numeric standard for dissolved oxygen because of natural conditions. This is known as the "0.1 rule" (South Carolina Department of Health and Environmental Control, 1993). Although the municipal water-reclamation facilities utilize land application of effluent to meet the restrictive permit limits based on the "0.1 rule," there are periods during wet weather when land application to saturated soil is not an option.

The Grand Strand Water and Sewer Authority is responsible for providing water and wastewater services to the majority of Horry County, and has been deeply involved in water-resource planning issues along coastal South Carolina. The Au-

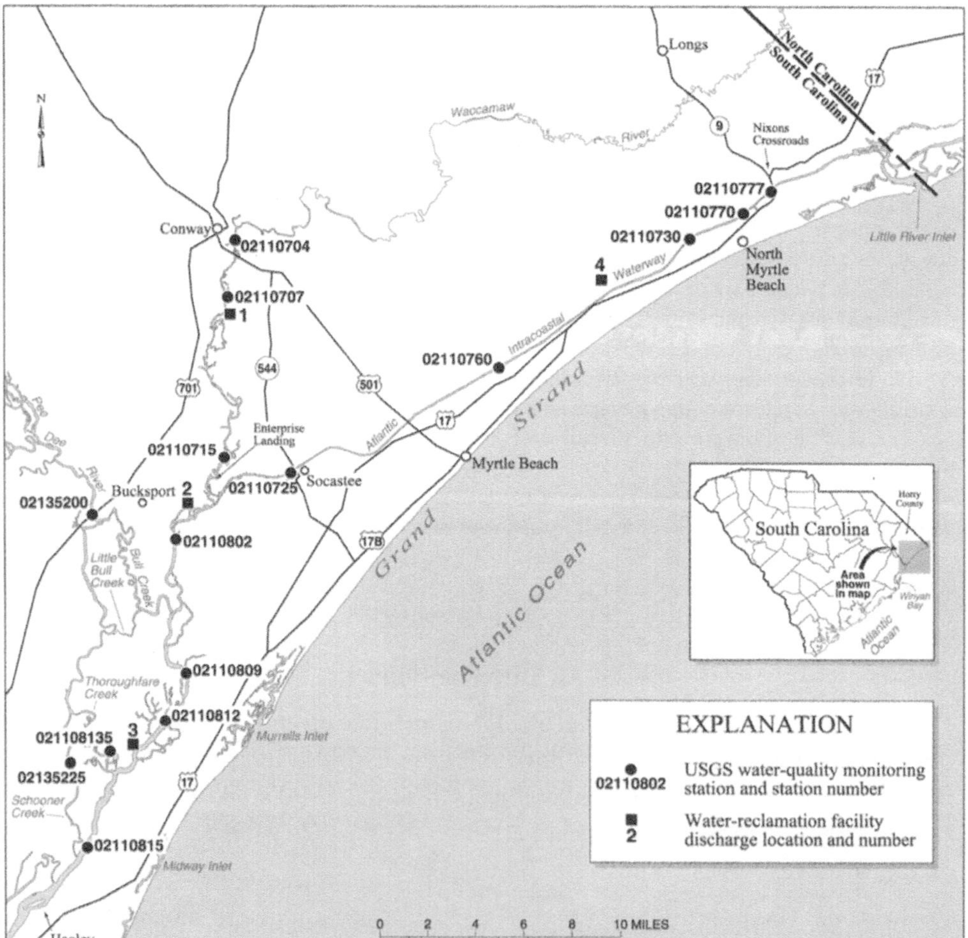

Figure 1. Location of real-time water-quality monitoring stations and water-reclamation facility discharge points.

thority, working with Jordan, Jones & Goulding, Inc., and the U.S. Geological Survey (USGS), developed a wastewater-release scheme that would maximize the increased assimilative capacity of the rivers during wet weather conditions when land application is not viable.

2. Description of Study Area

The Pee Dee and Waccamaw River Basins, approximately 13,000 and 1,300 square miles, respectively, supply freshwater inflow to the Grand Strand in the coastal zone, an increasingly developing strip of communities along the South Carolina coast from Little River Inlet to the north and Winyah Bay to the south. Inland and south of the U.S. Highway 701 Bridge, the Pee Dee River branches successively into Bull, Thoroughfare, and Schooner Creeks (Figure 1). These three creeks eventually flow into the Waccamaw River, with their net flows discharging 25 miles southward into Winyah Bay.

The Waccamaw River originates in North Carolina and enters the Atlantic Intracoastal Waterway about 10 miles north of the mouth of Bull Creek. Prior to the 1930's, the Waccamaw River flowed to the south toward Winyah Bay. In the 1930's, the U.S. Army Corps of Engineers constructed a canal to form the waterway from Enterprise Landing to the Little River Inlet, which altered the flow of the Waccamaw River north toward Little River Inlet along the Atlantic Intracoastal Waterway.

The majority of the freshwater flow to the segment of the Waccamaw River south of its junction with the waterway is from the Pee Dee River Basin and is carried by Bull Creek. The annual average streamflow from the Pee Dee Basin is about 14,100 cubic feet per second (ft^3/s), which is the combined streamflow of the three major rivers (Pee Dee, Little Pee Dee, and Lynches Rivers) (Carswell *et al.*, 1988). The Pee Dee River (downstream from the confluence with the Little Pee Dee River), Bull Creek, and Thoroughfare Creek are tidally affected during low and medium streamflows. The annual average streamflow of the Waccamaw River at Longs is 1,220 ft^3/s. The net streamflow of the Atlantic Intracoastal Waterway at the confluence with the Waccamaw River is north towards Little River Inlet from the Waccamaw River (Carswell *et al.*, 1988).

Saltwater enters the Atlantic Intracoastal Waterway through Winyah Bay to the south and Little River Inlet to the north. The Atlantic Intracoastal Waterway is affected by semidiurnal tides throughout the entire reach with a mean tide range of 4.0 feet (ft) at Nixons Crossroads and 3.5 ft at Hagley Landing (National Oceanic and Atmospheric Administration, 1995). The Pee Dee and Waccamaw Rivers are tidally affected during low and medium streamflows downstream of U.S. Highway 701 and U.S. Highway 501 bridges, respectively.

Point-source loads occur at four locations in the study area where (Figure 1) Grand Strand Water and Sewer Authority operate facilities in each of the locations. Prior to the recent TMDL, the total permitted loading to the system was 28,230 pounds per day of ultimate oxygen demand. The critical period analysis performed by South Carolina Department of Health and Environmental Control for the TMDL indicates that during low-flow regimes, a substantial reduction in total loading of ammonia and biochemical oxygen demand is required to maintain water-quality standards. Based on the flow period and critical condition boundary loadings for June 1, 1994, through July 15, 1994, a total assimilative capacity of 10,668 pounds per day of ultimate oxygen demand was determined. A summary of previous permit limits and the TMDL limits are shown in Table 1.

3. Approach

Models were used to simulate the unsteady streamflow, and the fate and transport of nutrients, biochemical oxygen demand, and dissolved oxygen in the Waccamaw and Pee Dee Rivers and in the Atlantic Intracoastal Waterway. The one-dimensional, dynamic flow BRANCH model (Schaffranek and others, 1981) was used to compute the two-way, tidal flow sand hydraulic properties of the riverine/estuarine system. Because the BRANCH model does not simulate the transport of constituents, it was necessary to use the one-dimensional dynamic Branched Lagrangian Transport Model (BLTM)(Jobson, 1997; Jobson and Schoelhamer, 1993) to simulate the mass transport and transformations of nutrients, biochemical oxygen demand, and dissolved oxygen.

As part of the TMDL development, the USGS, in cooperation with the Waccamaw Regional Planning and Development Council, applied the BRANCH and BLTM models to the Waccamaw River, Pee Dee River, Bull Creek, and Atlantic Intracoastal Waterway (South Carolina Department of Health and Environmental Control, 1999). As part of the model application, an extensive real-time data-collection network was established in the lower Waccamaw and Pee Dee River watersheds where continuous measurements of streamflow, water level, dissolved oxygen, temperature, and specific conductance are recorded (Figure 1). The two models were calibrated and validated to simulate the water quality and tidal dynamics of the

Table 1. Previous permit limits and TMDL limits for four discharge locations in the Grand Strand

[UOD, ultimate oxygen demand; TMDL, Total Maximum Daily Load]

Location	*Previous Permit Limits in UOD* *(pounds per day)*	*TMDL limits in UOD* *(pounds per day)*
Conway	1,873	303
Bucksport	228	84
Hagley	24,220	8,643
North Myrtle Beach	1,909	1,638
Total	28,230	10,668

system. Results from these models indicate that the assimilative capacity is dependent upon the selected flow conditions (Drewes and Conrads, 1995).

The flow period selected to analyze for critical conditions used for the TMDL determination was based on a review of several years of flow data collected by the USGS during model development; however, the selected critical flow period is considered analogous to the lowest 7-day average, 10-year recurrence streamflow typically used for assimilative capacity analyses. It is typical to further analyze the assimilative capacity of surface waters using higher flow regimes on either a seasonal or hydrograph-controlled release basis. Under a hydrograph-controlled release permit, dischargers are able to utilize the increased assimilative capacity during periods of higher streamflow. This seasonal analysis is consistent with the TMDL policy of the U.S. Environmental Protection Agency and is commonly utilized by South Carolina Department of Health and Environmental Control.

During the draft National Pollution Discharge Elimination System (NPDES) permitting process, South Carolina Department of Health and Environmental Control agreed to allow variable ultimate oxygen demand loadings based on actual streamflows, but specified that the variable loadings must be calculated using the dynamic model developed for the Waccamaw River and Atlantic Intracoastal Waterway system in order to be consistent with the previous analysis. To determine the assimilative capacity of various flow regimes, the BRANCH and BLTM models were used to simulate the streamflow and water quality for various flow conditions between 1990 and 1994. The hydraulic data for the BLTM were simulated with the BRANCH model by using measured water levels at the model boundaries. For consistency between model simulations for various flow periods, the only inputs to the model that were modified for each simulation were the historical period selected for the water-level time series used in the BRANCH model to simulate the flow and hydraulic characteristics of the system. All other inputs to the model, including water-quality conditions and meteorological conditions, remained unchanged.

Due, in part, to drainage from extensive wetlands and the relatively low flushing rate of this system, dissolved-oxygen concentrations tend to be naturally low, often lower than the 4.0 mg/L minimum required for these coastal waters in South Carolina. Additional point sources of ultimate oxygen demand loads may be permitted if they do not cause a cumulative impact to the dissolved-oxygen concentration of greater than 0.1 mg/L. The "0.1 rule" was applied consistent with the critical period analysis to determine the point-source loadings at varying flow periods. "Background" conditions were estimated by removing the point-source loadings (both pipe flow and effluent concentration) from the model and determining the resultant dissolved-oxygen profile. The next step was to include the point-source loadings in the model and to determine the associated dissolved-oxygen profile. The result of the proposed discharges is the difference between the background or no-load scenario and the load scenario.

The dynamic model BRANCH/BLTM developed for the Waccamaw and Atlantic Intracoastal Waterway system was used to analyze the tiered hydrograph-controlled release scheme. Programs were developed by USGS staff to analyze the flow data files, determine which streamflow tier would apply, input the load file, and then run the model. Model simulations were run for 45-day periods. The first 15 days were used as a "warm up" period and the assimilative capacity analysis was performed during the last 30 days. The load and no-load scenarios were then compared to verify that the "0.1 rule" was maintained. The process of determining the tier loading rates was an iterative process of selecting streamflow thresholds, determining loading amounts, and evaluating the water-reclamation facility operational feasibility of the hydrograph-controlled release.

Jordan, Jones, and Goulding performed an analysis of the USGS flow period records and associated statistical occurrences. The selection of higher flow levels for model analysis was based on meeting the following goals:

- Maintain all water-quality standards;
- Simplify South Carolina Department of Health and Environmental Control's review of permit compliance;
- Minimize complicated discharge decisions for water-reclamation facility operating staff; and
- Remain consistent with U.S. Environmental Protection Agency's TMDL policy (South Carolina Department of Health and Environmental Control, 1991).

4. Results

A three-tiered release schedule was determined to best meet all of the intended goals of the proposed hydrograph-controlled release scheme. These tiers are defined in Table 2. Specified minimum flow regimes must only occur in one of the streams to trigger an individual tier loading. For example, if flows in the Waccamaw River are less than 1,000 ft^3/s whereas flows in the Pee Dee River are 8,000 ft^3/s, then tier 1 loadings would apply.

Based on an analysis of the flow records from USGS gaging stations in the Waccamaw and Pee Dee River Basins, Tier 1 flows (and critical period loadings) would occur approximately 51 percent of the time. Tier 2 flows (and loadings) would occur approximately 27 percent of the time, and Tier 3 flows (and loadings) about 22 percent of the time.

5. Model Analysis

A tiered hydrograph-controlled release concept was evaluated using the Waccamaw and Atlantic Intracoastal Waterway model used for the TMDL development. This concept allows for additional loadings during flow regimes higher than the critical period while still protecting water-quality standards. Three separate flow periods

Table 2. Summary of tier streamflows and loadings

Tiers and Streamflow Levels	*Percent of Time Flow Occurs*	*Discharge Location*	*Loading (pounds per day UOD)*
Tier 1	51	Conway	303
Pee Dee River < 6,000 ft^3/s		Bucksport	84
or		Hagley	8,643
Wacamaw River < 1,000 ft^3/s		North Myrtle Beach	1,638
		Total	**10,668**
Tier 2	27	Conway	574
6,000 ft^3/s < Pee Dee River		Bucksport	121
< 10,000 ft^3/s		Hagley	12,955
or		North Myrtle Beach	2,446
1,000 ft^3/s < Waccamaw River		**Total**	**16,096**
< 2,500 ft^3/s			
Tier 3	22	Conway	956
Pee Dee River > 10,000 ft^3/s		Bucksport	201
or		Hagley	21,592
Waccamaw River > 2,500 ft^3/s		North Myrtle Beach	4,077
		Total	**26,827**

were selected to validate the tiered-concept approach. Each flow period exhibited unique flow characteristics and is considered appropriate for this analysis. A summary of the flow periods selected and the associated results is provided below.

December 1990 Flow Period

Flows in the Waccamaw River generally averaged less than 1,000 ft^3/s during December 1990 with periodic values greater then 1,000 ft^3/s (Tiers 1 and 2). Flows in the Pee Dee River generally averaged between 10,000 and 15,000 ft^3/s (Tier 3). The amount of permissible loading was controlled by the lower flows in the Waccamaw River, and varied between Tiers 1 and 2 during this simulation. Water-quality standards for dissolved oxygen were maintained during this flow scenario.

February 1992 Flow Period

Flows in the Waccamaw River averaged between 2,000 and 5,000 ft^3/s (Tiers 2 and 3) during February 1992 (Figure 2). Flows in the Pee Dee River generally ranged between 5,000 and 17,000 ft^3/s (Tiers 1, 2, and 3). The amount of permissible loading was controlled by the lower threshold flows of the Pee Dee River, and varied between Tiers 1 and 3 during this simulation. The differences in dissolved-oxygen concentration between the load and no-load scenarios ranged between 0.03

and 0.09 mg/L. Water-quality standards for dissolved oxygen were maintained during this flow scenario.

November 1992 Flow Period

Flows in the Waccamaw River varied between 100 and 4,500 ft^3/s during November 1992 (Tiers 1, 2, and 3). Flows in the Pee Dee River generally ranged between 12,000 and 32,000 ft^3/s (Tier 3). Permissible loading was controlled by the lower tier streamflows of the Waccamaw River which varied between Tiers 1, 2, and 3 during this simulation. Of the three flow periods, the November 1992 period was the most dynamic. Early in the simulation, the system was controlled by Tier 1 streamflow and loading conditions. The difference in dissolved-oxygen concentrations between the load and no-load scenarios was 0.09 mg/L. As streamflows increased in the Waccamaw River, the loading increased to Tiers 2 and 3. The resulting differences in dissolved-oxygen concentration decreased to 0.02 mg/L, well below the allowable dissolved-oxygen reduction of 0.1 mg/L (Figure 3).

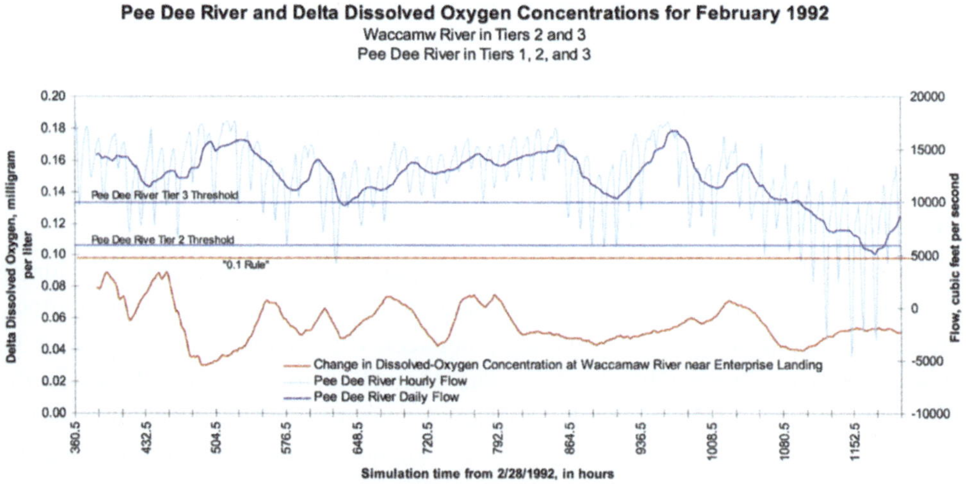

Figure 2. Simulated streamflow for the Pee Dee River and dissolved-oxygen differences for the Waccamaw River at Enterprise Landing for February 1992.

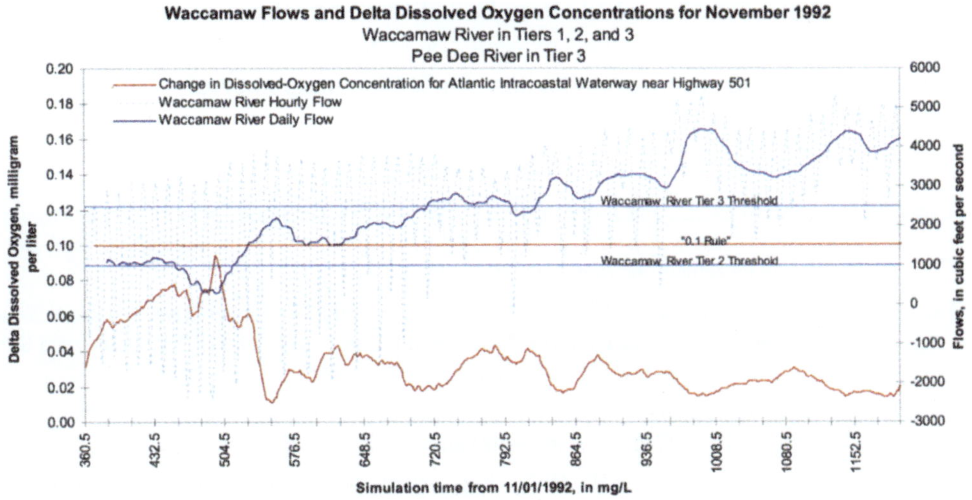

Figure 3. Simulated streamflow for the Waccamaw River and dissolved-oxygen differences for the AIW near Highway 501 November 1992.

6. Summary and Implementation

The tiered hydrograph-controlled release concept for the Pee Dee, Waccamaw, and Atlantic Intracoastal Waterway system was based on the variable flows typically experienced, which is consistent with U.S. Environmental Protection Agency's TMDL policy. Antidegradation rules in the South Carolina Department of Health and Environmental Control regulations allow a maximum deficit of 0.1 mg/L where waters do not meet the numeric standard for dissolved oxygen because of natural conditions. The "0.1 rule" was applied to the model simulations using the dynamic model developed by the USGS and approved by South Carolina Department of Health and Environmental Control for TMDL analyses. By selecting variable discharge loadings that were triggered on minimum flows in both the Pee Dee and Waccamaw River systems, water-quality standards were maintained.

The National Pollution Discharge Elimination System permits for the Grand Strand Water and Sewer Authority and other water-reclamation facilities operating in this basin are hydrograph-controlled release permits with loadings consistent with this analysis. The real-time data-collection network for the Grand Strand has been modified to support the operation of these hydrograph-controlled release permits. The network includes streamflow monitors entering the study area on the Pee Dee River and Waccamaw River and in Little River on the Atlantic Intracoastal Waterway. The streamflow monitoring stations are instrumented with acoustic velocity meters that record tidal streamflow on a 15-minute interval. Wastewater Reclamation Facilities access the real-time streamflow data, either through the internet or automated email, and adjust the effluent discharges to the system according to the streamflow conditions. In addition, water-quality monitors are located in the area of the dissolved-oxygen "sags" to record water temperature, dissolved oxygen, and specific conductance on a 60-minute interval.

References

Carswell, W.J., Sanders, C.L., and Johnson, D.M.: 1988, Freshwater supply potential of the Atlantic Intracoastal Waterway near Myrtle Beach, South Carolina: U.S. Geological Survey Water-Resources Investigations Report 88-4066, 45 pp.

Drewes, P.A., and Conrads, P.A.: 1995, Assimilative capacity of the Waccamaw River and the Atlantic Intracoastal Waterway near Myrtle Beach, South Carolina, 1989-92: U.S. Geological Survey Water-Resources Investigations Report 95-4111, 58 pp.

Jobson, H.E.: 1997, Enhancements to the Branched Lagrangian Transport Modeling System: U.S. Geological Survey Water-Resources Investigation Report 97-4050, 56 pp.

Jobson, H.E., and Schoelhamer, D.H.: 1987, Users manual for a Branched Lagrangian Transport Model: U.S. Geological Survey Water-Resources Investigation Report 87-4163, 73 pp.

National Oceanic and Atmospheric Administration: 1995, Tide tables 1995, high and low predictions—East Coast of North and South America, Including Greenland: U.S. Department of Commerce, National Ocean Service, 301 pp.

Schaffranek, R.W., Baltzer, R.A. and Goldberg, D.E.: 1981, A model for simulation of flow in singular and interconnected channels: U.S. Geological Survey Techniques of Water-Resource Investigation, book 7, chap. C3, 100 pp.

South Carolina Department of Health and Environmental Control: 1991, Agreement on the development of wasteload allocations/total maximum daily loads and NPDES permit limitations, 43 pp.

—— 1993, Water classifications and standards (Reg. 61-68) and classified waters (Reg. 61-69) for the State of South Carolina, 36 pp.

—— 1999, Total Maximum Daily Load Determination for the Waccamaw River and the Atlantic Intracoastal Water Way Near Myrtle Beach, SC., 80 pp.

A PROPOSED COAST-WIDE REFERENCE MONITORING SYSTEM FOR EVALUATING WETLAND RESTORATION TRAJECTORIES IN LOUISIANA

GREGORY D. STEYER[1*], CHARLES E. SASSER[2], JENNEKE M. VISSER[2]
ERICK M. SWENSON[2], JOHN A. NYMAN[3], AND RICHARD C. RAYNIE[4]

[1]*U.S. Geological Survey, National Wetlands Research Center, Baton Rouge, LA, U.S.A.;*
[2]*Louisiana State University Coastal Ecology Institute, Baton Rouge, LA, U.S.A.;*
[3]*Louisiana State University School of Forestry, Wildlife and Fisheries, Baton Rouge, LA, U.S.A.;*
[4]*Louisiana Department of Natural Resources, Baton Rouge, LA, U.S.A.*
[*]*(author for correspondence, email: gsteyer@usgs.gov)*

Abstract. Wetland restoration efforts conducted in Louisiana under the Coastal Wetlands Planning, Protection and Restoration Act require monitoring the effectiveness of individual projects as well as monitoring the cumulative effects of all projects in restoring, creating, enhancing, and protecting the coastal landscape. The effectiveness of the traditional paired-reference monitoring approach in Louisiana has been limited because of difficulty in finding comparable reference sites. A multiple reference approach is proposed that uses aspects of hydrogeomorphic functional assessments and probabilistic sampling. This approach includes a suite of sites that encompass the range of ecological condition for each stratum, with projects placed on a continuum of conditions found for that stratum. Trajectories in reference sites through time are then compared with project trajectories through time. Plant community zonation complicated selection of indicators, strata, and sample size. The approach proposed could serve as a model for evaluating wetland ecosystems.

Keywords: monitoring, wetland restoration, Louisiana, reference, HGM

1. Introduction

The Coastal Wetlands Planning, Protection and Restoration Act of 1990 (CWPPRA), also called the "Breaux Act", was enacted to create, restore, enhance and protect coastal wetlands, primarily those in Louisiana. The restoration plan developed pursuant to this act specifically requires: (1) "an evaluation of the effectiveness of each coastal wetland restoration project in achieving long-term solutions to arresting coastal wetland loss in Louisiana;" and (2) "a scientific evaluation of the effectiveness of the coastal wetlands restoration projects carried out under the plan in creating, restoring, protecting and enhancing coastal wetlands in Louisiana." A monitoring program was established to evaluate project-specific goals and objectives based on sound scientific procedures and relies heavily on the establishment of appropriate paired reference areas. Monitoring both project and reference areas provides a means to achieve statistically valid comparisons and is the most reliable way to assess project effectiveness.

Forty-seven Breaux Act projects have been constructed with another 75 authorized for funding through January 2001. These projects average over 5,700 hectares in size and utilize a variety of techniques ranging from simple shoreline protection projects to complex projects which manipulate hydrology (LCWCRTF, 1997). Many of the projects are close to each other, increasing the possibility of cumulative, indirect influences on landscape level processes. These cumulative effects are not being effectively evaluated at the project-specific level because they require a characterization of the landscape outside of the project areas.

The proposed Coast-wide Reference Monitoring System (CRMS) was developed to address Breaux Act Monitoring Program needs, and it has two objectives to address gaps in the current monitoring strategy. The first objective is to determine collectively the effectiveness of the projects carried out under the Breaux Act restoration plan by providing a network or "pool" of reference sites against which to evaluate project effectiveness, eliminating the need for one-on-one pairs of reference and project areas. Reference areas are fairly easy to find for small-scale projects such as shoreline protection but extremely difficult for large-scale, complex projects such as hydrologic restoration. The CRMS will provide an array of reference sites for the many projects for which no appropriate paired reference areas exist. Data collected under the CRMS will characterize typical conditions within various vegetation types for both project and nonproject areas and provide a basis of comparison (i.e., a standard) to evaluate differences in response to Breaux Act projects.

The second objective is to ensure that the strategic coastal plan for Louisiana (Coast 2050) is effective in re-creating a sustainable coastal ecosystem (LCWCRTF, 1998). This plan lays out a restoration strategy that depends on the cumulative effects of many individual projects that should collectively lead to the general restoration of Louisiana's coastal estuaries. Because these estuaries are open ecological systems, driven by complex hydrological circulation patterns that will be further altered by the combined effect of individual projects, it is imperative to develop a program to evaluate these cumulative effects. Therefore the second CRMS objective is to determine the ecological condition of the coastal wetlands based on the variables measured. The CRMS will provide an avenue to evaluate the effectiveness of the restoration plan and determine whether whole coastal ecosystems are being restored, not just the areas directly affected by individual projects.

The programmatic questions that CRMS will address are: (1) Did the projects carried out under the restoration plan contribute to reducing coastal wetland loss in Louisiana as well as within each hydrologic basin?; (2) Did the Breaux Act sustain a diversity of vegetation types in the hydrologic basins?; (3) Is the Breaux Act effective in reducing the effects of major stressors on wetlands (i.e., flooding regime, salinity, marsh elevation change)?; and (4) Which project types are the most effective in creating, restoring, protecting, and enhancing wetlands?

2. Approach

2.1 REFERENCE SITE ISSUES

The selection of reference sites has philosophical and practical implications. In studies that address the condition of landscapes or ecosystems, an attempt is often made to find "pristine" sites for comparison (i.e., areas that show little human impact). In the absence of natural sites, "ideal" conditions are sometimes assigned as reference standards. Reference standards are typically expected to be biotically diverse and to exhibit high productivity, two conditions that are usually considered to identify "valuable" ecosystems. Given the degree of human activity in Louisiana estuaries, there are no pristine sites.

An alternative approach in determining reference sites is to select multiple sites ecologically similar to a project area (rather than some ecologically ideal site). From a statistical point of view, although project performance can be evaluated against ideal goals, the evaluation of project performance may be best made against nonproject areas of comparable condition. Then changes in both project and reference areas can be compared over time.

Thus, for a coast-wide monitoring system in Louisiana, two options for reference sites are: (1) sites that reflect some ideal ecological condition against which project site restoration can be gauged (reference standards); or (2) sites similar to preproject conditions (controls). On the Louisiana coast there are few, if any, reference areas that are sufficiently similar to project areas to qualify as controls. Individual wetland characteristics vary independently in different wetlands because conditions of local sites are determined by an extremely complex spatial pattern of structure and process. Further, the prevalence of human impacts everywhere in the coastal estuaries means that there are no "natural" reference areas. Any local site is probably significantly influenced, either directly or indirectly, by human actions.

The design proposed here contains features of both alternatives. It is based primarily on recommendations for the hydrogeomorphic (HGM) approach (Brinson, 1993; Brinson *et al.*, 1994; Brinson and Rheinhardt, 1996) to wetland functional assessment and the EPA Ecological Monitoring and Assessment Program (EMAP) (Leibowitz *et al.*, 1991; Novitzki, 1994; Thornton *et al.*, 1994). EMAP classifications and HGM ideas for reference areas are used extensively, including: (1) the classification of the coastal wetlands into relatively homogeneous classes (which are blocked in the statistical design); (2) the selection of reference sites that span the whole range of ecological response characteristics of each class; and (3) the identification of those reference sites that most closely approach the sustainable functional potential of each class, as reference standards (Figure 1). Reference sites might be paired with project sites, but that is not necessary for the statistical design.

2.2 Statistical Design

The basic statistical design is an asymmetric analysis derived from BACI (Underwood, 1991; 1992; 1994), with the addition of multiple reference sites as well as multiple project sites. Underwood compared the power between BACI design and multiple-reference site design to detect impacts of various types. He found the multiple-reference site design is much more powerful than the traditional BACI, which is currently used to monitor wetland restoration on the Louisiana coast. He also found that increased replication of reference sites increased the power of the analysis.

We have adapted the classification and reference ideas, but the functional assessment in Louisiana is driven by the objectives of the Breaux Act projects. The proposed design of the monitoring program entails classifying the coast into a discrete number of similarly functioning regions and basins, and into project and nonproject areas. Marsh classes (also referred to as vegetation types) are also identified within each basin. A strength of the design is that the division of coastal habitats into vegetation types by basin allows analysis at a meaningful watershed scale, yet it also allows regrouping of the data to a more regional scale when desirable. The population sample within a basin encompasses all project and reference sites monitored through time. The design will provide: (1) an instantaneous "snapshot" that places project sites within the functional continuum of all reference sites; (2) the trajectory of change of project sites through time with respect to changes in functionally equivalent reference sites; and (3) comparisons to a group of reference standards (see Figure 2).

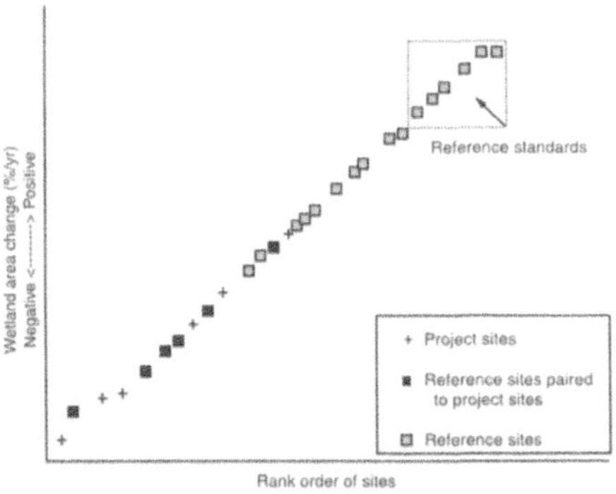

Figure 1. Monitoring sites showing range in response, paired project/reference sites, and reference standards.

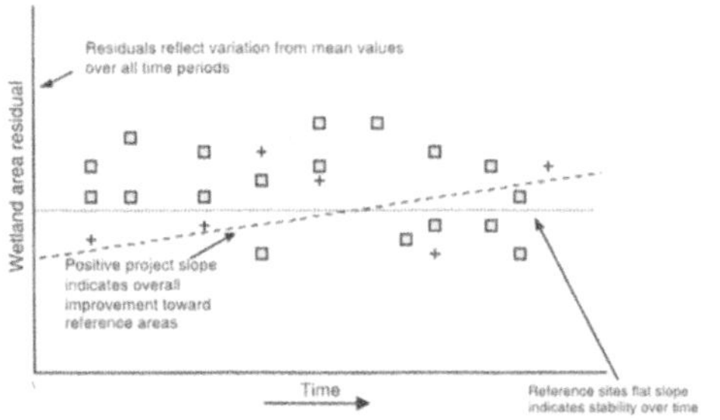

Figure 2. Trajectories of reference sites compared to project sites over time.

3. Design Attributes

3.1 MONITORING VARIABLES

The variables measured at each site are those necessary to address the objectives of the CRMS. These variables are crucial in determining the effectiveness of the Breaux Act program as well as those considered most important in affecting growth of vegetation. The frequency of sampling is that which is minimally required to evaluate long-term trends, while maintaining information on seasonal trends. At a minimum, the recommended variables to be monitored include land:water ratio, vegetation composition and cover, elevation, duration and frequency of flooding, salinity and sedimentation/erosion. Sampling methodologies for these variables are described in Steyer *et al.* (1995).

3.2 SAMPLE POPULATION AND STRATIFICATION

Chabreck and Linscombe (1997) assembled a vegetation data base that provides a grid of stations located along 157 transect lines that cover the majority of the Louisiana coastal zone from Mississippi to Texas. These transects were extended north to cover bottomland hardwood and swamp vegetation types by using the 1988/90 National Wetlands Inventory habitat data. The transects were also extended across saline bays and associated saline marsh islands in the Terrebonne, Barataria, and Pontchartrain Basins to complete polyhaline oystergrass coverages in the Louisiana coastal zone. The data base of coastal vegetation at 6,298 stations is evenly distributed over the entire coast (Figure 3) and provides a representative sample of the vegetation communities in the Louisiana coastal zone; therefore, random selection of sample sites from this sample population should be representative of the whole coast.

Traditionally, the Louisiana coastal marshes have been divided based on salinity into four vegetation zones: fresh, intermediate, brackish, and saline (Chabreck, 1972). A recent analysis by Visser and Sasser (1998) of coast-wide vegetation data obtained by Chabreck and Linscombe in 1997 identified 12 vegetation types that occur under five salinity regimes in the Louisiana coastal zone. These vegetation types, plus the swamp and bottomland hardwood types, are distributed over nine hydrologic basins as shown in Figure 3. Hydrologic basin and project/nonproject were selected as the strata for purposes of analysis and reporting. Samples were also proportionally allocated by vegetation type. Using vegetation type as an allocation strata reduces the number of samples needed in the overall design, while at the same time ensuring that all the major vegetation types in a basin are represented. This allows for comparisons within vegetation type among basins as needed.

3.3 SAMPLE SIZE

The number of reference sites monitored affects the quality of the analyses. The number of replicates reflects a trade-off between the increased cost of additional sample sites and improved confidence in results. We investigated three approaches in the preparation of our recommendation for the number of reference sites: central limit theorem, a power analysis of water salinity data, and a Monte-Carlo type resampling of existing vegetative composition data.

The Central Limit Theorem suggests using 30 samples per stratum as a starting point for a sound statistical design. With 18 strata (9 basins and project/nonproject), this approach would suggest a total of 540 reference sites in CRMS. Power analysis of salinity data was explored by Malaeb (1997) to evaluate the number of reference sites required. This investigation was of limited utility because the data were not complete enough and were spatially and serially correlated. Resampling techniques following Crowley (1992) were utilized to evaluate the sample size needed to describe the vegetative composition of the coast, and to detect changes in the vegetative composition over time.

This resampling technique evaluated the number of samples needed to avoid two types of potential error in statistical analyses. We used the coastal vegetation data base (Figure 3) consisting of observations of current vegetation types at 6,298 stations in coastal Louisiana. The number of reference sites needed to avoid a Type I Error is also the minimum number of sites required to accurately represent the current vegetation type distribution. We evaluated sample size and Type I Error by randomly selecting stations from the 6,298 stations, determining the vegetation type composition of that sample, and then using a chi-square statistic to test if the vegetation type composition of the sample differed from the vegetation type composition of the full data set. The sample sizes used to estimate vegetation type composition were 25, 50, 100, 150, 250, 300, 350, 400, 450, 500, 600, 800, 1000, 1200, 1400, 1600, and 2500. Each sample size was tested with 5,000 different

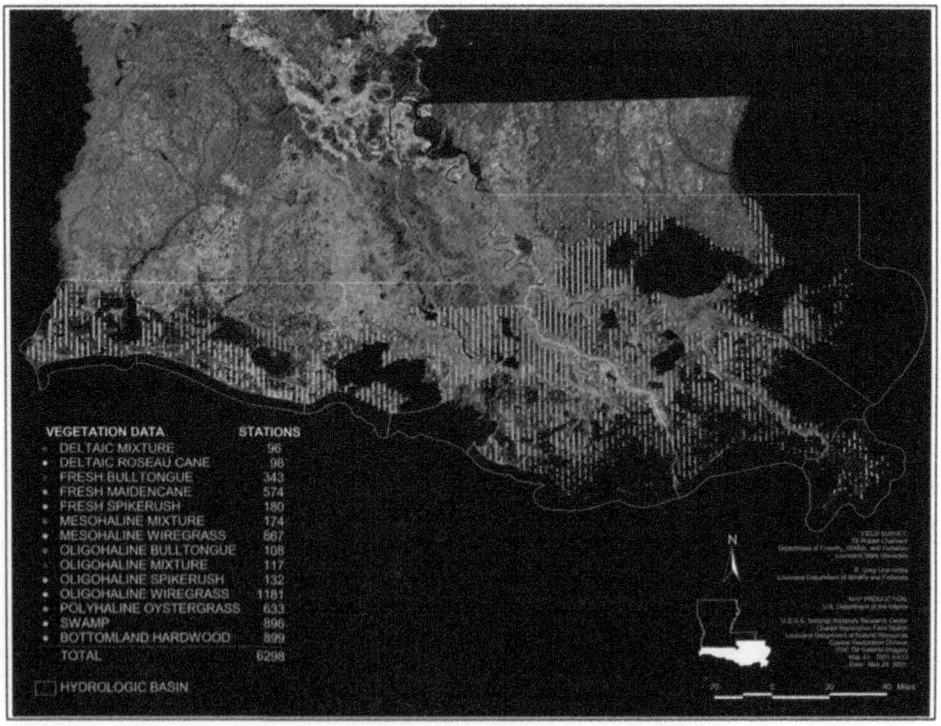

Figure 3. The 1997 geographic distribution of vegetation sites as established by Chabreck and Linscombe (1997) and classified according to Visser and Sasser (1998).

random samples. For instance, vegetation type composition was calculated by using 25 randomly selected stations 5,000 times. For each sample, the vegetation type composition was tested to determine if it was significantly different from the vegetation type composition of the full data set. For each sample size, we calculated an accuracy as the percent of times that the sample was not significantly different from the full data set. This approach indicated that 100 reference sites would indicate the true coastwide vegetation type composition 95% of the time (Figure 4). In addition to coastwide sampling, we evaluated the sample number needed to avoid a Type I Error on two of the eight hydrologic basins within the Louisiana coastal zone: Barataria Basin and Mermentau Basin. Based on these analyses, 100 reference sites were required to indicate the true basin-wide vegetation type composition with 95% confidence. The sample sizes for coastwide and basin are similar because this procedure only determines the sample size necessary to detect a statistical difference in composition. This difference is independent of area sampled but dependent on the complexity of the vegetation composition. Barataria Basin contains 13 and Mermentau 7 of the 14 vegetation types found coast-wide. Based on these analyses, the other hydrologic basins were assigned to two groups (one with a similar number of vegetation types to Barataria and the other similar to Mermentau). This assignment estimated that 650 reference sites are needed for the CRMS.

We evaluated the sample size needed to avoid a Type II Error by using the chi-square statistic to test if samples from a "time-one" were different from a "time-two" (Green, 1979). The vegetation type of each of the 6,298 stations was used as the time-one vegetation type. Each of the 6,298 stations was assigned a time-two vegetation type based on the most likely vegetation type shift anticipated by regional wetland experts using a marine intrusion scenario. The time-two vegetation type was the same as the time-one vegetation type for 80% randomly selected stations. The vegetation type was changed in the remaining 20% of the stations. A number of stations were then randomly selected and the time-one and time-two vegetation type compositions were compared with the chi-square statistic. This process was repeated 5,000 times for each sample size evaluated. We defined accuracy as the percentage of those tests that indicated a statistically significant difference between time-one and time-two. These analyses indicated that approximately 800 reference sites were needed to detect this difference 95% of the time (Figure 5).

We further investigated a possible reduction in sample size required by reducing the number of vegetation types using the change detection (Type II Error) approach. We collapsed the vegetation types into six vegetation classes: forested wetland, fresh, intermediate, brackish, saline, and deltaic marsh; however, very little reduction in the number of sites could be achieved in this manner. More importantly, combining classes would make the analyses less sensitive to one of the most important changes believed to proceed wetland loss in Louisiana: stress

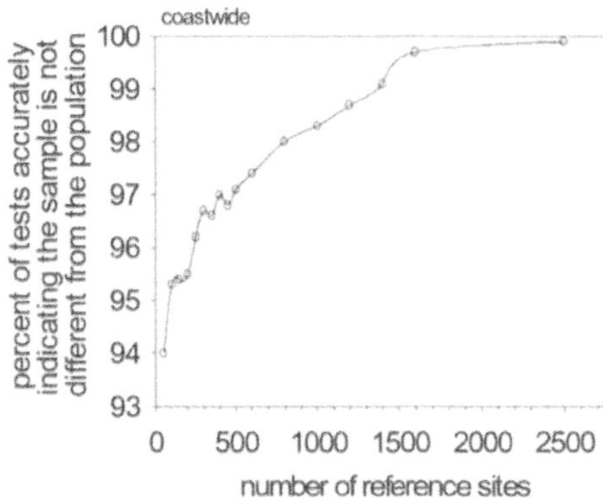

Figure 4. Resampling methodology indicating that 100 randomly selected samples (reference sites) would represent the true coastwide vegetative composition 95% of the time.

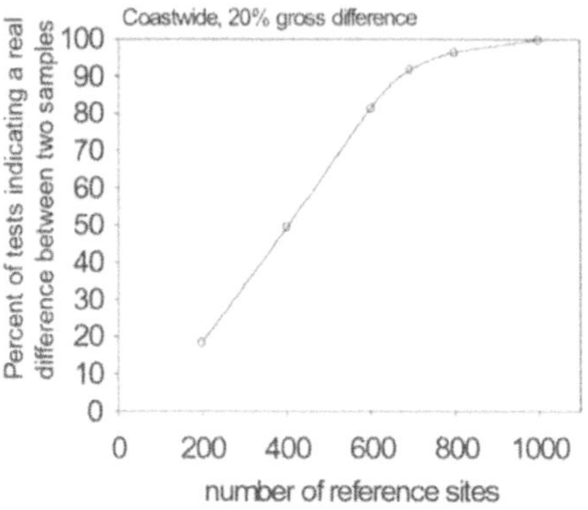

Figure 5. Resampling methodology indicating that 800 randomly selected samples (reference sites) would detect a 20% change in marsh type between time-one and time-two 95% of the time.

induced changes in vegetative composition. We therefore rejected combining classes of vegetation as a means of reducing sample size.

Based on the statistical approaches described above which estimated the number of reference sites from 540 to 800, and on the extensive professional experience of participating wetland scientists, we recommend a total of 700 reference sites in this sampling design. This level of replication should be sufficient to represent the functional variation of the various response variables measured.

Our methods for sample size determination can be adapted by other programs depending on the amount of existing information and the questions that the monitoring program is trying to address. Monitoring planners should consider resampling existing data to explore the consequences of various sample sizes on statistical power if existing data are available. The ability to avoid a Type I Error was achieved with relatively small sample sizes. Avoiding Type II Error was the more important consideration. Future analyses should evaluate the effect of the number of vegetation classes on statistical power, and should evaluate the effect of sample size on variables other than vegetative composition.

4. Sampling Locations and Distribution

The 700 reference sites were selected from the 6,298 station sample population based on the location of a random sample point in the space defined by the strata (Figure 6). This unequal probability design ensures representation of all vegetation types within the basin strata. The 700 reference sites will be spatially fixed to increase the likelihood of detecting temporal variability and correlations, which are critical to restoration evaluations. The random distribution of sites will be defined as project sites or nonproject sites for reporting purposes. There is no defined sample allocation to project sites because over time, new projects will be added and some existing project sites may revert to nonproject sites. Based on the current distribution of projects, 190 reference sites fall within project boundaries.

To provide a powerful "backbone" of sample stations, we suggest that 200 (28%) of the stations will be sampled every year. The 200 annual stations should represent the true coastwide vegetation composition with greater than 95% confidence. Of the remaining 500 stations, approximately one third will be sampled each year without replacement. In this arrangement, approximately 367 (52%) of the stations will be measured each year. Overall, this schedule should greatly enhance the coverage of temporal variation in the design, provide reference sites that are measured in the same year as the project sites, and at the same time provide efficiency in use of personnel and equipment.

5. Discussion

The proposed plan adequately addresses the twin reference area monitoring problems embodied in the two objectives set forth in the beginning of this document. The plan provides an adequate network or "pool" of reference sites that can be

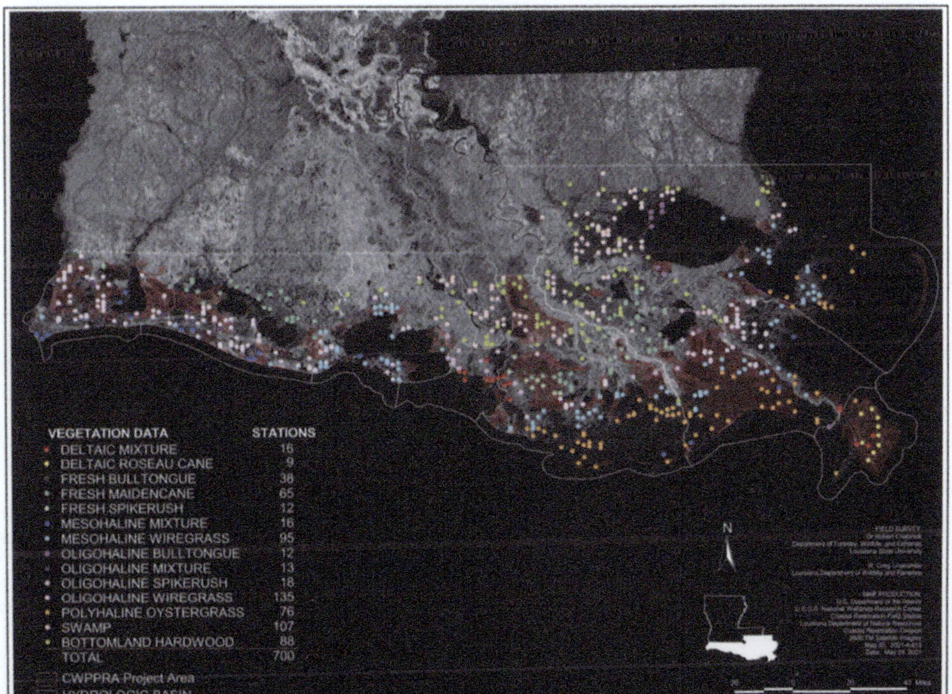

Figure 6. Distribution of 700 randomly selected reference sites to be sampled in the CRMS design.

used to address the serious problem of lack of suitable paired reference areas associated with most Breaux Act projects, and it also provides a framework of data collection on a large enough scale that evaluates the effectiveness of the Breaux Act restoration plan in creating a sustainable coastwide ecosystem.

The 700 recommended reference sites will provide adequate spatial representation of the Louisiana coastal marshes and adjacent, small waterbodies. The base of 200 annual sampling sites will provide sufficient replication for within year comparisons and the three year sampling period for the remaining stations will provide resource and cost efficiencies while not reducing our ability to compare long term trends in reference versus project sites.

Remaining design activities include: (1) identifying an additional indicator that exhibits quick response to environmental change in order to rapidly gauge effectiveness of projects; (2) testing the optimal power of the design in characterizing temporal versus spatial variability; and (3) developing ecological functional models following HGM methodologies.

The plan presented allows an evaluation of the cumulative effects of the many individual projects that should collectively lead to the general restoration of Louisiana's coastal estuaries.

Acknowledgments

We extend our appreciation to the Louisiana Department of Natural Resources who funded this work and to Darin Lee, Ralph Libersat, and Larry Reynolds who reviewed earlier versions of this paper. We also thank James Gosselink, Thomas Fontaine, Kevin Summers, Dennis Whigham, and Richard Novitzki, who provided valuable critical comments while serving on a panel to review the proposed approach. Additional appreciation is given to the peer reviewers for their comments.

References

Brinson, M. M.: 1993, 'A Hydrogeomorphic Classification for Wetlands', U.S. Army Corps of Engineers, Waterways Experiment Station, Vicksburg, MS, Wetlands Research Program Technical Report WRP-DE-4.

Brinson, M. M., Kruczynski, W., Lee, L. C., Nutter, W. L., Smith, R. D. and Whigham, D. F.: 1994, 'Developing an Approach for Assessing the Functions of Wetlands', in: Mitsch, W. J. (ed), Global Wetlands: Old World and New, Elsevier Science, NY, pp. 615–624.

Brinson, M. M. and Rheinhardt, R.: 1996, 'The Role of Reference Wetlands in Functional Assessment and Mitigation', *Ecological Applications* **6**, 69–76.

Chabreck, R. H.: 1972, 'Vegetation, Water, and Soil Characteristics of the Louisiana Coastal Region', Louisiana State University, Agricultural Experiment Station Bulletin 664.

Chabreck, R. H. and Linscombe, J.: 1997, 'Vegetation Type Map of the Louisiana Coastal Marshes', New Orleans, LA, Louisiana Department of Wildlife and Fisheries.

Crowley, P. H.: 1992, 'Resampling Methods for Computation Intensive Data Analysis in Ecology and Evolution', *Annual Review of Ecology and Systematics* **23**, 405–447.

Green, R. H.: 1979, Sampling Design and Statistical Methods for Environmental Biologists, Wiley Interscience, Chichester, England.

Leibowitz, N. C., Squires, L. and Baker, J. P.: 1991, Research Plan for Monitoring Wetland Ecosystems, EPA/600/3-91/010, Environmental Research Laboratory, U.S. Environmental Protection Agency, Corvallis, OR.

Louisiana Coastal Wetlands Conservation and Restoration Task Force (LCWCRTF): 1997, 'The 1997 Evaluation Report to the U.S. Congress on the Effectiveness of Louisiana Coastal Wetland Restoration Projects', Baton Rouge, LA, 76 pp.

Louisiana Coastal Wetlands Conservation and Restoration Task Force (LCWCRTF): 1998, 'Coast 2050: Toward a Sustainable Coastal Louisiana', Baton Rouge, LA, 161 pp.

Malaeb, Z. A.: 1997, 'CWPPRA Sample Size Determination and Sampling Design Issues', 2 Volume Report to Louisiana Department of Natural Resources, Baton Rouge, LA.

Novitzki, R. P.: 1994, 'EMAP-Wetlands: A Program for Assessing Wetland Condition', in: Mitsch, W. J. (ed), Global Wetlands: Old World and New, Elsevier Science, NY, pp. 691-709.

Steyer, G. D., Raynie, R. C., Steller, D. L., Fuller, D. and Swenson, E.: 1995, 'Quality Management Plan for Coastal Wetlands Planning, Protection, and Restoration Act Monitoring Program', Open-file Report 95-01, Louisiana Department of Natural Resources, Coastal Restoration Division, Baton Rouge, 97 pp.

Thornton, K. W., Saul, G. E. and Hyatt, D. E.: 1994, 'Environmental Monitoring and Assessment Program Assessment Framework', EPA/620/R-94/016, U.S. Environmental Protection Agency, Office of Research and Development, EMAP Center, Research Triangle Park, NC.

Underwood, A. J.: 1991, 'Beyond BACI: Experimental Designs for Detecting Human Environmental Impacts on Temporal Variation in Natural Populations', *Australian Journal of Marine and Freshwater Restoration* **42**, 569–587.

Underwood, A. J.: 1992, 'Beyond BACI: The Detection of Environmental Impacts on Populations in the Real, but Variable World', *Journal of Experimental Marine Biology and Ecology* **161**, 145–178.

Underwood, A. J.: 1994, 'On Beyond BACI: Sampling Designs that Might Reliably Detect Environmental Disturbances', *Ecological Applications* **41**, 3–15.

Visser, J. M. and Sasser, C. E.: 1998, '1997 Coastal Vegetation Analysis', Report to Louisiana Department of Natural Resources, Baton Rouge, LA.

STORMWATER TOXICITY IN CHOLLAS CREEK AND SAN DIEGO BAY, CALIFORNIA

KENNETH SCHIFF[1*], STEVEN BAY[1], AND DARIO DIEHL[1]

[1]Southern California Coastal Water Research Project, 7171 Fenwick Lane, Westminster, CA, U.S.A.
*(author for correspondence, e-mail: kens@sccwrp.org)

Abstract. Stormwater discharges from Chollas Creek, a tributary of San Diego Bay, have been shown to be toxic to aquatic life. The primary objective of this study was to provide the linkage between in-channel measurements and potential impairments in the receiving waters of San Diego Bay. This study addressed this objective within the context of four questions: (1) How much area in San Diego Bay is affected by the discharge plume from Chollas Creek during wet-weather conditions?; (2) How much of the wet-weather discharge plume is toxic to marine aquatic life?; (3) How toxic is this area within the wet-weather discharge plume?; and (4) What are the constituent(s) responsible for the observed toxicity in the wet-weather plume?

The stormwater plume emanating from Chollas Creek was dynamic, covering areas up to 2.25 km^2. Approximately half of the plume was estimated to be toxic to marine life, based upon the results of purple sea urchin (*Strongylocentrotus purpuratus*) fertilization tests. The area nearest the creek mouth was the most toxic (NOEC = 3 to 12% plume sample), and the toxicity decreased with distance from the creek mouth. The toxicity of plume samples was directly proportional to the magnitude of plume mixing and dilution until, once outside the plume margin, no toxicity was observed. Trace metals, most likely zinc, were responsible for the observed plume toxicity based upon toxicity identification evaluations (TIEs). Zinc was also the constituent identified from in-channel samples of Chollas Creek stormwater using TIEs on the storms sampled in this study, and in storms sampled during the previous storm season.

Keywords: stormwater, toxicity, TIE, *Strongylocentrotus*, San Diego Bay

1. Introduction

Stormwater inputs are a large source of pollutants discharged to receiving waters around the country (U.S. EPA, 1995a). In southern California, stormwater inputs are among the largest of all sources that discharge pollutants to our coastal water bodies (Schiff *et al.*, 2000). Runoff in the southern California region is exacerbated by the area's expansive urbanization, which increases the number of potential non-point sources and promotes runoff due to a larger proportion of impervious surfaces (e.g., cement). The problem is compounded further as a result of the area's infrequent, but intense rainfall events, which promotes the build-up of potentially toxic constituents.

Previous monitoring of urbanized watersheds in San Diego demonstrated that stormwater runoff discharges significant loads of pollutants and was toxic to aquatic life (Schiff and Stevenson 1996, Skinner *et al.*, 1998). One such watershed is Chollas Creek, a heavily urbanized (> 83% developed) tributary to San Diego Bay. Samples collected near the end of the Chollas Creek channel, approximately 5 kilometers upstream of San Diego Bay, were exposed to both marine and freshwater organisms (Schiff *et al.*, 2001). Chollas Creek runoff was toxic to both the freshwater

Environmental Monitoring and Assessment **81**: 119–132, 2003.

and marine organisms, however, the marine organisms were more sensitive (i.e., their response indicated more toxicity). To determine which constituent(s) was responsible for the observed toxicity, toxicity identification evaluations (TIEs) were also conducted on samples of wet-weather discharges from Chollas Creek. Trace metals, most likely zinc, were the constituents responsible for the toxicity to the purple sea urchin. Managers have added Chollas Creek to the state's list of impaired waterbodies, the 303(d) list.

Although in-channel samples of stormwater discharge were shown to be toxic from Chollas Creek, scientists and managers were unsure what potential effects may exist in the marine receiving waters of San Diego Bay. This is a common problem nationwide for many stormwater monitoring programs that conduct whole effluent toxicity tests. A link between in-channel measurements and measurements in the receiving water environment needed to be established. The primary objective of this study was to provide the linkage between in-channel measurements and potential impairments in the receiving waters of San Diego Bay.

This study attempted to make the linkage between in-channel measurements and receiving waters by answering four questions: 1) How much area in San Diego Bay is affected by the discharge plume from Chollas Creek during wet-weather conditions?; 2) How much of the areal extent of the wet-weather discharge plume is toxic to marine aquatic life?; 3) How toxic is the area within the wet-weather discharge plume?; and 4) What are the constituent(s) responsible for the observed toxicity in the wet-weather plume?

2. Methods

2.1 General Approach

This project was completed in four integrated tasks, each associated with one of the study questions. First, the discharge plume was mapped based upon physical water quality parameters including salinity, temperature, and turbidity that are tracers of wet-weather discharges. The second task was to assess the extent of the toxicity within the discharge plume by sampling multiple locations along the gradient of plume influence. The purple sea urchin (*Strongylocentrotus purpuratus*) was used as the test species. The magnitude of toxicity was then assessed by conducting toxicity tests with dilution series at a site near the most concentrated part of the plume and at the in-channel site upstream of the bay. Finally, toxicity identification evaluations (TIEs) were conducted on samples from the most concentrated portion of the plume and from the in-channel sampling site upstream of the bay during the same storm events to determine the toxic constituents of concern.

2.2 Field Sampling

Plume mapping and receiving water sampling occurred during or immediately following rainfall, attempting to capture the maximum extent of the plume. A Sea-

Bird Electronics pumped SBE 19 CTD (2 Hz internal recording) was used to make all physical water quality measurements. Sensors on the instrument were temperature, conductivity, pressure (Paine strain gauge, 100 psia), and a SeaTech transmissometer (660 nm, 25 cm path length). The pump delivered constantly flowing (18 mL/s) water over the temperature and conductivity sensors, so that slow descent or tow speeds had minimal bias on sensor measurements.

Surface water measurements were made with the CTD hung in the water using a downrigger. The downrigger submerged and stabilized the instrument at an average depth of 0.4 meter while the research vessel cruised at slow speeds (< 5 knots). This configuration minimized boat turbulence and plume mixing. A differential global positioning system (dGPS) was connected to the Sea-Bird Electronics data collection system so that latitude and longitude were appended to the CTD data stream. The CTD data were collected at half-second intervals and dGPS positions were updated every two seconds.

Water column profiles were also taken across and along shore of the Chollas Creek mouth. At each water column profile site, surface water samples were taken for analysis of toxicity, trace metals, and total suspended solids (TSS). All water samples were immediately placed on Blue Ice in a hard plastic ice chest.

Composited storm samples were collected from within the channel of Chollas Creek at a site upstream of San Diego Bay. These samples were collected using an automated, flow-paced sampler as described in Schiff *et al.* (2001).

2.3 Toxicity Testing

All samples of stormwater and receiving water were evaluated for toxicity using the purple sea urchin fertilization test (U.S. EPA, 1995b). Details of the test method are given in Schiff *et al.* (2001). In brief, the test consisted of a 20-minute exposure of sperm to the samples, eggs were then added and given 20 min for fertilization to occur. Toxic effects are expressed as a reduction in fertilization percentage relative to controls. The purple sea urchins used in the tests were collected from the intertidal zone in northern Santa Monica Bay. The tests were conducted in glass shell vials containing 10 mL of solution at a temperature of 15°C. All stormwater samples were adjusted to a salinity of 34 g/kg with the addition of hypersaline brine prepared by freezing and partially thawing seawater. Three parameters were used to describe the magnitude of toxicity: highest concentration not producing toxicity (No effect concentration, NOEC), concentration producing a 50% response (EC50), and toxic units (TU, 100/EC50).

2.4 Toxicity Identification Evaluations

2.4.1 Toxicant Characterization (Phase I TIEs)

A modified Phase I TIE (U.S. EPA, 1991; 1996) was conducted simultaneously with baseline testing of stormwater and receiving water samples to minimize holding time and any possible associated change in toxicity. Test conditions were the same as for the baseline test, except that a reduced number of replicates (from five to three) were tested as recommended by U.S. EPA guidance. The salinity of each water sample was adjusted to 34 g/kg before the application of the treatments.

Four treatments were applied to each sample. Ethylenediaminetetraacetic acid (EDTA), a chelator of metals, was added to the test samples at a concentration of 60 mg/L. Sodium thiosulfate (STS), a treatment that reduces oxidants such as chlorine and also decreases the toxicity of some metals, was added to separate portions of each sample at a concentration of 50 mg/L. The EDTA and STS treatments were begun at least 1h prior to the addition of the test organisms to allow interaction with the sample. Samples were centrifuged for 30 min to remove particle-borne contaminants. A portion of the centrifuged sample (200 to 600 mL) was passed through a 1 or 2 g Varian Bond Elut C-18 solid phase extraction column in order to remove non-polar organic compounds. The filtrate was retained for toxicity testing. The C-18 columns were sealed and stored under refrigeration for later elution during Phase II testing. A control sample (laboratory dilution water) was included with each type of treatment to verify that the manipulation itself was not causing toxicity.

2.4.2 Toxicant Identification and Confirmation (Phase II and III TIEs)

The Phase II TIE tests with sea urchin sperm focused upon obtaining information to account for the effectiveness of the EDTA and C-18 treatments during the Phase I studies. These tests had three objectives: (1) to determine whether metal concentrations in the samples were sufficient to cause toxicity, (2) to investigate the effect of C-18 column treatment on trace metal toxicity, and (3) to determine whether non-polar compounds were present in toxic amounts.

The influence of trace metals was assessed by measuring the concentration of dissolved metals in the stormwater and plume samples analyzed for toxicity. Dissolved fractions of trace metals were extracted from the samples with ammonium pyrrolidine dithiocarbamate (APDC) after Bloom and Crecelius (1984). The chelated precipitate was captured on a membrane filter, which was then digested in nitric acid and subsequently analyzed using a high resolution inductively coupled plasma mass spectrometer (ICP/MS) (US EPA, 1983). The resulting data were compared to previously established toxicity thresholds for selected metals.

An experiment was conducted to determine whether extraction using a C-18 column altered the amount of toxicity in a sample contaminated with trace metals. Samples of filtered laboratory seawater were spiked with toxic concentrations of either copper (40 μg/L) or zinc (30 μg/L). Each sample was split into two portions, and one portion was passed through a C-18 column using the same procedure applied in previous Phase I TIE experiments. A sea urchin fertilization test was then conducted to compare the toxicity of the C-18 extracted and unaltered samples.

The presence of toxic concentrations of non-polar compounds was investigated by measuring the toxicity of solvent eluates from the Phase I C-18 columns. Each C-18 column was eluted with methanol (MeOH) followed by dichloromethane (DCM) for the February sample or with MeOH only for the March sample. The solvent eluates were exchanged into isopropanol and adjusted in volume so that when a 0.5% solution was made in seawater, a concentration factor of 2 to 3 times the original sample was achieved. Isopropanol was used since MeOH and DCM introduces toxicity to control samples. This maximum concentration and two additional dilutions (50 and 25%) were tested for toxicity using the sea urchin fertilization test.

Confirmation was achieved by first comparing the estimated TUs based upon concentration in each sample to the observed TUs. A second confirmation was achieved by plotting the concentration of the identified toxicants versus reduction in fertilization.

3. Results

3.1 Storm Event Characteristics and Sampling

Plume mapping and toxicity studies were conducted during three storm events that covered a range of rainfall and tidal conditions. The first event occurred on January 25, 2000 and was relatively small; rainfall amounts were less than 0.1 cm. Plume mapping required 3.47 hours to complete and occurred across both the ebb and flood cycle of a neap tide; tidal heights increased 0.3 meter during surface water measurements. Only two samples were collected for toxicity. The first sample was very near the creek mouth, and the second was collected in the bay in an area outside of the runoff influence. An in-channel sample was not collected upstream of the bay. The second event occurred on February 12, 2000 and was larger with rainfall amounts of nearly 1.0 cm. The plume mapping during this event occurred exclusively during flood tide; tidal heights increased 0.7 meter during the surface water measurements. Plume mapping required 4.72 hours to complete. Nine samples were collected for toxicity analysis. One sample was collected from the in-channel site upstream of the bay, and eight samples were collected from receiving water sites that ranged from nearest the creek mouth to the open bay outside of the plume influence. The third event occurred on March 5, 2000 and was the largest event sampled. Rainfall quantities exceeded 1.6 cm and plume mapping occurred exclusively during ebb tide; tidal heights decreased 1.8 meters during the surface water measurements. Plume mapping activities required 5.12 hours to complete. Nine samples were also collected for toxicity analysis during this event. One sample was collected from the in-channel site upstream of the bay, and eight samples were collected from receiving water sites that ranged from nearest the creek mouth to the open bay outside of the plume influence.

3.2 PLUME MAPPING

The extent of the stormwater plume varied by size of storm, but was large enough to extend across the bay beyond navigable waters (Figure 1). The area of decreased salinity, defined as 32 practical salinity units (psu), and increased turbidity, relative to open bay water, extended from less than 0.02 km^2 during the smallest storm event to 2.25 km^2 during the second runoff event. The most concentrated portions of the plume were always located nearest the creek mouth, within the channel leading to the bay, and decreased away from the creek mouth as mixing and dispersion with open bay water occurred. During larger events, sufficient runoff volumes were discharged so that plumes were advected more than 2 km away from the Chollas Creek mouth out into the open bay.

Freshwater runoff plumes floated over the denser bay water and formed lenses that were thickest near the creek mouth and thinnest out near the margins (Figure 2). The depth of the plume nearest the creek mouth ranged from 2 to 4 meters, depending upon the amount of rainfall (i.e., discharge volume) and overall water depth. The vertical water column structure showed that complete mixing occurred within the first 400 meters of the creek mouth during the first and smallest event, and extended more than 1,500 meters offshore during the second runoff event. The second storm event also penetrated deepest into the water column.

Turbidity (transmissivity in beam C units) measurements also clearly defined the margins of the plume (Figure 2). Transmissivity results were similar to the salinity measures for the cross-shore water column structure; the thickest plume penetration was nearest the creek mouth, and the freshwater lens thinned near the plume margin out in the open bay. Transmissivity at the margins changed up to 20% within a distance of meters. These changes were so dramatic they were visible to field crews aboard the research vessel during surveys.

3.3 SPATIAL EXTENT OF TOXICITY

Tests of San Diego Bay surface water samples collected during three runoff events revealed the presence of toxicity offshore of Chollas Creek. Toxicity to sea urchin sperm was detected in all samples collected within the portion of the plume containing 10% runoff (as determined from salinity measurements). Fertilization (normalized to the control response) in these samples ranged from 5 to 74%. The highest magnitude of toxicity was found in samples taken near the mouth of Chollas Creek. Reference samples of San Diego Bay water collected from 0.3 to 1.5 km outside of the plume were non-toxic, with fertilization values of 88 to 100%. Changes in toxicity corresponded to changes in salinity across the plume gradient emanating from Chollas Creek.

The boundaries of the discharge plume and the toxicity plume differed (Figure 3). In the second runoff event, the toxic portion of the plume (the region producing (80% fertilization), extended across 1.03 km^2 of the bay and comprised approxi-

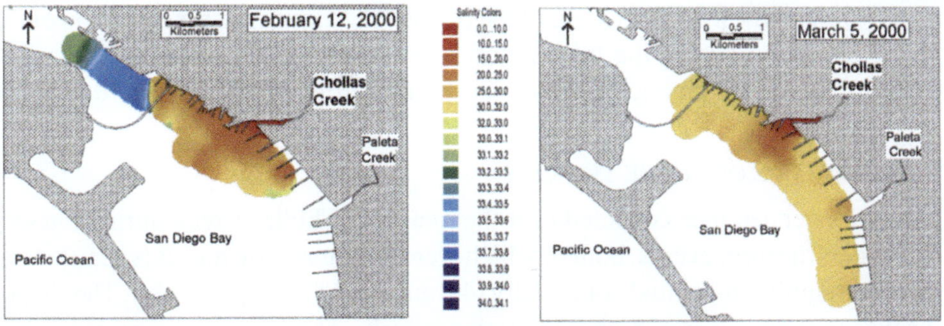

Figure 1. Surface plume mapping of salinity (in practical salinity units, psu) offshore Chollas Creek during two separate storm events (February 12 and March 5, 2000).

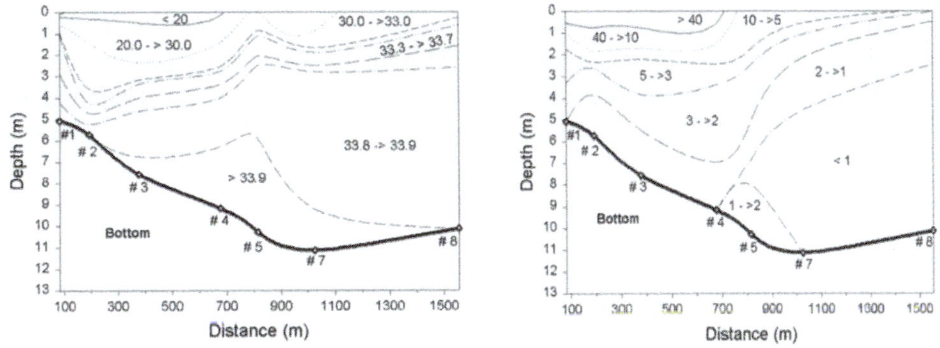

Figure 2. Salinity (practical salinity units, psu) and turbidity (Beam C attenuation in meters) in water column cross-section moving offshore from the Chollas Creek mouth into San Diego Bay following the second storm event (February 12, 2000).

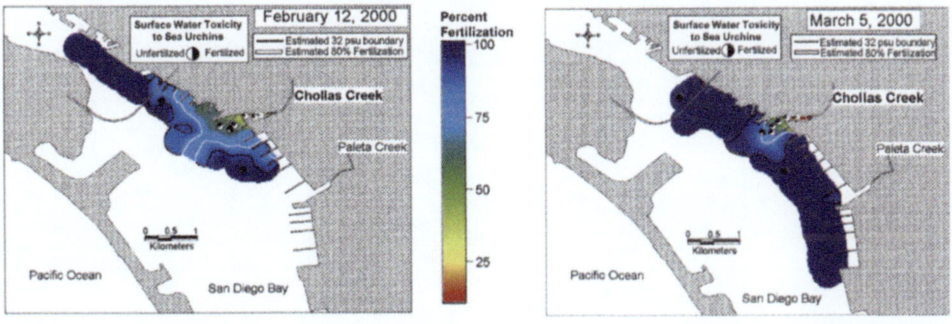

Figure 3. Map of surface layer water toxicity to sea urchins in San Diego Bay associated with Chollas Creek discharges following two separate storm events. The colors show the estimated percent of sea urchin fertilization based upon the toxicity results for water samples (indicated by pie diagrams) and measured salinity in the surface layer.

mately 50% of the physical extent of the plume. The size of the toxic portion of the plume measured during the third runoff event was smaller, covering an area of 0.24 km^2, as was the physical extent of the plume during the third event (0.98 km^2).

3.4 MAGNITUDE OF PLUME TOXICITY

Surface water samples collected near the mouth of Chollas Creek during the second and third storm events, both of which were dominated by runoff (salinity <1 g/kg), had a similar magnitude of toxicity (4.0 and 3.9 TUs, respectively). The EC50s for these samples were 25 and 26%, respectively. The no effects concentrations (NOECs) for these samples were 3 and 12% for the second and third events, respectively. These data indicate that strong toxic effects are expected to occur in portions of the plume containing 25% runoff.

The overall magnitude of toxicity in the plume samples was similar to that measured from Chollas Creek in-channel samples during each runoff event. The channel composite was slightly more toxic (5.9 TUs) than the plume sample during the second event, while the creek composite sample for the third event contained 3.4 TUs, which was similar to the plume sample. The NOECs for the creek composites (3 and 12%) were the same as those measured for the plume samples, indicating that the threshold concentration for toxicity was similar in both types of samples.

Similarities in the magnitude of toxicity were also evaluated by comparing the dose-response curves of the creek composite sample to the toxicity of San Diego Bay surface water samples collected during each event (Figure 4). The toxicity of surface water samples was similar to that measured in the corresponding creek composite sample diluted to contain a similar concentration of runoff. This indicates that most of the plume toxicity could be accounted for by dilution of creek discharge.

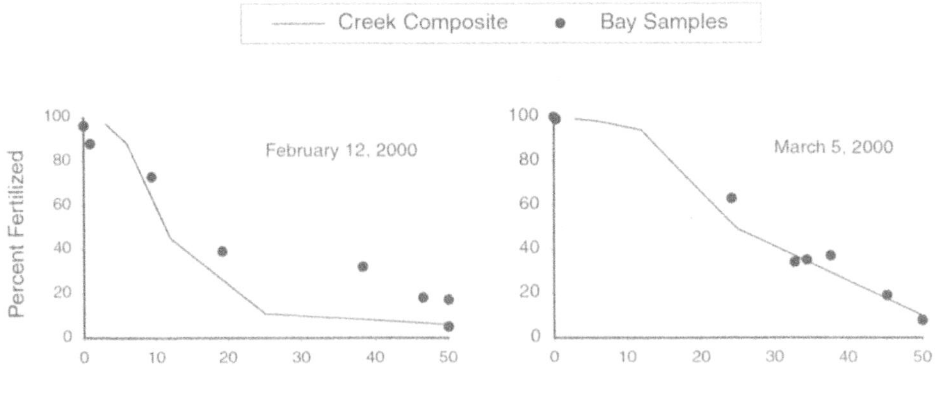

Figure 4. Comparison of sea urchin toxicity in San Diego Bay surface water samples to the dose-response curve obtained from in-channel samples obtained from Chollas Creek during storms two (February 12, 2000) and three (March 5, 2000).

3.5 IDENTIFICATION OF TOXICANTS

Phase I TIE tests consistently demonstrated that EDTA effectively eliminated toxicity and STS was ineffective for both in-channel and both plume samples (Table 1). Particle removal by centrifugation was only partially effective in samples from one event. Treatment using a C-18 solid phase extraction column removed the toxicity in three of four samples, including both in-channel samples. Elimination of toxicity by EDTA is an indication that trace metals are likely constituents responsible for toxicity.

Table 1. Summary of Phase I toxicity identification evaluation (TIE) results for stormwater samples taken from Chollas Creek and bay surface water samples taken from the runoff plume near the mouth of the creek. Each sample tested contained 50% runoff.

Date	*Sample Type*	*Baseline Toxicity (% fertilization)*	*TIE Treatment (% fertilization)*			
			Centrifuge	*C-18 Column*	*EDTA*	*STS*
02/12/00	Creek	6	8	98	100	6
02/12/00	Plume	5	12	5	91	5
03/05/00	Creek	10	41	95	85	15
03/05/00	Plume	8	46	98	88	7

No toxicity was detected in solvent eluates of the C-18 columns that removed toxicity during the Phase I tests. The DCM or MeOH eluates of the columns tested at 2 to 3 times the original concentration in the aqueous sample were non-toxic to sea urchin sperm; fertilization in these samples was 93 to 100% of the control value. These results indicate that the toxicants removed by the C-18 column either had polar characteristics or had degraded during storage prior to elution.

A Phase II experiment was conducted in order to investigate the effectiveness of the C-18 column treatment on dissolved trace metals. The toxicity of seawater solutions spiked with toxic concentrations of copper or zinc was eliminated after C-18 column treatment (Figure 5). The effect was particularly noteworthy for zinc, where a solution causing a 55% reduction in fertilization was rendered non-toxic after passage through the column. These tests used the same type of column and procedures that were used in the Phase I experiments.

Eight metals were identified as having elevated concentrations in the composite sample relative to those in surface water (Table 2). Aside from iron, dissolved zinc and copper were present in the highest concentrations in samples collected from within Chollas Creek. Zinc levels from the most concentrated portion of the plume (essentially 100% stormwater discharge) ranged from 92 to 152 μg/L; these concentrations were above the zinc EC50 determined from previous fertilization experiments (29 μg/L). Copper concentrations in the same samples ranged from 37

Table 2. Salinity (practical salinity units), sea urchin toxicity (% of control fertilization), and dissolved concentration of selected metals in Chollas Creek and San Diego Bay water samples. Plume samples were collected from the portion of the plume having the lowest salinity while the bay samples were collected outside of the plume. Median effects concentration (EC50) and no effects concentration (NOEC) values for selected metals were calculated from independent fertilization experiments.

Date	Sample Type	Salinity (psu)	Fertilized (%)	Metal Concentration (µg/L) Sb	Cr	Cu	Fe	Mo	Ni	V	Zn
	EC50					31					29
	NOEC					20–44					5–10
01/25/00	Plume	32.90	74	0.19	0.12	2.46	6	8.92	1.24	2.03	16.50
01/25/00	Bay	33.80	94	0.14	0.16	3.21	2	8.61	0.91	2.04	11.40
02/12/00	Creek	0	6	17.56	2.68	51.20	2,184	25.80	10.72	11.14	150.80
02/12/00	Plume	0	18	3.45	1.93	36.88	1,752	14.60	8.75	12.57	92.00
02/12/00	Bay	32.40	92	0.18	0.15	3.08	1	8.95	0.94	2.06	13.80
03/05/00	Creek	0	10	8.32	3.36	63.00	3,080	25.20	12.44	11.28	146.00
03/05/00	Plume	0	8	6.20	3.30	65.60	3,260	26.40	13.78	13.44	152.20
03/05/00	Bay	32.96	100	0.16	0.14	3.22	3	9.44	10.42	2.00	14.30

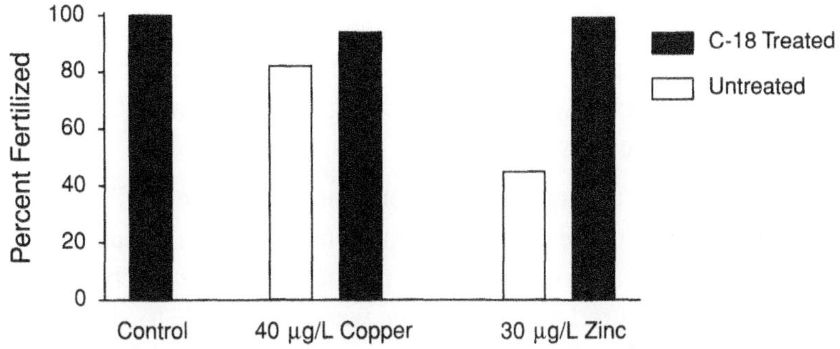

Figure 5. Effect of C-18 column treatment on the toxicity of metal-spiked seawater to sea urchins.

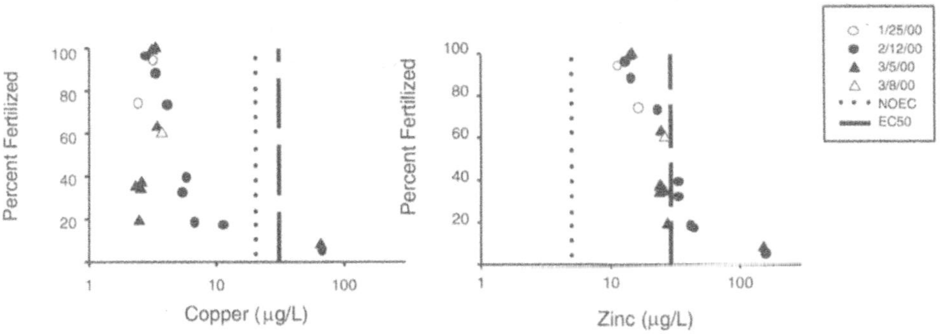

Figure 6. Comparison of sea urchin toxicity to dissolved zinc concentrations in San Diego Bay surface water samples collected during three storm events. Reference lines indicate the no effect concentration (NOEC) and median effect concentration (EC50), which was determined from laboratory experiments by Bay et al. (1997).

to 66 µg/L, which also exceeded the EC50 for copper (31 µg/L) measured in reference toxicant tests conducted during this study. Background concentrations of zinc and copper in the bay during this study were approximately 14 µg/L and 3 µg/L, respectively. Little is known about the sensitivity of sea urchin sperm to the other metals in Table 2.

Gradients in dissolved zinc concentration among the San Diego Bay surface water samples showed a dose response relationship with sea urchin fertilization (Figure 6). Fertilization percentage declined as zinc concentration increased, and the magnitude of effect was consistent with the dose response relationship for zinc determined from prior laboratory experiments with spiked seawater. For example, large reductions in fertilization were observed in samples containing dissolved zinc concentrations similar to or greater than the EC50 value. A similar plot for dissolved copper also indicated a correspondence between toxicity and concentration, but the data were more variable and did not correspond as well with the NOEC and EC50 for copper-spiked seawater (Figure 6). Many of the toxic samples contained less copper than the NOEC, indicating that the toxicity was not caused by copper. The two samples containing copper concentrations above the EC50 were highly toxic, however.

Toxic unit calculations based upon zinc and copper EC50s confirmed that these two metals were present in sufficient concentrations to cause most of the toxicity measured in the heart of the runoff plume and in the in-channel samples. Most of the toxicity measured in each sample was attributable to zinc, and the total amount of toxicity calculated from zinc plus copper accounted for 72 to 109% of the total toxicity measured in the samples.

4. Discussion

The wet-weather events sampled from Chollas Creek generated large volumes of stormwater runoff that extended across more than 2 km^2 of San Diego Bay. These plumes were turbid and, because the freshwater discharge was less dense than seawater, they formed thin lenses of 2 to 5 meters thickness on the surface of the bay. Thin stormwater plumes of large areal extent have been observed offshore other urban watersheds in southern California. Plumes extending over 4 km^2 and 3 meters thickness were measured offshore Ballona Creek following storm events of 7 cm precipitation and greater (Bay *et al.*, 1999). Similarly, thin plumes were measured in upper Newport Bay following storm events discharging runoff from urban areas of the San Diego Creek watershed (Lee *et al.*, 1999).

Several limitations apply to the plume measurements made as part of this study. The first limitation is attempting to describe a fixed plume extent when plumes are dynamic water quality features. For example, our plume surveys lasted up to 6 hours, yet plumes can shift and change in size or shape over the course of a tidal cycle. A second limitation to our plume mapping abilities was the difficulty of attempting to extrapolate our measurements to unmeasured areas. We used a kriging method to construct our plume maps. However, kriging techniques work best under steady state conditions, an assumption that was violated at times during our surveys because plumes, particularly during flooding events, are dynamic. On the other hand, we were able to find the plume margins using distinct salinity and turbidity signatures. Hence, our estimates of plume extent were verified using empirical data. Moreover, the repetitiveness of the plume extent over multiple storm events adds confidence to our assessment. In fact, the storms we measured were not extraordinarily large and a greater extent of impact may occur during larger-sized events.

Plume toxicity was directly related to toxicity measured in the Chollas Creek discharge. The magnitude of toxicity near the creek mouth was similar to the toxicity measured from within the creek. The magnitude of toxicity then decreased as the plume mixed away from the creek mouth until, reaching outside the plume margin, the toxicity was not observed. In fact, the toxicity could be predicted as a function of salinity. Further, the magnitude of toxicity in the creek discharge was similar to that of the discharge plume after accounting for dilution of the plume with bay waters. These findings are consistent with plume toxicity measurements taken offshore of other creek mouths in southern California (Bay *et al.*, 1997). Offshore of Ballona Creek, toxicity observed using the purple sea urchin test was in direct proportion to the amount of stormwater present in that plume.

Not only was the plume toxicity proportional to the amount of freshwater mixing, but the same constituents were identified as the toxic agents from the creek and from the plume. Zinc was implicated as the primary toxicant to the purple sea urchin in Chollas Creek discharges during the 1997/1998 wet season (Schiff *et al.*, 2001) and again during this study. Zinc was also implicated as the primary toxicant to the purple sea urchin in the stormwater plumes in San Diego Bay that emanated from Chollas Creek during this study. However, this finding was different from TIE testing with other species, such as the freshwater daphnid *Ceriodaphnia*, which was sensitive to other constituents (e.g., organophosphate pesticides). We do not expect all species to be affected similarly by stormwater exposures. While we observed some very compelling evidence that Chollas Creek stormwater affected purple sea urchins, this finding may not be applicable to other marine species. We continue to conduct research on the linkages between single species toxicity tests and ecological consequences at the population and community levels.

Stormwater plumes are not the only source of pollutants that generate water column toxicity in San Diego Bay. The toxicity observed in the wet-weather plume offshore of Chollas Creek was greater than the toxicity observed in other 303(d) listed portions of San Diego Bay (RWQCB unpublished data). For example, purple sea urchins exposed to receiving waters sampled from the Shelter Island Yacht Harbor in San Diego Bay yielded NOECs of 20% whereas samples taken from the Chollas Creek wet-weather plume yielded NOEC values as low as 3%. Similarly, dissolved zinc and copper concentrations were as much as an order of magnitude higher in the wet-weather plume than in the yacht harbor. However, the toxicity in the Shelter Island waters persists every day of the year while the toxicity in the plume persists for at least 3 days following runoff events. When prioritizing areas for remediation, managers need to decide whether increased toxicity for shorter time periods is more important than persistent low-level toxicity.

Acknowledgments

This project was partially funded through the State Water Resources Control Board. The authors are grateful for the review provided by the City of San Diego, Port of

San Diego, US Navy SPAWAR, and San Diego BayKeeper. In-channel sampling was conducted by URS-Woodward Clyde America. Field sampling was accomplished using the R/V Fisherette. Many thanks to David Tsukada.

References

Bay, S., B. Jones, and K. Schiff. 1999. Study of the impact of stormwater discharge on Santa Monica Bay. SeaGrant Program, University of Southern California. Los Angeles, CA USCSG-TR-02-99

Bay, S., D. Greenstein, A. Jirik, and A. Zellers. 1997. Toxicity of stormwater from Ballona and Malibu Creeks. pp. 96-104 in: S. Weisberg, C. Francisco, and D. Hallock (eds.), Southern California Coastal Water Research Project Annual Report 1995–1996. Westminster, CA.

Bloom, N.S., and B.A. Crecelius. 1984. 'Determination of silver in seawater by coprecipitation with cobalt pyrrolidine dithiocarbamate and Zeeman graphite-furnace atomic absorption spectrometry,' *Analytica Chemica Acta* **156**:139–145.

Lee, G.F., S. Taylor, and D. Neiter. 1999. Review of Existing Water Quality Characteristics of Upper Newport Bay, Orange County, California, and its Watershed and Results of Aquatic Life Toxicity Studies Conducted during 1997–98 in the Upper Newport Bay Watershed. Prepared for California State Water Resources Control Board, Santa Ana Regional Water Quality Control Board, and Orange County Public Facilities and Resources Department. Lee and Associates. Madero, CA.

Schiff, K., M.J. Allen, S. Bay, and E. Zeng. 2000. Seas at the millennium: the Southern California Bight. pp. 467-472 in: C. Sheppard (ed.), Seas at the Millennium. Elsevier Press. London, UK.

Schiff, K., S. Bay, and C. Stransky. 2001. Characterization of stormwater toxicity from Chollas Creek, San Diego. pp 71-84 In Weisberg, S. and D. Elmore (eds) Southern California Coastal Water Research Project Annual Report 1999–2000. Westminster, CA.

Skinner, L., A. dePeyster, and K. Schiff. 1998. 'Developmental effects of urban stormwater in San Diego County, California in medaka (*Oryzias latipes*) and inland silverside (*Menidia beryllina*),' *Archives of Environmental Toxicology* **37**:227–235

U.S. Environmental Protection Agency, Environmental Monitoring and Systems Laboratory (US EPA). 1983. Methods for the chemical analysis of water and wastes. EPA-600/4-79-020. Cincinnati, OH.

U.S. Environmental Protection Agency (U.S. EPA). 1991. Methods for Aquatic Toxicity Identification Evaluation: Phase I Toxicity Characterization Procedures. Second Edition. EPA/600/6-91/003. U.S. Environmental Protection Agency, Environmental Research Laboratory. Duluth, MN.

U.S. Environmental Protection Agency (U.S. EPA). 1995a. National Water Quality Inventory; 1994 Report to Congress. EPA/841/R-95/005. Washington, DC.

U.S. Environmental Protection Agency (U.S. EPA). 1995b. Short-term Methods for Estimating the Chronic Toxicity of Effluents and Receiving Waters to West Coast Marine and Estuarine Organisms. First Edition. EPA/600/R-95/136. U.S. Environmental Protection Agency, National Exposure Research Laboratory. Cincinnati, OH.

U.S. Environmental Protection Agency (U.S. EPA). 1996. Marine Toxicity Identification Evaluation (TIE): Phase I guidance document. EPA/600/R-96/054. U.S. Environmental Protection Agency, National Health and Environmental Effects Research Laboratory. Narragansett, RI.

MANAGING TROUBLED DATA: COASTAL DATA PARTNERSHIPS SMOOTH DATA INTEGRATION

STEPHEN S. HALE[1*], ANNE HALE MIGLARESE[2], M. PATRICIA BRADLEY[3], THOMAS J. BELTON[4], LARRY D. COOPER[5], MICHAEL T. FRAME[6], CHRISTOPHER A. FRIEL[7], LINDA M. HARWELL[8], ROBERT E. KING[9], WILLIAM K. MICHENER[10] DAVID T. NICOLSON[11], AND BRUCE G. PETERJOHN[12]

[1]U. S. Environmental Protection Agency (USEPA), Office of Research and Development, National Health and Environmental Effects Research Laboratory, Atlantic Ecology Division, Narragansett, RI; [2]NOAA Coastal Services Center, Charleston, SC; [3]USEPA, Mid-Atlantic Integrated Assessment, Ft. Meade, MD; [4]NJ Department of Environmental Protection, Trenton, NJ; [5]Southern California Coastal Water Research Project, Westminster, CA; [6]U. S. Geological Survey (USGS), Reston, VA; [7]Florida Marine Research Institute, St. Petersburg, FL; [8]USEPA, Gulf Breeze, FL; [9]USEPA, Washington, DC; [10]University of New Mexico, Albuquerque, NM; [11]Smithsonian Institution, Washington, DC; [12]USGS, Laurel, MD

[*](author for correspondence, email: hale.stephen@epa.gov)

Abstract. Understanding the ecology, condition, and changes of coastal areas requires data from many sources. Broad-scale and long-term ecological questions, such as global climate change, biodiversity, and cumulative impacts of human activities, must be addressed with databases that integrate data from several different research and monitoring programs. Various barriers, including widely differing data formats, codes, directories, systems, and metadata used by individual programs, make such integration troublesome. Coastal data partnerships, by helping overcome technical, social, and organizational barriers, can lead to a better understanding of environmental issues, and may enable better management decisions. Characteristics of successful data partnerships include a common need for shared data, strong collaborative leadership, committed partners willing to invest in the partnership, and clear agreements on data standards and data policy. Emerging data and metadata standards that become widely accepted are crucial. New information technology is making it easier to exchange and integrate data. Data partnerships allow us to create broader databases than would be possible for any one organization to create by itself.

Keywords: coastal monitoring, coastal databases, information management, data partnerships, data integration

1. Introduction

The 1990 National Research Council report *Managing Troubled Waters: The Role of Marine Environmental Monitoring* (Boesch *et al.*, 1990) called for improved monitoring of coastal waters to better address environmental problems. One of the stated criteria for success was "Not only must data be gathered, but attention must be paid to their management, synthesis, interpretation, and analysis." Ten years later, monitoring data are still "troubled" because individual monitoring programs often manage their data in individual ways that make it difficult to find and bring data together for syntheses. No single database can hold all the data needed for assessing the marine environment. Nevertheless, we are making progress. The Internet has made it easier to publish, find, and access data. Standards for data and metadata, essential for data exchange and integration, are emerging. Many new developments are being driven by disciplines other than environmental monitor-

Environmental Monitoring and Assessment **81**: 133–148, 2003.

ing, such as biodiversity and genetic databases. Partnerships of monitoring programs lead to better data and data products.

A workshop panel (Managing Troubled Waters: Partnerships Smooth Data Integration), 26 April 2001, Pensacola Beach, FL, discussed various ways that partnerships can improve information management systems. Questions posed to the panelists were:

1. Why do we need to share and integrate data?
2. What barriers exist to data sharing and integration?
3. How can data partnerships help overcome these barriers?
4. How can we make data partnerships work?
5. What are some examples of successful data partnerships?
6. What data standards can be widely adopted?
7. What new technology would be helpful?

The workshop considered the current state of coastal data systems, particularly environmental monitoring databases, and how partnerships have improved data systems, assessments, and management decisions. This paper summarizes the workshop findings.

2. How Partnerships Lead to Better Data

2.1 WHY DO WE NEED TO SHARE AND INTEGRATE DATA?

2.1.1 Environmental Problems Require Data from Many Sources

The questions currently being addressed by environmental scientists transcend historic disciplinary, jurisdictional, spatial, and temporal boundaries and deal with problems that cannot be solved by individuals or small teams from single disciplines. For example, the National Science Foundation's Long-Term Ecological Research (LTER) program addresses multidecadal phenomena at population to ecosystem scales (Franklin *et al.*, 1990), and the Biocomplexity in the Environment Program (Colwell, 1998) searches for unifying principles that govern complex biological phenomena. Successfully addressing these questions requires approaches that are holistic, interdisciplinary, broad-scale, and long-term.

Environmental monitoring programs on regional and national scales need multiple sources of data to assess regional environments (NRC, 1995; CENR, 1997). No single organization can conduct wall-to-wall multidisciplinary studies or maintain a comprehensive data management system with all the necessary data. Assessments of watersheds and ecoregions often need data from multiple political jurisdictions. States need to share data because they contain broadly divergent physiographic provinces and ecoregions that connect with other states.

We need to further develop generalized global data for taxa and natural communities in order to guide conservation actions and assess environmental impacts. The nation's natural resources continue to be impacted by residential and urban growth and by invasions of exotic species. Sharing and integrating existing biological information is paramount to improved decision making.

Sharing environmental monitoring data maximizes its usefulness. Users include resource managers, agency regulators, researchers, educators, and the general public, each of whom needs a particular type of information. Scientists may need raw data for their own analyses, while other users may prefer synthesized results. Data may need to be summarized at different geographic scales for different users.

2.1.2 Sharing Data Leads to Better Information

Shared data can improve research by providing greater spatial, temporal, and disciplinary coverage than individual organizations can offer. They can address cumulative impacts, detect emerging properties, and offer new perspectives on old problems.

Some recent advances in data partnerships are in bioinformatics. Several biodiversity database efforts are underway, such as Species 2000, the Census of Marine Life, and the Ocean Biogeographic Information System (Bisby, 2000; Edwards *et al.*, 2000; Grassle, 2000; Malakoff, 2000; Pennisi, 2000; Vieglais *et al.*, 2000). Some of these consolidate information from distributed databases. Gathering knowledge about all species is driven by the need to protect species and habitats, assess effects of environmental change, and understand ecosystems and evolutionary biology.

Sharing data may minimize redundancy and save money that could be better spent on collecting supplementary data. Integrating data provides a way for organizations to build repositories of cohesive, high-quality environmental data that may help break down "data empires."

Environmental managers face a wide array of challenges. Regulations and laws are produced at many levels of political organizations and often conflict or overlap. Balancing societal needs for employment, transportation, and other things with environmental concerns is a complex task. Managers need diverse and detailed information for issues that can shift with the political winds. One day we worry about global warming; the next, we have an energy crisis. The same diverse pool of data is needed to respond to both.

Finally, issues can emerge for which we are not prepared. Sharing data may allow us to address these more quickly than if a new program has to be started. For example, when the toxic dinoflagellate *Pfiesteria* became a problem in U.S. Mid-Atlantic estuaries in 1997, agencies were able to work together to combine existing data and design appropriate new monitoring. Without these partnerships, the issue would have been much more difficult.

2.2 WHAT BARRIERS EXIST TO DATA SHARING AND INTEGRATION?

2.2.1 Technical Barriers

General lack of consistency—in resources, technology, data collection methods, experimental design, data quality standards, laboratory procedures—among potential partners hinders efforts to share and integrate data. First, data must be comparable before they can be integrated. Lack of a standard sampling protocol can make data integration difficult. If one group uses a 0.5-mm mesh to sieve benthic invertebrate grabs while another group uses a 1.0-mm mesh, their data are incompatible. Data may come from incomparable sampling designs—sampling targeted at known problem areas versus random sampling. Scale is another important consideration; for example, when federal programs collect data at scales too small to be used for local questions (hydrography at 1:100,000 instead of the 1:12,000 needed by states). As analytical procedures become more sensitive, new data may not be comparable with old data. Knowledge of data quality is another necessary condition for data integration.

Different programs may use widely differing data formats, codes, database design, directories, systems, and metadata. Common variables may lack standard names. Lack of standardization can inhibit data integration (NRC, 1995). Conversely, adoption of standard methods helps compare results from different studies, helps document methods for publications and metadata, and can reduce costs (Michener, 2000a).

Data discovery remains challenging, particularly for data that are not "registered" in directories such as the Global Change Master Directory and have not been cited in peer-reviewed publications (as may be the case with monitoring data). Inadequate and incomplete metadata often prevent correct interpretation of data previously collected by others and retard subsequent data integration efforts. Comprehensive metadata are critical for slowing "information entropy," the normal degradation in informational content of data with time (Michener *et al.*, 1997).

The vast amount of information made available by the explosive growth of the Internet presents another challenge. Data producers are now able to easily deliver this information worldwide. To take advantage of this capacity, however, takes data standards, consistent tools and processes, and good quality assurance.

Data archival is a problem that needs more attention. Environmental data must be kept accessible, even as hardware, software, and storage media change. Storage media degrade with time. It is difficult for small data groups to have an effective long-term archival system. Data archival (indexing, documentation) takes a significant effort; often there is inadequate funding. Too little attention is given to maintaining data beyond the lifetime of the project.

2.2.2 Social, Cultural, and Organizational Barriers

Sociological aspects of collaborative studies can be as important as technical aspects (NRC, 1995; Farrey *et al.*, 1999). Each organization has its own rules, regulations, and management practices. The differences can seem insurmountable as organizations begin to collaborate. Organizational interactions are central to the success or failure of data interfacing; problems revolve around rewards, priorities, ingrained attitudes, data ownership, and cooperation (NRC, 1995). Programs are generally focused towards specific disciplines, each with a different jargon. Partnering becomes complex when combining biologists, chemists, geologists, economists, political scientists, and managers. As in diplomacy, one must learn the other's terminology and frame of reference.

Agency policies can complicate things. States may get conflicting messages from different federal agencies. Timely processing of data can become an issue, because data from multiple sources often need to be from the same periods in order to be compared statistically. Policies on releasing data can also be problematic if they reduce access to essential information.

Other barriers to sharing data include concerns over publication rights, credit and other "rewards" for developing databases (Porter and Callahan, 1994). A lack of full and open data sharing can create long time lapses between data collection and distribution.

2.2.3 Level of Effort Barriers

Inadequate funding for data centers is a common problem (Hale, 1999). That forces data centers to focus on their own data management needs rather than interacting with others to address broader needs. Many organizations that collect environmental monitoring data do not have adequate staff or expertise to run their own database management systems, servers, and networks. They may not have staff to create metadata, which is more important for shared data than it is for local data. Preparation of metadata for large field sampling partnerships may take as much time as the preparation of the data (Hale *et al.*, 2000).

Data integration and synthesis, as well as the collaboration necessary to make them possible, requires time, money, and personnel that are often overlooked in project budgets (Michener, 2000b). Data translation costs in compiling common analytical databases from multi-organizational field data can be high (Hale *et al.*, 2000).

Not all valuable monitoring data are available in electronic format, nor are all electronic data readily available on the Internet. Some groups who do not have money and technological expertise to create a firewall may be reluctant to open their databases to the Internet.

2.3 How can data partnerships help overcome these barriers?

2.3.1 Overcoming Technical Barriers

A partnership may be able to sample more parameters with wider spatial and temporal coverage than organizations could individually. It also allows studies to be more comprehensive and thereby satisfy more constituencies, reduce redundancy, lead to more nested scalable datasets, and help aggregate local data at regional scales.

Partnerships can reduce many technical barriers by agreeing on core data standards and the flow of data. For example, participants may decide to standardize their taxonomy and to use the same chemical analytical methods. Data can have a longer life span when collected by two or more groups. Some biodiversity databases (e.g., FishNet) have been able to link disparate databases by getting all partners to agree on a few data standards and by developing common software (Pennisi, 2000).

Partners can create inventories of monitoring programs and data directories. Centralized metadata can point to distributed data. Partnerships can help groups share tools and transfer technology. Making new, costly technology such as supercomputers available to partners provides incentives to cooperate and share data. Advanced data centers can help smaller centers develop standards, design databases, archive their data, and construct metadata.

States are concerned about issues of scale (e.g., bathing beaches versus ecoregions) and regulatory endpoints (e.g., health of humans versus ecosystems) in designing coastal and estuarine monitoring networks and interpreting their results. State programs nested within regional sampling designs may allow local data to be interpreted more robustly.

Following are some examples of how partnerships have overcome technical barriers. The National Coastal Assessment (Coastal 2000) is a partnership between the U.S. EPA's Environmental Monitoring and Assessment Program (EMAP) and the 24 coastal states to develop a national coastal monitoring program (EMAP, 2001). It integrates state-level data to regional and national scales by using a core set of sampling methods and data standards for chemistry and bottom communities. The Long-Term Ecological Research (LTER) Network Information System is a cooperative, federated database system at 24 sites that range from urban watersheds to tropical rainforests to marine and dry-valley sites in Antarctica (Briggs and Su, 1994; Ingersoll *et al.*, 1997; Brunt, 1998; Baker *et al.*, 2000). Two of its key technical elements are: (1) modularity in software design; and (2) development of independent prototypes that are tested rigorously for scalability and sustainability prior to broad use by all sites (Baker *et al.*, 2000).

The National Biological Information Infrastructure (NBII, 2001) addresses the complexity of species and ecosystems on national and regional scales by developing new standards and tools that make it easier to find and integrate biological resources information. The Integrated Taxonomic Information System (ITIS, 2001) has developed a reference system on the scientific names of the biota of North America. The success of this multi-agency effort is rooted in the development of common standards, procedures, and software.

STORET is a repository for water quality, biological, and physical data collected by federal, state and local agencies, Indian Tribes, and others (STORET, 2001). The rigorous database schemata and intricate database management system ensures that data from individual sites are compatible. STORET has succeeded because there is a common need by all 50 states to store the data collected in support of the Clean Water Act. LabNet, at the other end of the standardization spectrum, is a set of tools developed by the National Association of Marine Laboratories to provide a means to locate, retrieve, and process data that reside at participating marine laboratories (LabNet, 2001).

The Gulf of Mexico Aquatic Mortality Network (GMNET) is a joint effort by state agencies, EPA, USGS, NOAA, and the Gulf of Mexico Program to integrate aquatic mortality events across the five Gulf states (GMNET, 2001). GMNET achieves uniformity with a standard approach and data format shared by all states.

2.3.2 Overcoming Social, Cultural, and Organizational Barriers

Successful data partnerships are marked by equal levels of commitment, ownership, influence, and perception of benefits among the partners. Partners share a common interest in group success. They contribute from their strengths while learning from the other partners.

The level of detail in a partnership ranges from single sites to entire regions. Small organizations contribute detailed data that big ones need but lack. Larger organizations provide smaller ones with data that would otherwise be too expensive for them to acquire and too difficult to manage.

Organizations that have to exchange data tend to work out data standards and thereby ease the pressure on all. Pride and peer pressure encourage organizations to give high-quality data to their partners and to do it on time. Having to exchange data makes groups better manage their own data and metadata. In short, joint ownership promotes stewardship.

Sharing and integrating data are also helped by the broadened base of technology, resources, and personnel brought into a partnership. Partnerships bridge the gaps between groups of environmental professionals and establish platforms for sharing innovative ways to manage data, create data standards, and manage people and money. Partners who work towards shared goals and products come to understand each other's culture, mission, and discipline.

Questions of data 'ownership' or 'stewardship' can often be addressed by agreeing early on data flow and data use. Journals such as *Ecological Archives* provide reasons for academics to publish databases.

Partnerships can also stabilize long-term monitoring programs. Just as interwoven biological communities are resilient to perturbations, interactions among organizational communities may offer resilience to the waxing and waning of individual budgets. The partners' investments reduce the chance of the partnership's work being dissolved.

Several partnerships have overcome sociological barriers. Coastal 2000 works because states get useful data they would not have gotten otherwise and because EMAP gets state data that can be aggregated regionally and nationally. NSF's LTER program has succeeded by creating well-defined partnerships, a central focus, and explicit responsibilities for its individual sites.

The NBII partnership works because federal agencies and private industry cooperated to deliver bioinformatics tools and products that were mutually beneficial and needed by its customers. ITIS was formed by four federal agencies (NOAA, EPA, USGS, USDA) and the Smithsonian Institution's National Museum of Natural History. Because each partner has a mission to inventory, monitor, research, or manage biological resources and information, all members need a common vocabulary of scientific names of organisms. ITIS has a distributed group of taxonomic experts to lend scientific credibility to its database.

STORET is operated independently by states that control access and set their own rules for its use. This autonomy from EPA is both the largest change from historical STORET usage and, in the eyes of many users, its most desirable feature. In fact, the states have provided many ideas for the structure of the database. LabNet laboratories are motivated to disseminate their data because that gives them access to data from the other laboratories. GMNET has succeeded by integrating data to understand larger-scale environmental changes.

The Mid-Atlantic Integrated Assessment (MAIA) formed partnerships among federal, state, and local agencies in the U. S. Mid-Atlantic region to integrate research and monitoring information to provide answers to policy and management questions (Bradley *et al.*, 2000; MAIA, 2001). An Inventory of Environmental Monitoring Programs includes maps of sampling locations, lists of measurements, and design information for hundreds of federal environmental monitoring programs (Jackson and Gant, 1998; NEMI, 2001). All of the above examples keep data management close to the data collectors.

Sometimes the demand for an integrated data product is strong enough to drive the formation of successful partnerships. The New Jersey Atmospheric Deposition Network is a partnership with a broad-based need for deposition data from several data sources (NJDEP, 2001). It works because all partners agreed to a format and structure for sharing data. Florida's Marine Resources GIS combines LANDSAT data with aerial photography to map estuarine and marine fisheries habitat. The widespread demand for these data has led to dozens of mutually beneficial data-sharing agreements (Friel, 1994). The Bight '98 regional survey of the southern

California Bight brought together 62 different agencies (possibly a record number for a collaborative venture). The only way that this could have succeeded was to develop a common data format for data exchange. Bight '98 got agreement on such a format by actively engaging all the partners, so that they each had a voice in creating the data standard, and then openly discussing differences of opinion. Partners were motivated to participate because planned joint analyses could not proceed with incomplete data (SCCWRP, 2001).

2.3.3 Overcoming Level of Effort Barriers

The combined resources of partnerships may be great enough to eliminate problems that their individual partners could not handle. Similarly, management groups such as steering committees can decide how and where to apply their limited resources for the greatest good. The shared voice of a partnership may be more effective in getting attention, resources, and money than its individual members could.

When resources can be effectively shared, partnerships can reduce hardware and software costs. A larger partner may donate the expertise and consulting that make it possible to aggregate the group's data. Data from small organizations who lack firewalls may be housed by a larger group with a firewall. Partnerships can give data sources with paper files an incentive to computerize their data. Partnerships also provide an impetus to standardize data procedures.

Level of effort barriers have been overcome by many programs. Coastal 2000 was started with funds from EPA to bolster routine state monitoring. Standards established by ITIS partners, along with a scientifically credible system, save time and money, avoid redundancy, and foster multiagency support. STORET helps individual sites use its software by offering extensive training, a listserver, and user conferences. MAIA provides an economy of scale by reducing redundancy and helping small programs get access to better data. The Chesapeake Information Management System (CBP, 2001) succeeds because of high-level commitment, strong partners, and reliable annual funding. LabNet has taken a simpler approach. It requires little from its data sources, who control their own data. The prototype application uses flat text files that can be created by almost any application. All of the above programs have invested energy, money, and collaborative leadership to start their partnership and keep it going.

2.4 HOW CAN WE MAKE DATA PARTNERSHIPS WORK?

A data partnership needs more than just the removal of barriers; it also needs active promotion. Keys to successful collaboratives in general (Chrislip and Larsen, 1994) include: good timing and clear need, strong partners, broad-based involvement, credibility and openness of process, commitment of high-level and visible leaders, support of established authorities, diminution of mistrust and skepticism, strong leadership, interim successes, and a shift to broader concerns. When integrating diverse environmental data, as much consideration be given to organizational issues as to technical ones (NRC, 1995).

Scurlock *et al.* (in press) suggest that the environmental research community take the following actions to facilitate collaboration and data sharing: "provide incentives for sharing and archiving data; recognize data with metadata as valuable research products; establish a universal citation policy for data; establish guidelines or expectations for metadata; develop data distribution, analysis, and archive centers; and ensure long-term financial and institutional support."

2.4.1 Create Shared Need for Data

Data partnerships work best by fostering dependence on the highest-quality data and information needed to address problems. This dependence requires a commitment from data providers, decision-makers, and funding sources; adequate resources; and an atmosphere that promotes teamwork. Projects with data product goals that require the researchers to have access to each other's data can motivate people to share data (NRC, 1995).

The Distributed Oceanographic Data System (DODS, 2001) relies on a joint need for physical oceanographic data. The Ocean Biogeographic Information System (Grassle, 2000) and the Census of Marine Life (Malakoff, 2000) succeed because their collaborative leadership ties solutions to a common need for data.

2.4.2 Use Collaborative Leadership and Create Open Communication

Among ten keys to success, a case study of data management practices of six environmental monitoring programs (NRC, 1995) listed four sociological items: plan for and build on success, use a collaborative approach, account for human behavior and motivation, and consider the needs of participants as well as users. Partnerships require open interactions, sincere effort, and shared goals and responsibilities. Strong collaborative leadership is important (Chrislip and Larsen, 1994), as is interaction among scientists, information managers, and program managers. The objectives and expectations of the partnership should be clearly defined and communicated. Workshops may help create standards and build consensus. Decision-making should be shared, and executive managers must agree on the important questions. Partnerships do not work because of altruistic motives, nor do they work by top-down mandates. They work only if each party believes that its own objectives will be enhanced by its participation.

Written documents are useful. Data policies (e.g., USGLOBEC, 1994; NRC, 1995; USGCRP, 1998; Hale *et al.*, 1999) adopted by the partners can eliminate many problems. Such statements can lay out standards and expectations for sharing data. The Chesapeake Bay Program (CBP, 2001) has shown that a written agreement on cooperative data management, signed by high-level authorities, can provide significant force. An information management plan (e.g., Hale *et al.*, 1999) can help outline a clear path. McLaughlin *et al.* (2001), based on their experience with integrating data from several organizations, suggest that database developers need to have clear-cut project objectives and to let data contributors know if the objectives change.

2.4.3 Invest Energy and Money

Partners have to be willing to invest the extra work and communication that it takes to make the partnership work. Partnerships need to be promoted from within each organization. Successful partnerships ultimately require organizations to commit to or jointly fund a common initiative. Paid partners may be more likely to submit their data on time, but partnerships without committed resources depend solely on individual champions to make them succeed. The energy needed to form and maintain partnerships is best spent when there is a likelihood of a long-term relationship. Just like chemical reactions, which stop unless the reagents are replenished, partnerships need a continual supply of reagents (energy and money) to sustain the reaction.

Data management should be funded as a continuing core element of monitoring programs (Boesch *et al.*, 1990). Data management and integration need priority, visibility, and adequate funding, even if that means less sampling.

2.5 WHAT DATA STANDARDS AND NEW TECHNOLOGY WOULD BE HELPFUL?

2.5.1 Data Standards

The best data standards are those that will be maintained by organizations willing to make long-term commitments. The two most common pitfalls are developing standards independent of the users (e.g., scientists and others who will be asked to comply with standards) and developing comprehensive standards without the user-friendly software needed to use them.

Most potential partners agree on the need for certain federal standards such as state names and abbreviations, ZIP codes, and others. They also agree in principle to use standard names for chemical substances and biological organisms, although complete lists do not yet exist. Some common standards used in environmental monitoring include: National Spatial Data Infrastructure (FGDC, 2001), Biological Data Profile and the Biological Names and Taxonomy Data Standards (NBII, 2001), the Integrated Taxonomic Information System for species (ITIS, 2001), and Chemical Abstracts Service codes for chemicals (CAS, 2001). Core data elements for reporting water-quality monitoring results have been proposed (USGS, 2001a).

Different levels of standardization exist. STORET is an example of maximum standardization (strict data dictionary and database schema for distribute databases). Another model uses more (CBP, 2001) or less (EMAP, 2001; GMNET, 2001) centralized metadata to point to partially-standardized distributed data. LabNet and DODS use minimum standardization. Yet another approach is to use common fields only for things that all partners record, such as date, location, and environmental conditions (USGLOBEC, 1994).

Uniform standards for metadata ensure that essential information about the data will be provided to users. The Federal Geographic Data Committee's content standard for digital geospatial metadata (FGDC, 1998, 2001) is mandatory for U.S. federal agencies. The NBII Biological Profile to the FGDC standard on geospatial metadata provides a foundation for biological data. Meeting the FGDC content standard does not preclude using other metadata formats, such as LabNet's Meta-Lite (LabNet, 2001), STORET's metadata fields (STORET, 2001), and nongeospatial metadata (Michener *et al.*, 1997). The Darwin Core metadata standard is being used to describe specimen collections in museums and herbariums (Bisby, 2000; Vieglais *et al.*, 2000). Metadata standards are essential for extending the useful life of archived data.

2.5.2. New Technology

Rapid increases in environmental data holdings, along with increasing demand for the data and information needed for informed environmental decision-making, will require additional advances in data mining and knowledge discovery tools (Michener, 2000b). Also needed are middleware applications for translating among major databases, collaborative computing tools to improve communication, and software tools for developing metadata. Additional mechanisms and outlets are needed for peer-reviewed publication of data to provide the professional credit and quality publications that can facilitate data partnerships.

Portals offer great potential for creating and promoting partnerships. A portal is a customized electronic gateway that provides a virtual workspace for individuals to interact and share information. Portals provide collaborative technologies such as video conferencing tools, online discussion forums, and document sharing. These allow researchers, program managers, and educators to access real-time information, collaborate, and discuss issues with fellow portal members. An example portal for migratory bird data that allows users to get bird population and habitat data at various geographic scales is being developed by the USGS Patuxent Wildlife Research Center (USGS, 2001b).

New interest in biological databases (bioinformatics), resulting from genetic and biodiversity databases, coupled with fast-paced developments in information technology, are opening up many possibilities for sharing data. The Species Analyst system provides interoperability among museum and herbarium specimen collections by using the Z39.50 protocol to locate data from several different databases and return records that are transformed into Extensible Mark-up Language

(XML) for use by Web browsers (Bisby, 2000; Edwards *et al.*, 2000; Vieglais *et al.*, 2000). Species 2000 uses Common Object Request Broker Architecture (CORBA) to poll an array of up to 200 global species databases on the Internet to provide a functional virtual catalog of all known species (Bisby, 2000).

BioBot is an example of a robot application that helps users discover biological information on the Internet (NBII, 2001). Commercial search engines such as AltaVista and Yahoo are currently indexing and harvesting information from a number of biological sites. BioBot can simultaneously search that existing biological meta-information, as well as NBII information that has already been indexed by the NBII Search Engine.

3. Recommendations

- Form coastal data partnerships to help surmount technical, organizational, and level of effort barriers to data integration;
- Create shared need for data among partners to encourage better data stewardship;
- Use collaborative leadership to start the partnership and keep it going;
- Maintain open communication to make sure all partners are satisfied with direction and progress;
- Invest enough money and effort to achieve results;
- Adopt uniform data standards wherever possible to make data exchange easier; and
- Embrace relevant new technology to improve data management, communication, and data sharing.

Acknowledgments

We thank W. Galloway, R. Haebler, D. Cobb, B. Melzian, K. Rahn, and three anonymous reviewers for useful suggestions. Although the research described in this article has been funded in part by the U.S. Environmental Protection Agency, it has not been subjected to Agency review. Therefore, it does not necessarily reflect the views of the Agency. Mention of trade names or commercial products does not constitute endorsement or recommendation for use. This is contribution number AED-01-083 of the U.S. Environmental Protection Agency, Office of Research and Development, National Health and Environmental Effects Research Laboratory, Atlantic Ecology Division, Narragansett, RI.

References

Baker, K.S., Benson, B.J., Henshaw, D.L, Blodgett, D., Porter J.H. and Stafford, S.G.: 2000, 'Evolution of a Multisite Network Information System: The LTER Information Management Paradigm', *BioScience* **50**, 963–978.

Bisby, F.A.: 2000, 'The Quiet Revolution: Biodiversity Informatics and the Internet'. *Science* **289**, 2309–2312.

Boesch, D.F., Schubel, J.R., Bernstein, B.B., Eichbaum, W.M., Garber, W., Hirsch, A., Holland, A.F., Johnson, K.S., O'Connor, D.J., Speer, L. and Wiersma, G.B.: 1990, Managing Troubled Waters: The Role of Marine Environmental Monitoring, Committee on a Systems Assessment of Marine Environmental Monitoring, National Research Council. National Academy Press, Washington, DC, USA. 125 pp.

Bradley, M.P., Brown, B.S., Hale, S.S., Kutz, F.W., Landy, R.B., Shedlock, R., Mangold, R., Morris, A., Galloway, W., Rosen, J.S., Pepino, R. and Wiersma, B.: 2000, 'Summary of the MAIA Working Conference', *Environmental Monitoring and Assessment* **63**, 15–29.

Briggs, J.M. and Su, H.: 1994, 'Development and Refinement of the Konza Prairie LTER Research Information Management Program', in: Michener, W.K, Brunt, J.W. and Stafford, S.G (eds), Environmental Information Management and Analysis: Ecosystem to Global Scales, Taylor & Francis, London pp. 87–100.

Brunt, J.W.: 1998, 'The LTER Network Information System: A Framework for Ecological Information Management', in: Aguirre-Bravo, C. and. Franco, C.R (eds), North American Science Symposium: Toward a Unified Framework for Inventorying and Monitoring Forest Ecosystem Resources, 2–6 Nov 1998, Guadalajara, Mexico, Proceedings RMRS-P-12US, Department of Agriculture, Forest Service, Rocky Mountain Research Station, Fort Collins, CO, pp. 435–440.

CAS [Chemical Abstracts Service]: 2001, CAS Web site, [http://www.cas.org].

CBP [Chesapeake Bay Program]: 2001, CBP Web site, [http://www.chesapeakebay.net].

CENR [Committee on Environment and Natural Resources]: 1997, Integrating the Nation's Environmental Monitoring and Research Networks and Programs: A Proposed Framework, CENR, National Science and Technology Council, Washington, DC, USA. 102 pp.

Chrislip, D.D. and Larson, C.E.: 1994, Collaborative Leadership, Jossey-Bass Publishers, San Francisco, CA, USA. 192 pp.

Colwell, R.: 1998, 'Balancing the Biocomplexity of the Planet's Living Systems: A Twenty-first Century Task for Science', *BioScience* **48**, 786–787.

DODS [Distributed Oceanographic Data System]: 2001, DODS Web site, [http://www.unidata.ucar.edu/packages/dods].

Edwards, J.L., Lane, M.A and Nielsen, E.S.: 2000, 'Interoperability of Biodiversity Databases: Biodiversity Information on Every Desktop', *Science* **289**, 2312–2314.

EMAP [Environmental Monitoring and Assessment Program]: 2001, EMAP Web site, [http://www.epa.gov/emap].

Farrey, P.M., Mooney-Seuss, M.L. and Tausig, H.C. (eds.): 1999, Out of the Fog: Furthering the Establishment of an Electronic Environmental Information Exchange for the Gulf of Maine, New England Aquarium, Aquatic Forum Series, Report 99-1, Boston, MA, USA.

FGDC [Federal Geographic Data Committee]: 2001, FGDC Web site, [http://www.fgdc.gov].

FGDC [Federal Geographic Data Committee]: 1998, Content Standard for Digital Geospatial Metadata, version 2.0, FGDC-STD-001-1998, Washington, DC, USA.

Franklin, J.F., Bledsoe, C.S. and Callahan, J.T.: 1990, 'Contributions of the Long-term Ecological Research Program', *BioScience* **40**, 509–523.

Friel, C.A.: 1994, 'Nautical Information as a Component of a Marine Geographic Information System', Chapter 5 in: Charting a Course into the Digital Era, National Research Council, National Academy Press, Washington, D.C., pp. 37–43.

GMNET [Gulf of Mexico Aquatic Mortality Network]: 2001, GMNET Web site, [http://pelican.gmpo.gov].

Grassle, J.F.: 2000, 'The Ocean Biogeographic Information System (OBIS): An On-line, Worldwide Atlas for Accessing, Modeling and Mapping Marine Biological Data in a Multidimensional Geographic Context', *Oceanography* **13**, 5–7.

Hale, S. S.: 1999, 'How to Manage Data Badly (Part 1)', *Bulletin of the Ecological Society of America* **80**(4), 265–268.

Hale, S.S., Bahner, L.H. and Paul, J.F.: 2000, 'Finding Common Ground in Managing Data Used for Regional Environmental Assessments', Environmental Monitoring and Assessment 63, 143–157.

Hale, S., Rosen, J., Scott, D., Paul, J. and Hughes, M.: 1999, EMAP Information Management Plan: 1998–2001, EPA/620/R-99/001, U.S. Environmental Protection Agency, Office of Research and Development, Research Triangle Park, NC, 164 pp.

Ingersoll, R.C., Seastedt, T.R and Hartman, M.: 1997, 'A Model Information Management System for Ecological Research', *BioScience* **47**, 310–316.

ITIS [Integrated Taxonomic Information System]: 2001, ITIS Web site, [http://www.itis.usda.gov].

Jackson, L.E. and Gant, M.P.: 1998, 'An Interactive, Spatial Inventory of Environmental Data in the Mid-Atlantic Region', *Environmental Monitoring and Assessment* **51**, 325–329.

LabNet: 2001, LabNet Web site, [http://www.mbl.edu/labs/NAML/labnet].

MAIA [Mid-Atlantic Integrated Assessment]: 2001. MAIA Web site, [http://www.epa.gov/maia].

Malakoff, D.: 2000. 'Grants Kick Off Ambitious Count of All Ocean Life', *Science* **288**, 1575–1576.

McLaughlin, R.L., Carl, L.M., Middel, T., Ross, M., Noakes, D.L.G., Hayes, D.B. and Baylis, J.R. : 2001, 'Potentials and Pitfalls of Integrating Data from Diverse Sources: Lessons from a Historical Database of Great Lakes Stream Fishes', *Fisheries* **26**, 14–23.

Michener, W.K.: 2000a, 'Research Design: Translating Ideas to Data', in: Ecological Data: Design, Management and Processing, Michener, W.K. and Brunt, J.W. (eds), Blackwell Science, Oxford, Great Britain, pp. 1–24.

Michener, W.K.: 2000b, 'Ecological Knowledge and Future Data Challenges', in: Ecological Data: Design, Management and Processing, Michener, W.K. and Brunt, J.W. (eds), Blackwell Science, Oxford, Great Britain, pp. 162–174.

Michener, W.K., Brunt, J.W, Helly, J., Kirchner, T.B. and Stafford, S.G.: 1997, 'Non-geospatial Metadata for the Ecological Sciences', *Ecological Applications* **7**, 330–342.

NBII [National Biological Information Infrastructure]: 2001, NBII Web site, [http://www.nbii.gov].

NEMI [National Environmental Monitoring Initiative]: 2001, NEMI Web site, [http://www.epa.gov/monitor].

NJDEP [New Jersey Department of Environmental Protection]: 2001, NJDEP Web site, [http://www.state.nj.us/dep].

NRC [National Research Council]: 1995, Finding the Forest in the Trees: The Challenge of Combining Diverse Environmental Data, NRC, National Academy Press, Washington, DC, USA. 129 pp.

Pennisi, E.: 2000. 'Taxonomic Revival', *Science* **289**, 2306–2308.

Porter, J.H. and Callahan, J.T.: 1994, 'Circumventing a Dilemma: Historical Approaches to Data Sharing in Ecological Research', in: Environmental Information Management and Analysis: Ecosystem to Global Scales, Michener, W.K., Brunt, J.W and Stafford, S.G. (eds), Taylor and Francis, Ltd., London, pp. 193–202.

SCCWRP [Southern California Coastal Water Research Project]: 2001, SCCWRP Web site, [http://www.sccwrp.org].

Scurlock, J.M.O., Olson, R.J., McCord, R.A. and Michener, W.K.: In press, 'Data Banks: Archiving Ecological Data and Information', in: Encyclopedia of Global Environmental Change, John Wiley & Sons, New York.

STORET: 2001, STORET web site, [http://www.epa.gov/OWOW/STORET].

USGCRP [U.S. Global Change Research Program]: 1998, Data Management for Global Change Research. Policy Statements for the National Assessment Program, July 1998. National Science Foundation, Washington, DC, USA.

USGLOBEC [U. S. Global Ocean Ecosystems Dynamics]: 1994, U. S. GLOBEC Data Policy, Report No. 10. Woods Hole, MA, [http://globec.whoi.edu].

USGS [U. S. Geological Survey]: 2001a, 'Data Elements for Reporting Water Quality Results of Chemical and Microbiological Analytes', *Federal Register* **66**(52), 15273–15275.

USGS [U. S. Geological Survey]: 2001b, Migratory bird research Web site, [http://www.mbr-pwrc.usgs.gov].

Vieglais, D., Wiley, E.O., Robins, C.R. and Peterson, A.T.: 2000, 'Harnessing Museum Resources for the Census of Marine Life: The FISHNET project', *Oceanography* **13**, 10–13.

INCIDENCE OF STRESS IN BENTHIC COMMUNITIES ALONG THE U.S. ATLANTIC AND GULF OF MEXICO COASTS WITHIN DIFFERENT RANGES OF SEDIMENT CONTAMINATION FROM CHEMICAL MIXTURES

JEFFREY L. HYLAND[1*], W. LEONARD BALTHIS[1], VIRGINIA D. ENGLE[2]
EDWARD R. LONG[3], JOHN F. PAUL[4], J. KEVIN SUMMERS[2], AND ROBERT F. VAN DOLAH[5]

[1]NOAA National Ocean Service, Charleston, SC, U.S.A.; [2]U.S. Environmental Protection Agency, Gulf Breeze, FL, U.S.A.; [3]ERL Environmental, Salem, OR, U.S.A.; [4]U.S. Environmental Protection Agency, Narragansett, RI, U.S.A.; [5]SC Department of Natural Resources, Charleston, SC, U.S.A.

[*](author for correspondence, email: jeff.hyland@noaa.gov)

Abstract: Synoptic data on concentrations of sediment-associated chemical contaminants and benthic macroinfaunal community structure were collected from 1,389 stations in estuaries along the U.S. Atlantic and Gulf of Mexico coasts as part of the nationwide Environmental Monitoring and Assessment Program (EMAP). These data were used to develop an empirical framework for evaluating risks of benthic community-level effects within different ranges of sediment contamination from mixtures of multiple chemicals present at varying concentrations. Sediment contamination was expressed as the mean ratio of individual chemical concentrations relative to corresponding sediment quality guidelines (SQGs), including Effects Range-Median (ERM) and Probable Effects Level (PEL) values. Benthic condition was assessed using diagnostic, multi-metric indices developed for each of three EMAP provinces (Virginian, Carolinian, and Louisianian). Cumulative percentages of stations with a degraded benthic community were plotted against ascending values of the mean ERM and PEL quotients. Based on the observed relationships, mean SQG quotients were divided into four ranges corresponding to either a low, moderate, high, or very high incidence of degraded benthic condition. Results showed that condition of the ambient benthic community provides a reliable and sensitive indicator for evaluating the biological significance of sediment-associated stressors. Mean SQG quotients marking the beginning of the contaminant range associated with the highest incidence of benthic impacts (73-100% of samples, depending on the province and type of SQG) were well below those linked to high risks of sediment toxicity as determined by short-term toxicity tests with single species. Measures of the ambient benthic community reflect the sensitivities of multiple species and life stages to persistent exposures under actual field conditions. Similar results were obtained with preliminary data from the west coast (Puget Sound).

Keywords: Benthic indicators, chemical contaminants, sediment quality, predicting benthic stress, U.S. Atlantic and Gulf of Mexico estuaries.

1. Introduction

Although concentrations of some chemical contaminants have shown decreasing trends in estuarine sediments in recent years (Valette-Silver, 1992), potential biological impacts from sediment contamination remain a key environmental concern in many coastal areas around the U.S. Numerical sediment quality guidelines (SQGs) have been developed to provide interpretive tools for assessing the biological significance of individual chemical contaminants in sediments (e.g., see review by Chapman, 1989). Combining (e.g., by averaging) the ratios of individual chemical concentrations in a sample relative to corresponding SQG values

Environmental Monitoring and Assessment **81**: 149–161, 2003.

also has been used to quantify potentially harmful mixtures of chemical contaminants present at varying concentrations (Long *et al.*, 1998).

An empirical framework for estimating the probability of observing sediment toxicity within different ranges of mean SQG quotients was introduced recently, based on results of amphipod toxicity tests conducted with samples from estuaries nationwide (Long *et al.*, 1998, 2000; Long and MacDonald, 1998). These analyses demonstrated a relatively low incidence of toxicity in samples with the lowest amount of chemical contamination (mean SQG quotients < 0.1) and an increasing incidence of toxicity with increasing contamination. The incidence of highly toxic samples (i.e., mean survival < 80% of controls, and significantly different at $p < 0.05$) was highest at mean ERM quotients > 1.5 and mean PEL quotients > 2.3 (Long and MacDonald, 1998). Hyland *et al.* (1999) applied a similar approach in efforts to develop a basis for evaluating risks of benthic community-level effects in relation to sediment contamination for estuaries of the southeastern U.S. This latter study demonstrated that condition of the ambient benthic community can serve as a reliable and sensitive indicator of potential disturbances resulting from these chemical stressors. Mean SQG quotients associated with a high incidence of benthic impacts (i.e., mean ERM quotients > 0.058 and mean PEL quotients > 0.096) were well below those associated with a similar high incidence of sediment toxicity in the prior laboratory studies.

In the present study, we extend the initial analysis of benthic-contaminant relationships in southeastern estuaries by Hyland *et al.* (1999) to additional data sets from mid-Atlantic and Gulf of Mexico estuaries. The purposes of the study were to determine bioeffect ranges for evaluating risks of chemical contaminant effects on benthic communities in these other coastal regions, and ultimately to expand our understanding of the relationships between chemical exposure levels and risks of adverse effects on resident fauna over broader spatial scales. The use of the term "risk" here means the likelihood of an effect occurring based on its observed incidence (percentage of occurrence) in the sample population.

2. Methods

Data were collected as part of the estuaries component of the U.S. EPA's Environmental Monitoring and Assessment Program (EMAP), conducted in collaboration with the National Status and Trends (NS&T) program of NOAA and multiple other federal, state, academic, and private research institutions. The data were derived from sediment samples collected at 1,389 stations sampled throughout three EMAP estuarine provinces (Figure 1): (1) the Virginian Province (Cape Cod to mouth of Chesapeake Bay) represented by 511 stations sampled during summers 1990–93; (2) Carolinian Province (mouth of Chesapeake Bay to St. Lucie Inlet, FL) represented by 290 stations sampled during summers 1994–97; and (3)

Louisianian Province (Anclote Key, FL to Rio Grande, TX) represented by 588 stations sampled during summers 1991–94. Though many environmental variables were sampled as part of EMAP (Holland, 1990), the present analysis was performed using data on concentrations of chemical contaminants in sediments and condition of benthic macroinfaunal communities measured synoptically at each site.

Concentrations of organic chemicals and metals were measured in a single sample of composited surface sediment (upper 2–3 cm) collected at each station with a 0.04-m^2 Young grab sampler. Typically, 16 inorganic elements, four butyltins, 27 aliphatic hydrocarbons, 44 polynuclear aromatic hydrocarbons (PAHs), 20 polychlorinated biphenyl (PCB) congeners, and 24 pesticides were measured at each station using methods consistent with general EMAP and NS&T protocols (U.S. EPA, 1995; Lauenstein and Cantillo, 1993). Specific methods for the Virginian, Carolinian, and Louisianian Provinces are summarized in Paul *et al.* (1999), Hyland *et al.* (1999), and Summers *et al.* (1992), respectively.

Total chemical contamination in each sediment sample was expressed as the mean of the ratios of individual chemical concentrations relative to corresponding SQG values. Thus, of the total number of chemical contaminants that were measured, only those for which SQGs have been developed were included in this computation (Table 1). These "concentration-to-SQG quotients" were computed using two different types of SQGs: Effects Range-Median (ERM) values (Long *et*

Figure 1. Map of EMAP-Estuaries Provinces showing location of the three provinces included in the present study (larger bolded font). Data supporting this study were from 1,389 stations sampled as follows: 511 stations in the Virginian Province from 1990–93; 290 stations in the Carolinian Province from 1994–97; and 588 stations in the Louisianian Province from 1991–94.

al., 1995) and Probable Effects Level (PEL) values (MacDonald *et al.*, 1996). ERM and PEL values both represent mid-range concentrations of chemicals above which adverse effects on a wide variety of benthic organisms are likely to occur. Both types of guidelines are listed in Table 1 by each corresponding chemical contaminant. Methods used here to compute mean ERM and PEL quotients are the same as those used by Long *et al.* (1998; 2000) and Long and MacDonald (1998) to evaluate the incidence of sediment toxicity within different chemical contaminant ranges. Concentrations of chemical contaminants below analytical detection limits were treated here as zeros and included in the computations.

The condition of benthic macroinfaunal communities was assessed using diagnostic, multi-metric benthic indices specific for each region:

- Virginian Province: EMAP Multivariate Benthic Index (Paul *et al.*, 1999). Component variables are the Gleason diversity index and abundances of tubificid oligochaetes and spionid polychaetes.
- Louisianian Province: EMAP Multivariate Benthic Index (Engle and Summers, 1999). Component variables are Shannon-Wiener diversity (H'); abundances of tubificids; and proportions of capitellid polychaetes, bivalves, and amphipods.

Table 1. Chemicals and corresponding sediment quality guideline (SQG) values used in the calculation of mean SQG Quotients. NA = Not Available.

Chemical	*ERM*[a]	*PEL*[b]	*Chemical*	*ERM*[a]	*PEL*[b]
Metals (μg/g or ppm)			*PAHs (ng/g or ppb)*		
Arsenic	70	41.6	Acenaphthene	500	88.9
Cadmium	9.6	4.21	Acenaphthylene	640	128
Chromium	370	160	Anthracene	1100	245
Copper	270	108	Benzo[a]anthracene	1600	693
Lead	218	112	Benzo[a]pyrene	1600	763
Mercury	0.71	0.7	Chrysene	2800	846
Silver	3.7	1.77	Dibenz[a,h]anthracene	260	135
Zinc	410	271	Fluoranthene	5100	1494
			Fluorene	540	144
Pesticides (ng/g or ppb)			2-Methylnaphthalene	670	201
4,4'-DDD (p,p'-DDD)	NA	7.81	Naphthalene	2100	391
4,4'-DDE (p,p'-DDE)	27	374	Phenanthrene	1500	544
4,4'-DDT (p,p'-DDT)	NA	4.77	Pyrene	2600	1398
Dieldrin	NA	4.3			
Gamma BHC (Lindane)	NA	0.99	*PCBs (ng/g or ppb)*		
Total Chlordane[c]	NA	4.79	Total PCBs	180	189
Total DDTs[d]	46.1	51.7			
			Total No. SQGs Used	24	29

[a]From Long *et al.* (1995)
[b]From MacDonald *et al.* (1996)
[c]Includes alpha-, gamma-, and oxychlordane
[d]Includes 4,4'-DDD, 4,4'-DDE, 4,4'-DDT, 2,4'-DDD, 2,4'-DDE, and 2,4'-DDT

- Carolinian Province: Multi-metric Benthic Index of Biotic Integrity, B-IBI (Van Dolah *et al.*, 1999). Component variables are species richness, total faunal abundance, dominance, and % abundance of pollution-sensitive taxa.

Different approaches were used to develop the three benthic indices and each is based on a different combination of biological attributes as indicated above. However, they are all similar in that multiple, benthic attributes were combined into a single measure designed to maximize the ability to distinguish between degraded versus non-degraded benthic condition. Also, all three indices were developed with methods included to help account for biological variability attributable to key, natural controlling factors (e.g., latitude, salinity, sediment particle size). Efforts to account for such variability focused on factors shown, through multivariate statistical analysis, to have the strongest influence on benthic distributions under non-degraded (reference) conditions within each specific province (e.g., latitude and salinity in the case of the Carolinian Province). Further details on these procedures are included in the references cited above.

Data used in the calculation of benthic indices were derived from replicate macroinfaunal samples collected at each station with a 0.04-m^2 Young grab sampler. Benthic and chemistry samples were collected at each station from separate grabs on the same day (usually < 2 hours apart) and within close proximity to one another. Because sampling was conducted from an anchored boat, both types of samples generally were taken within a 10-m radius of the station coordinates (i.e., within the anchor swing). Benthic samples were collected to a maximum sediment depth of 10 cm (the penetration limit of the grab) and were rejected if < 5 cm (a precautionary measure used to prevent significant loss of fauna concentrated in this upper sediment layer). Contents of the grabs were live-sieved in the field with a 0.5-mm mesh screen, fixed in 10%-buffered formalin with rose bengal, and transferred to the laboratory for further processing. Animals were sorted from sample debris under a dissecting microscope and identified to the lowest possible taxon (usually to species).

The cumulative frequencies (percentages) of stations with degraded vs. non-degraded benthic communities were plotted in relation to ascending values of the two mean SQG quotients. Plots for the Carolinian Province are included here for illustration purposes (Figure 2). For each individual region, 10th-, 50th-, and 90th-percentile points from the cumulative-frequency plots of degraded stations were used as critical points to separate mean SQG quotients into four ranges: those associated with either a low incidence of benthic impacts (values below the corresponding 10th-percentile point), moderate incidence of impacts (values between the 10th- and 50th-percentile points), high incidence of impacts (values between the corresponding 50th- and 90th-percentile points), or very high incidence of impacts (values above the 90th-percentile point).

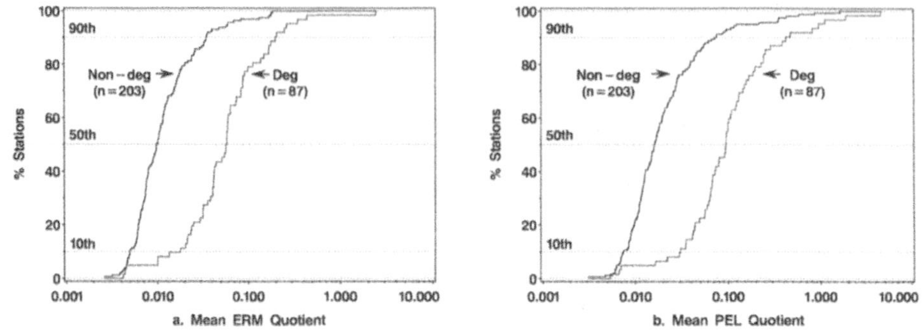

Figure 2. Cumulative percentage of stations with degraded vs. non-degraded benthic assemblages plotted in relation to (a) mean ERM and (b) mean PEL quotients for the EMAP Carolinian Province. Similar plots were generated for the other EMAP provinces (though not shown here).

Data also were merged across provinces to support the derivation of a single set of critical points for all three provinces combined. For this larger combined data set, 5th- and 95th-percentile points were used in place of the 10th and 90th to enhance predictability in the lower and upper contaminant ranges. With the larger merged data set, these narrower limits could be used while still retaining sufficient sample sizes at either end.

In the present analysis, stations with low dissolved oxygen (DO) were excluded to reduce the chance of misinterpreting any benthic responses that may have been influenced by this factor. Stations were excluded if DO concentrations were < 2 mg/L. This range is often associated with a high risk of adverse effects on benthic fauna (Diaz and Rosenberg, 1995). Although this precaution was taken, it is important to recognize the possibility that some of the remaining stations included in the analysis may have experienced low-DO conditions prior to the sampling date. Similarly, confounding effects of other unmeasured stressors (e.g., high concentrations of ammonia or sulfide in sediments), although unknown, may have contributed to the condition of benthic communities at some of these stations as well.

3. Results and Discussion

3.1 Chemical-Contaminant Bioeffect Ranges for the Benthos

Relationships between the incidence of degraded benthic communities and increasing chemical contamination are illustrated in Figure 2. Values of mean SQG quotients along the x-axis were always higher for degraded than non-degraded cases at corresponding cumulative-frequency points along the y-axis. For example, 10% of Carolinian Province stations with a degraded benthos had mean ERM quotients ≤ 0.018, while 10% of those with a healthy benthos had values ≤ 0.005 (Figure 2a). Fifty percent of the stations with a degraded benthos had mean ERM quotients ≤ 0.057, while 50% of those with a healthy benthos had values ≤ 0.010. Hence, the proportion of degraded to non-degraded benthic communities increased with increasing chemical contamination.

Similar patterns were observed for mean PEL quotients (Figure 2b), though individual values of the two different SQG quotients varied for any given sample. Mean ERM quotients were always less than mean PEL quotients (typically within a factor of two). This difference is due to the fact that PEL guidelines are lower than corresponding ERM guidelines for most chemicals (Table 1). Thus, concentration-to-SQG quotients based on ERM values (as a denominator) will be lower than those based on PEL values.

Lowest risks of benthic impacts were observed at mean SQG quotients below the 10th-percentile critical points (Tables 2 and 3). Six to 32% of stations in these lower ranges, depending on the location or type of SQG, had degraded benthic communities. Higher probabilities occurred at mean SQG quotients between the 50th- and 90th-percentile points, with about 54-86% of stations in these ranges having degraded benthic communities. The incidence of degraded benthic communities was highest (73-100%) at stations with quotients above the 90th-percentile points.

Critical points separating chemical contamination into the four concentration ranges varied from region to region (Tables 2 and 3). Such differences were fairly small (generally within a factor of 2) for critical points separating the first two ranges and the largest (up to a factor of 8) for those defining the beginning of the uppermost range. Generally, the critical points were the lowest for the Louisianian Province, highest for the Virginian Province, and in-between for the Carolinian Province. Reasons for these differences are not known but may be related to variations in magnitude of sediment contamination, susceptibility of benthic fauna, or confounding effects of other unmeasured environmental factors across the provinces. Nevertheless, all three provinces showed a consistent response of a very high incidence of degraded benthic condition (73–100% of samples, depending on the location or type of SQG) beginning at relatively moderate chemical-contaminant concentrations (upper ERM critical points of 0.062 to 0.473, and upper PEL critical points of 0.118 to 0.969).

Similar results were obtained when data from all provinces were merged (Table 4). The incidence of degraded benthic communities was lowest (18%) at mean SQG quotients below the 5th-percentile critical points (mean ERM and PEL quotients < about 0.01). The incidence of these effects increased with increasing chemical contamination. For example, a much higher incidence of degraded benthic communities (62–63%) occurred at mean SQG quotients between the 50th and 95th percentile points (i.e., mean ERM quotients of 0.044 to 0.361, and mean PEL quotients of 0.076 to 0.781). The highest incidence of degraded benthic communities (74–77%) occurred at mean SQG quotients above the 95th-percentile points (i.e., mean ERM quotients > 0.361 and mean PEL quotients > 0.781).

Table 2. Benthic risk levels and the corresponding observed incidence of effects within different ranges of mean ERM quotients. VA = EMAP Virginian Province; CAR = EMAP Carolinian Province; LA = EMAP Louisianian Province.

Risk Level	*EMAP Province*	*Mean ERM Quotient Range*	*Number of Stations*	*% Stations with Degraded Benthos*
Low	VA	≤ 0.022	148	9%
	CAR	≤ 0.018	158	7%
	LA	≤ 0.013	118	30%
Medium	VA	> 0.022–0.098	220	31%
	CAR	> 0.018–0.057	70	43%
	LA	> 0.013–0.036	267	52%
High	VA	> 0.098–0.473	123	54%
	CAR	> 0.057–0.196	51	69%
	LA	> 0.036–0.062	166	86%
Very High	VA	> 0.473	20	85%
	CAR	> 0.196	11	100%
	LA	> 0.062	37	92%

Table 3. Benthic risk levels and the corresponding observed incidence of effects within different ranges of mean PEL quotients. VA = EMAP Virginian Province; CAR = EMAP Carolinian Province; LA = EMAP Louisianian Province.

Risk Level	*EMAP Province*	*Mean PEL Quotient Range*	*Number of Stations*	*% Stations with Degraded Benthos*
Low	VA	≤ 0.040	160	10%
	CAR	≤ 0.030	155	6%
	LA	≤ 0.022	119	32%
Medium	VA	> 0.040–0.179	209	32%
	CAR	> 0.030–0.094	69	43%
	LA	> 0.022–0.062	265	52%
High	VA	> 0.179–0.969	120	55%
	CAR	> 0.094–0.420	51	71%
	LA	> 0.062–0.118	166	85%
Very High	VA	> 0.969	22	77%
	CAR	> 0.420	15	73%
	LA	> 0.118	38	95%

Table 4. Benthic risk levels and the corresponding observed incidence of effects within different ranges of mean SQG quotients for all three EMAP provinces combined.

SQG Type	*Risk Level*	*Percentile*	*Mean SQG Quotient Range*	*Number of Stations*	*% Stations with Degraded Benthos*
ERM	Low	≤ 5th	≤ 0.007	165	18
	Medium	5th – 50th	> 0.007–0.044	752	36
	High	50th – 95th	> 0.044–0.361	430	63
	Very High	> 95th	> 0.361	42	74
PEL	Low	≤ 5th	≤ 0.012	177	18
	Medium	5th – 50th	> 0.012–0.076	737	37
	High	50th – 95th	> 0.076–0.781	435	62
	Very High	> 95th	> 0.781	40	77

3.2 Comparison to Chemical Contaminant Ranges Associated with Sediment Toxicity

Long and MacDonald (1998) reported a relatively high probability of observing sediment toxicity in samples with mean ERM quotients > 1.5 and mean PEL quotients > 2.3. These predictions were based on matching sediment chemistry and toxicity data from multiple surveys of sediment quality performed in coastal areas nationwide. Toxicity was based on standard, 10-day, survival tests conducted with adult amphipods (*Ampelisca abdita* and *Rhepoxynius abronius*). The incidence of highly toxic samples (i.e., mean survival in sample < 80% of control, and significantly different at $p < 0.05$) in these upper, chemical contamination ranges was 74% and 76% for mean ERM and PEL quotients, respectively.

Probable, chemical-contaminant bioeffect ranges observed here for benthic communities were well below the chemical contamination ranges associated with an equivalent high incidence of sediment toxicity in the above laboratory toxicity studies. This pattern was repeated across all three EMAP provinces. We hypothesize that benthic community-level responses should be observed at lower, chemical concentration ranges because they reflect the varying sensitivities of multiple species and life stages to longer-term exposures that may be persisting over several generations. Such measures provide a realistic snapshot of how the ambient benthic ecosystem is responding to chemical exposure.

These comparisons are not meant to imply that the laboratory toxicity data are invalid or of lesser value. Rather, they help to show that the biological significance of chemical contamination varies with different receptors, endpoints, and exposure regimes. If other toxicological indicators (e.g., those based on sublethal endpoints, longer-term chronic exposures, earlier life stages) were evaluated in relation to mean SQG quotients, one might find toxicity thresholds that are much closer to these benthic community-level responses. For example, Chung (1999) found 50% mortality in young juvenile clams (*Mercenaria mercenaria*) at a mean ERM quotient of about 0.4, which is very similar to the mean-ERM-quotient, upper-range critical point of 0.36 reported here for the merged benthic data set (Table 4). Also, acute:chronic ratios from results of aqueous-phase toxicity tests are typically within a factor of 10 (sometimes up to 100 or more) for many chemicals, which is similar to the magnitude of variation observed here between benthic community-level responses and results of acute toxicity tests conducted with adult amphipods.

3.3 PRELIMINARY COMPARISON WITH BENTHIC DATA FROM OTHER REGIONS

Preliminary data from a recent survey of Puget Sound conducted by NOAA and the State of Washington provided an additional opportunity to see if similar patterns of benthic response extended to other regions. Figure 3 shows the relationship between the mean values (and 95% confidence intervals) of two benthic variables, H' diversity (Shannon and Weaver, 1949) and Swartz Dominance Index (Swartz *et al.*, 1985), across different ranges in mean ERM quotients, based on data from 297 stations sampled throughout Puget Sound, from 1997–99.

Note that for these data, unlike the East Coast and Gulf of Mexico data sets, we did not have the benefit of a diagnostic benthic index that could be used to classify samples as having degraded vs. non-degraded benthic condition, and which was designed to account for the influence of natural controlling factors on benthic distributions. Thus, these data could not be subjected to the same kind of analysis used to determine chemical-contaminant bioeffect ranges for Atlantic and Gulf of Mexico benthic communities. Nevertheless, Figure 3 shows clear reductions in both benthic indicators at mean ERM quotients beginning at about the 0.1 to 0.2 range. This is the same range in which we observed a high incidence of benthic stress for the Atlantic and Gulf of Mexico coasts and again well below the range associated with an equivalent high incidence of sediment toxicity. Similarly, Long

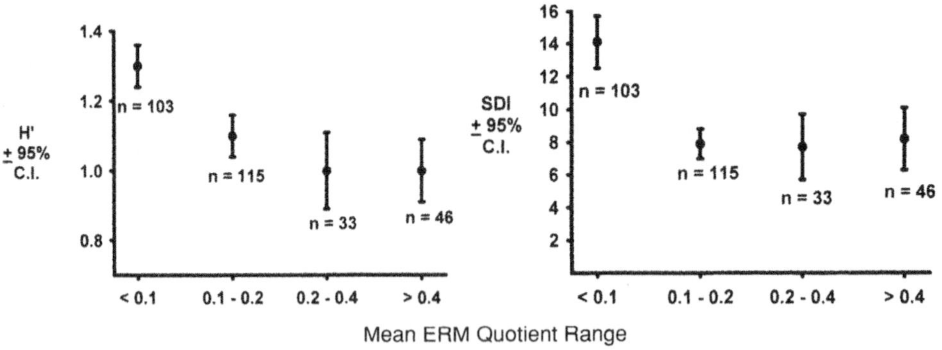

Figure 3. Plots of mean values of H' and Swartz Dominance Index (± 95% confidence intervals) within different ranges of mean ERM quotients at 297 stations sampled throughout Puget Sound from 1997–99.

et al. (in review) found evidence of benthic impacts in Biscayne Bay, FL at mean ERM quotients beginning in this same range (0.1-0.2), and that were lower than those associated with acute toxicity in laboratory survival tests performed on samples from the same locations (i.e., mean ERM quotients > 1.25).

Among the 297 stations sampled in Puget Sound, only one had sediments that were highly toxic in amphipod assays (mean survival in sample < 80% of control, and significantly different at $p < 0.05$), and this site had no chemical contaminants present at concentrations above the SQGs. Thus, an important observation is that adverse effects in these benthic communities showed a higher concordance with increases in chemical contamination (and possibly other unmeasured stressors) than did sediment toxicity based on the amphipod assay.

The Puget Sound data appear to show patterns consistent with those observed in the other EMAP provinces on the Atlantic and Gulf of Mexico coasts. However, these results must be regarded as preliminary until the data can be subjected to more rigorous analyses, including efforts to determine what natural controlling factors may be influencing benthic responses. It would be helpful as well to have the ability to apply a benthic condition index, or some other diagnostic method for classifying samples as degraded versus non-degraded, and which takes into account the influence of these other controlling factors on benthic responses.

3.4 DATA UNCERTAINTIES

We have attempted here to provide an empirical framework for predicting relative risks of benthic community-level effects within different ranges of sediment contamination by chemical mixtures. Predictive ability within these ranges, though high, was not perfect. As shown in Tables 2–4, it is possible that some sites in the lowest risk range will have degraded benthic communities and that some sites in the highest risk range will have non-degraded ones. A variety of factors may contribute to such uncertainty. A degraded benthic community might occur in sediments with low mean SQG quotients (false positives) due to potential effects from: (1) unmeasured chemical contaminants and other stressors (e.g., ammonia, sulfide); (2) measured chemical contaminants not included in the computation of mean SQG quotients (due to lack of bioeffect guidelines); or (3) other physical and biological sources of disturbance (e.g., sediment erosion and deposition due to bottom currents and storm events; predation and other biological interactions). Possible explanations for observing a non-degraded benthic community in sediments with high mean SQG quotients (false negatives) include: (1) contaminants were present at high concentrations but not in bioavailable forms (e.g., due to binding with sediments and organic complexes); (2) the fauna were not exposed to toxic concentrations of contaminants due to small-scale spatial variations; or (3) actual bioeffect levels for individual chemical contaminants present at the site are higher than the corresponding SQGs used to compute the concentration-to-SQG quotients.

Future research should focus on reducing such uncertainties. Of particular importance is the need to differentiate between the relative contributions of measured chemical contaminants to observed benthic effects vs. other possible stressors (e.g., ammonia, sulfide, unmeasured contaminants) that may be co-varying in relation to common environmental factors (such as % sediment fines). Condition of the ambient benthic community appears to be a sensitive indicator of the biological significance of sediment contamination. However, the incidence of a degraded benthic community at the high chemical-contamination range does not preclude the possibility that the observed bioeffects were due partly to other co-varying stressors. Where co-occurrences of high sediment contamination and degraded benthic-community condition are observed, follow-up studies are recommended to determine exact causes and stressor sources.

Acknowledgments

This study was sponsored by the U.S. EPA's Gulf and Atlantic Ecology Divisions of the National Health and Environmental Effects Research Laboratory and by NOAA's Center for Coastal Monitoring and Assessment of the National Centers for Coastal Ocean Science. The study was made possible through coordination of EPA's Environmental Monitoring and Assessment Program (EMAP) and NOAA's National Status and Trends (NS&T) program. Appreciation is extended to the hundreds of people involved in various aspects of these programs and who contributed to the generation of the invaluable data set on which the present analysis is based. We also thank the Washington State Department of Ecology for helping us to access preliminary data from Puget Sound.

References

Chapman, P.M.: 1989, 'Current Approaches to Developing Sediment Quality Criteria,' *Environmental Toxicology and Chemistry* **8**, 589–599.

Chung, K.: 1999, 'Toxicity of Cadmium, DDT, and Fluoranthene to Juvenile *Mercenaria mercenaria* in Aqueous and Sediment Bioassays,' Masters Thesis, University of Charleston South Carolina, Charleston, SC, 123 pp.

Diaz, R.J. and Rosenberg, R.: 1995, 'Marine Benthic Hypoxia: A Review of its Ecological Effects and the Behavioural Responses of Benthic Macrofauna,' *Oceanography and Marine Biology: an Annual Review*, **33**, 245–303.

Engle, V.D. and Summers, J.K.: 1999, 'Refinement, Validation, and Application of a Benthic Condition Index for Gulf of Mexico Estuaries,' *Estuaries* **22**, 624–635.

Holland, A.F. (ed.): 1990, 'Near Coastal Program Plan for 1990: Estuaries,' EPA/600/4-90/033, U.S. EPA, Atlantic Ecology Division, Narragansett, RI.

Hyland, J.L., Van Dolah, R.F. and Snoots, T.R.: 1999, 'Predicting Stress in Benthic Communities of Southeastern U.S. Estuaries in Relation to Chemical Contamination of Sediments,' *Environmental Toxicology and Chemistry* **18**, 2557–2564.

Lauenstein, G.G. and Cantillo, A.Y.: 1993, 'Sampling and Analytical Methods of the National Status and Trends Program: National Benthic Surveillance and Mussel Watch Projects, 1984–1992, Volumes III and IV,' NOAA Technical Memorandum NOS ORCA 71, NOAA National Ocean Service, Silver Spring, MD, 219 pp (Volume III) and 182 pp (Volume IV).

Long, E.R., Field, L.J. and MacDonald, D.D.: 1998, 'Predicting Toxicity in Marine Sediments With Numerical Sediment Quality Guidelines,' *Environmental Toxicology and Chemistry* **17**, 714–727.

Long, E.R., Hameedi, J.J., Sloane, G.M. and Read, L.: In review, 'Chemical contamination, toxicity, and benthic community indices in sediments of the lower Miami River and adjoining portions of Biscayne Bay, Florida, *Estuaries* (submitted March 2001).

Long, E.R., and MacDonald, D.D.: 1998, 'Recommended Uses of Empirically Derived, Sediment Quality Guidelines for Marine and Estuarine Ecosystems,' *Human and Ecological Risk Assessment* **4**, 1019–1039.

Long, E.R., MacDonald, D.D.., Severn, C.G., and Hong, C.B.: 2000, 'Classifying the probabilities of acute toxicity in marine sediments with empirically-derived sediment quality guidelines,' *Environmental Toxicology and Chemistry* **19**, 2598-2601.

Long, E.R., MacDonald, D.D. , Smith, S.L. and Calder, F.D.: 1995, 'Incidence of Adverse Biological Effects Within Ranges of Chemical Concentrations in Marine and Estuarine Sediments,' *Environmental Management* **19**, 81–97.

MacDonald, D.D, Carr, R.S., Calder, F.D., Long, E.R. and Ingersoll, C.G.: 1996, 'Development and Evaluation of Sediment Quality Guidelines for Florida Coastal Waters,' *Ecotoxicology* **5**, 253–278.

Paul, J.F., Gentile, J.H., Scott, K.J., Schimmel, S.C., Campbell, D.E. and Latimer, R.W.: 1999, 'EMAP-Virginian Province Four-Year Assessment Report (1990–93),' EPA/600/R-99/004, U.S. EPA, Atlantic Ecology Division, Narragansett, RI, 119 pp.

Shannon, C.E. and Weaver, W.: 1949, The Mathematical Theory of Communication, University of Illinois Press, Urbana, IL, USA.

Summers, J.K., Macauley, J.M., Heitmuller, P.T., Engle, V.D., Adams, A.M. and Brooks, G.T.: 1992, 'Annual Statistical Summary: EMAP-Estuaries Louisianian Province—1991,' EPA/600/R-93/001, U.S. EPA, Gulf Ecology Division, Gulf Breeze, FL, 101 pp. and 3 Appendices.

Swartz, R.C., Shultz, D.W., Ditsworth, G.R., DeBen, W.A. and Cole, F.A.: 1985, 'Sediment Toxicity, Contamination, and Macrobenthic Communities Near a Large Sewage Outfall,' in: Boyle, T.T. (ed.), *Validation and Predictability of Laboratory Methods for Assessing the Fate and Effects of Contaminants in Aquatic Ecosystems*, ASTM STP 865, American Society for Testing and Materials, Philadelphia, PA, pp. 152–175.

U.S. Environmental Protection Agency: 1995, 'Environmental Monitoring and Assessment Program: Laboratory Methods Manual—Estuaries. Volume 2: Chemical Methods,' EPA/620/R-95/009, U.S. EPA, ORD, Washington, D.C.

Valette-Silver, N.J.: 1993, 'The Use of Sediment Cores to Reconstruct Historical Trends in Contamination of Estuarine and Coastal Sediments,' *Estuaries* **16**, 577–588.

Van Dolah, R.F., Hyland, J.L., Holland, A.F., Rosen, J.S. and Snoots, T.R.: 1999, 'A Benthic Index of Biological Integrity for Assessing Habitat Quality in Estuaries of the Southeastern USA,' *Marine Environmental Research* **48**, 1–15.

APPLICATION OF THE BENTHIC INDEX OF BIOTIC INTEGRITY TO ENVIRONMENTAL MONITORING IN CHESAPEAKE BAY

ROBERTO J. LLANSÓ[1*], DANIEL M. DAUER[2], JON H. VØLSTAD[1], AND LISA C. SCOTT[1]

[1]*Versar, Inc., 9200 Rumsey Rd., Columbia, MD 21045;* [2]*Department of Biological Sciences, Old Dominion University, Norfolk, VA 23529*

[*]*(author for correspondence, email: llansorob@versar.com)*

Abstract: The Chesapeake Bay benthic index of biotic integrity (B-IBI) was developed to assess benthic community health and environmental quality in Chesapeake Bay. The B-IBI provides Chesapeake Bay monitoring programs with a uniform tool with which to characterize bay-wide benthic community condition and assess the health of the Bay. A probability-based design permits unbiased annual estimates of areal degradation within the Chesapeake Bay and its tributaries with quantifiable precision. However, of greatest interest to managers is the identification of problem areas most in need of restoration. Here we apply the B-IBI to benthic data collected in the Bay since 1994 to assess benthic community degradation by Chesapeake Bay Program segment and water depth. We used a new B-IBI classification system that improves the reliability of the estimates of degradation. Estimates were produced for 67 Chesapeake Bay Program segments. Greatest degradation was found in areas that are known to experience hypoxia or show toxic contamination, such as the mesohaline portion of the Potomac River, the Patapsco River, and the Maryland mainstem. Logistic regression models revealed increased probability of degraded benthos with depth for the lower Potomac River, Patapsco River, Nanticoke River, lower York River, and the Maryland mainstem. Our assessment of degradation by segment and water depth provided greater resolution of relative condition than previously available, and helped define the extent of degradation in Chesapeake Bay.

Keywords: biological indicators, benthic communities, degradation, Benthic Index of Biotic Integrity, Chesapeake Bay, low dissolved oxygen.

1. Introduction

The Chesapeake Bay benthic index of biotic integrity (B-IBI) was developed to assess benthic community health and environmental quality in Chesapeake Bay. The B-IBI evaluates the ecological condition of a sample by comparing values of key benthic community attributes to reference values expected under non-degraded conditions in similar habitat types. It is therefore a measure of deviation from reference conditions. This approach has been applied successfully to estuaries of the mid-Atlantic region and the southeastern United States (Weisberg *et al.*, 1997; Van Dolah *et al.*, 1999; Ranasinghe *et al.*, In Review). It was first applied to Chesapeake Bay in 1994. The application of the B-IBI to Chesapeake Bay has been possible through a partnership among the U.S. EPA Chesapeake Bay Program, the Maryland Department of Natural Resources, and the Virginia Department of Environmental Quality. Academic institutions and the private sector facilitate this partnership.

The B-IBI provides Chesapeake Bay monitoring programs with a uniform tool for characterizing bay-wide benthic community condition. A probability-based sampling design (stratified simple random) is used to estimate the areal extent of conditions in Chesapeake Bay. Annually, twenty-five samples are allocated randomly to each of ten strata. The benthic community condition for each sample is

Environmental Monitoring and Assessment **81**: 163–174, 2003.

Table 1. B-IBI ranges used to classify samples in 3 categories used in the assessment of benthic community degradation by Chesapeake Bay Program segment and water depth. Ranges are based on 90% confidence limits established by analysis of the original reference and degraded index development datasets.

	Category		
Habitat	*Degraded*	*Non-Degraded*	*Intermediate/Indeterminate*
Tidal Freshwater	< 2.5	> 3.5	2.5–3.5
Oligohaline	< 2.5	> 3.7	2.5–3.7
Low Mesohaline	< 3.0	> 3.4	3.0–3.4
High Mesohaline Sand	< 2.7	> 3.0	2.7–3.0
High Mesohaline Mud	< 2.2	> 2.5	2.2–2.5
Polyhaline Sand	< 1.8	> 3.7	1.8–3.7
Polyhaline Mud	< 2.3	> 3.0	2.3–3.0

3. Results

Figure 1 shows benthic community condition for each of the random samples. Condition has been classified according to the convention currently used by the Chesapeake Bay monitoring programs (see Methods). The figure clearly shows areas with a greater concentration of samples with degraded or severely degraded benthos (e.g., lower Potomac River, Maryland mainstem, Patapsco River), and areas with predominantly good benthos (e.g., Virginia mainstem).

Table 2 lists the probabilities of observing degraded benthos, non-degraded benthos, or benthos of intermediate condition (indeterminate for low salinity habitats) by Chesapeake Bay Program segment. Condition has been classified according to Table 1. Regions with degraded benthos occurring in more than half of the segment area (> 50% probability) were considered cause of concern, and are shown in red in Figure 2. Segments with predominantly good benthos (> 50% probability) are also shown in this figure. The confidence of the estimate is expressed by the intensity of the color. High color intensity indicates high confidence (the 90% confidence interval is within the 50–100% probability range). Low color intensity indicates low confidence (the 67% confidence interval is within the 50–100% probability range). Benthic community condition estimates for which the lower 67% confidence limit was ≤ 50% were categorized as "other".

Depth-related changes in the probability of observing degraded benthos were statistically significant at the 90% confidence level for five segments: Patapsco River (PATMH), Maryland mainstem (Segments CB2OH and CB3OH), Nanticoke River (NANMH), and lower York River (YRKPH). Other segments had significant relationships with depth but the intersect of the regression model was not significant. Figure 3 illustrates depth relationships for 4 segments. Although the intersect was not significant, we have included the Potomac River (Segment

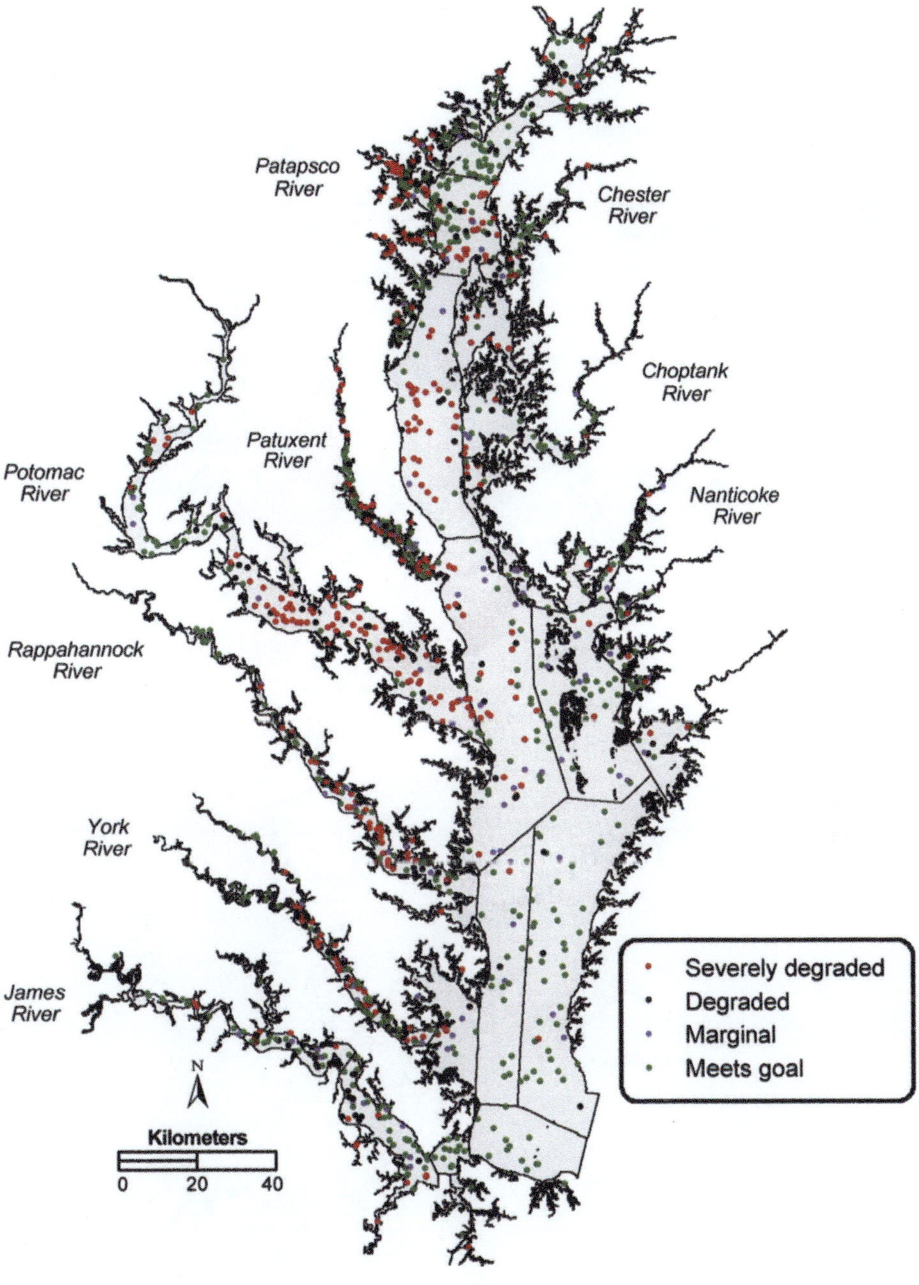

Figure 1. Benthic community condition for each of 1,466 samples collected 1994–2000 by the Maryland and Virginia benthic monitoring programs.

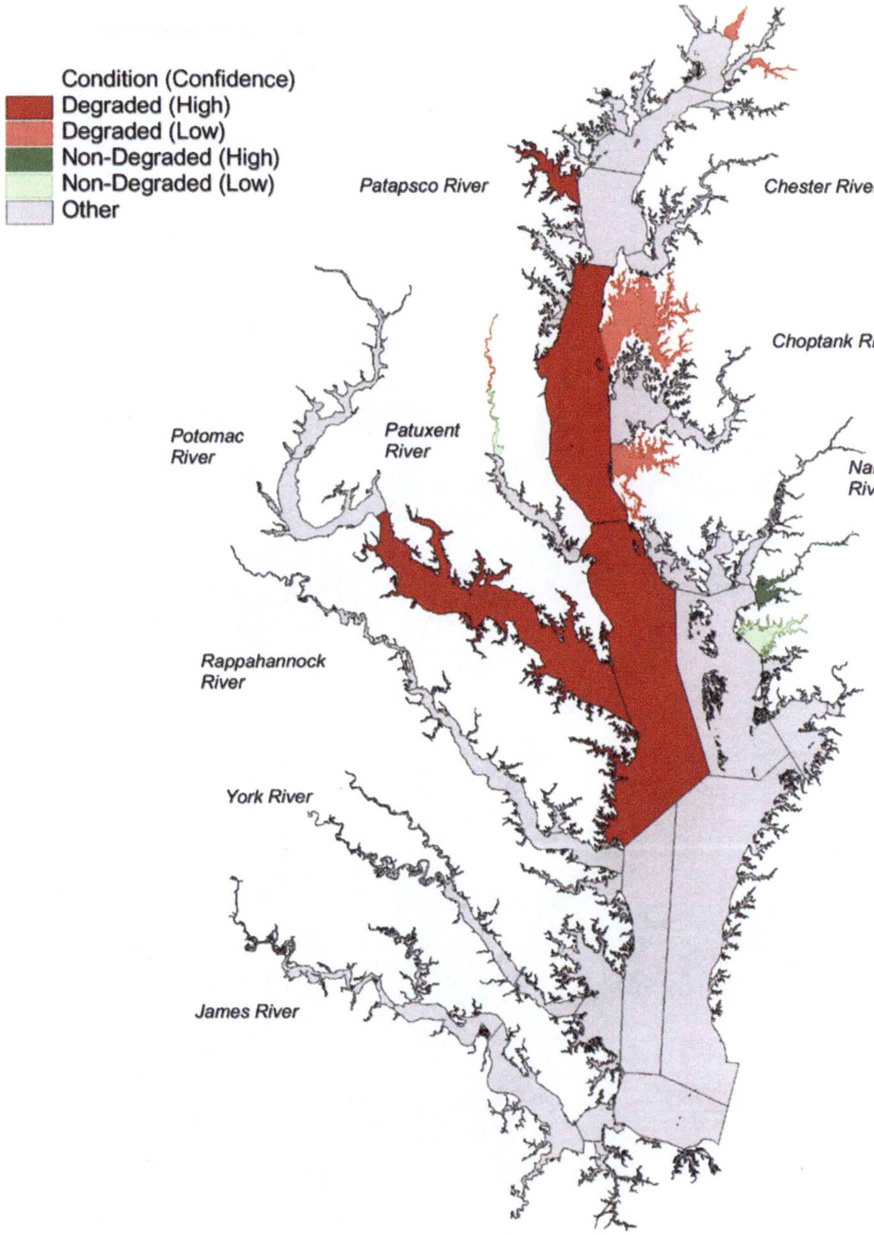

Figure 2. Chesapeake Bay Program segments with a >50% probability of observing degraded or non-degraded (good quality) benthos.

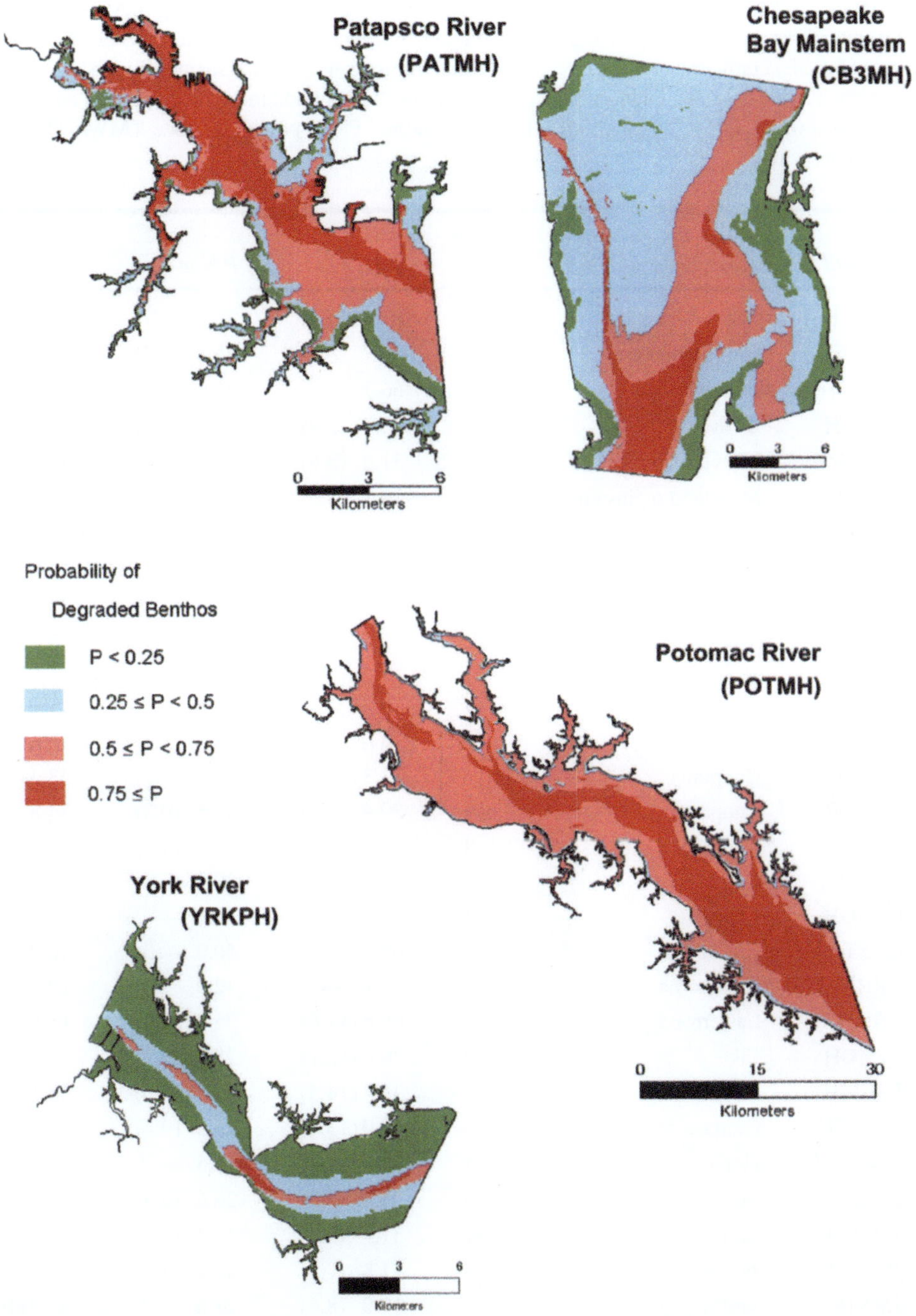

Figure 3. Depth-related changes in probability of observing degraded benthos for four segments for which regression models were statistically significant at the 90% confidence level.

Table 2. Probabilities (and SE) of observing degraded benthos, non-degraded benthos, or benthos of intermediate condition (indeterminate for low salinity habitats) for each of 67 Chesapeake Bay Program segments for which samples were collected by the Maryland and Virginia benthic monitoring programs, 1994–2000. Probabilities and standard errors were adjusted according to Agresti and Caffo (2000) for segments with ≥ 5 samples. Standard errors were used to calculate 67% (± SE) and 90% (± 1.65 x SE) confidence limits. Exact confidence limits were used for segments with < 5 samples, and are not shown in the table. Adjusted probabilities may not add to 100%. Segments codes: TF = tidal freshwater, OH = oligohaline, MH = mesohaline, PH = polyhaline. See TMWA (1999) for segment location.

Segment	*River/Water Body*	*No. of Sites*	*P Deg.*	*P Non-deg.*	*P Interm.*
BACOH	Back	9	61.5 (13.5)	23.1 (11.7)	30.8 (12.8)
BIGMH	Big Annemessex	14	16.7 (8.8)	44.4 (11.7)	50.0 (11.8)
BOHOH	Bohemia	2	66.7 —	33.3 —	33.3 —
BSHOH	Bush	10	35.7 (12.8)	21.4 (11.0)	57.1 (13.2)
CB1TF	Maryland mainstem	25	31.0 (8.6)	24.1 (7.9)	51.7 (9.3)
CB2OH	Maryland mainstem	44	20.8 (5.9)	35.4 (6.9)	47.9 (7.2)
CB3MH	Maryland mainstem	81	42.4 (5.4)	44.7 (5.4)	15.3 (3.9)
CB4MH	Maryland mainstem	49	76.6 (4.3)	11.1 (3.8)	14.8 (3.8)
CB5MH	Maryland mainstem	60	58.3 (4.7)	19.0 (4.1)	25.0 (4.5)
CB6PH	Virginia mainstem	22	11.5 (6.3)	42.3 (9.7)	53.8 (9.8)
CB7PH	Virginia mainstem	44	4.2 (2.9)	54.2 (7.2)	45.8 (7.2)
CB8PH	Virginia mainstem	12	12.5 (8.3)	37.5 (12.1)	62.5 (12.1)
CHKOH	Chickahominy	1	40.0 —	40.0 —	60.0 —
CHOMH1	Choptank	14	33.3 (11.1)	44.4 (11.7)	33.3 (11.1)
CHOMH2	Choptank	19	30.4 (9.6)	43.5 (10.3)	34.8 (9.9)
CHOOH	Choptank	8	16.7 (10.8)	33.3 (13.6)	66.7 (13.6)
CHSMH	Chester	38	45.2 (7.7)	33.3 (7.3)	26.2 (6.8)
CHSOH	Chester	4	50.0 —	37.5 —	37.5 —
CHSTF	Chester	1	60.0 —	40.0 —	40.0 —
CRRMH	Corrotoman	3	57.1 —	42.9 —	28.6 —
EASMH	Eastern Bay	12	68.8 (11.6)	25.0 (10.8)	18.8 (9.8)
ELKOH	Elk	10	28.6 (12.1)	28.6 (12.1)	57.1 (13.4)
FSBMH	Fishing Bay	7	36.4 (14.5)	45.5 (15.0)	36.4 (14.5)
GUNOH	Gunpowder	10	50.0 (13.4)	14.3 (9.4)	50.0 (13.4)
HNGMH	Honga	4	25.0 —	62.5 —	37.5 —
JMSMH	James	60	32.8 (5.9)	26.6 (5.5)	43.8 (6.2)
JMSOH	James	31	51.4 (8.4)	11.4 (5.4)	42.9 (8.4)
JMSPH	James	13	11.8 (7.8)	47.1 (12.1)	52.9 (12.1)
JMSTF	James	17	19.0 (8.6)	28.6 (9.9)	61.9 (10.6)
LAFMH	Lafayette	1	40.0 —	60.0 —	40.0 —

Table 2—Continued

Segment	*River/Water Body*	*No. of Sites*	*P Deg.*	*P Non-deg.*	*P Interm.*
LCHMH	Little Choptank	8	66.7 (13.6)	16.7 (10.8)	33.3 (13.6)
LYNPH	Lynnhaven Inlet	1	40.0 —	40.0 —	60.0 —
MAGMH	Magothy	17	57.1 (10.8)	28.6 (9.9)	23.8 (9.3)
MANMH	Manokin	12	25.0 (10.8)	62.5 (12.1)	25.0 (10.8)
MIDOH	Middle	12	18.8 (9.8)	12.5 (8.3)	81.3 (9.8)
MOBPH	Mobjack Bay	15	31.6 (10.7)	26.3 (10.1)	52.6 (11.5)
MPNOH	Mattaponi	7	27.3 (13.4)	54.5 (15.0)	36.4 (14.5)
MPNTF	Mattaponi	5	22.2 (13.9)	44.4 (16.6)	55.6 (16.6)
NANMH	Nanticoke	16	20.0 (8.9)	60.0 (11.0)	30.0 (10.2)
NANOH	Nanticoke	1	40.0 —	60.0 —	40.0 —
NANTF	Nanticoke	3	42.9 —	42.9 —	42.9 —
NORTF	Northeast	5	66.7 (15.7)	22.2 (13.9)	33.3 (15.7)
PATMH	Patapsco	56	61.7 (6.3)	21.7 (5.3)	20.0 (5.2)
PAXMH	Patuxent	134	37.0 (4.1)	42.8 (4.2)	21.7 (3.5)
PAXOH	Patuxent	14	27.8 (10.6)	66.7 (11.1)	16.7 (8.8)
PAXTF	Patuxent	3	71.4 —	28.6 —	28.6 —
PIAMH	Piankatank	1	60.0 —	40.0 —	40.0 —
PMKOH	Pamunkey	8	33.3 (13.6)	25.0 (12.5)	58.3 (14.2)
PMKTF	Pamunkey	3	28.6 —	42.9 —	57.1 —
POCMH	Pocomoke	9	30.8 (12.8)	38.5 (13.5)	46.2 (13.8)
POCOH	Pocomoke	1	60.0 —	40.0 —	40.0 —
POTMH	Potomac	124	72.7 (3.9)	9.4 (2.6)	19.5 (3.5)
POTOH	Potomac	37	24.4 (6.7)	34.1 (7.4)	46.3 (7.8)
POTTF	Potomac	19	43.5 (10.3)	17.4 (7.9)	47.8 (10.4)
RHDMH	Rhode	6	50.0 (15.8)	40.0 (15.5)	30.0 (14.5)
RPPMH	Rappahannock	106	49.1 (4.8)	23.6 (4.1)	29.1 (4.3)
RPPOH	Rappahannock	6	50.0 (15.8)	30.0 (14.5)	40.0 (15.5)
RPPTF	Rappahannock	10	14.3 (9.4)	50.0 (13.4)	50.0 (13.4)
SASOH	Sassafras	11	33.3 (12.2)	13.3 (8.8)	66.7 (12.2)
SBEMH	Elizabeth	2	50.0 —	50.0 —	33.3 —
SEVMH	Severn	13	52.9 (12.1)	29.4 (11.1)	29.4 (11.1)
SOUMH	South	11	33.3 (12.2)	46.7 (12.9)	33.3 (12.2)
TANMH	Tangier Sound	45	18.4 (5.5)	55.1 (7.1)	30.6 (6.6)
WICMH	Wicomico	13	17.6 (9.2)	76.5 (10.3)	17.6 (9.2)
WSTMH	West	7	36.4 (14.5)	45.5 (15.0)	36.4 (14.5)
YRKMH	York	69	41.1 (5.8)	23.3 (4.9)	38.4 (5.7)
YRKPH	York	31	28.6 (7.6)	31.4 (7.8)	45.7 (8.4)

POTMH) in Figure 3 because the regression was highly significant ($p = 0.003$) and a large proportion of samples in the main channel routinely fail the B-IBI. Contours are based on depth intervals in meters obtained from the regression model at four pre-determined probability ranges.

4. Discussion

Our B-IBI analysis shows greatest degradation in areas that are known to experience hypoxia. Hypoxia is most frequent and extensive in the mainstem of the Bay and the mesohaline Potomac River (Dauer *et al.*, 1992; Ranasinghe *et al.*, 1994; Dauer *et al.*, 2000). The Patapsco River also experiences hypoxia, but sediment contamination is a problem in this and other upper western shore tributaries (U.S. EPA, 1999; Dauer *et al.*, 2000; McGee *et al.*, 2001). Other western shore tributaries such as the Back and Severn Rivers also showed a > 50% probability of observing degraded benthos, but we had less than 67% confidence in the estimate.

Benthic community degradation clearly increased with water depth, principally in the mainstem of the Bay, and the Patapsco, Potomac, and York Rivers. Higher probabilities of observing degraded benthos with depth in these segments are associated with low dissolved oxygen conditions (Dauer *et al.*, 2000). Degradation in the upper portion of the Bay (Bohemia and Northeast Rivers) is probably associated with nutrient enrichment. Thirty percent of the sites sampled during 1996–2000 in the upper Bay by the Maryland long-term benthic monitoring program failed to meet the restoration goals due to excess abundance, excess biomass, or both (Llansó *et al.*, 2001). Excess abundance and biomass are phenomena usually associated with eutrophic conditions (Pearson and Rosenberg, 1978). Benthic community condition was best in two eastern shore tributaries (the Manokin and Wicomico Rivers) and in the oligohaline portion of the Patuxent River.

The application of the B-IBI to Chesapeake Bay provides an objective method for distinguishing degraded from non-degraded benthic communities bay-wide. The B-IBI provides context for evaluating the effectiveness of Bay management activities and a well-defined endpoint for restoration activities. Currently, Chesapeake Bay benthic monitoring programs use a B-IBI score of 3 as a quantitative benchmark against which to measure the health of benthos. While there are certain advantages to using a system for which samples are classified as meeting or not meeting the restoration goals, the identification of samples that are of intermediate quality improves the resolution of the interpretation of relative condition (Alden *et al.*, 2002). The classification approach of Alden *et al.* (2002) also has the advantage of using confidence limits to classify B-IBI scores. The use of confidence limits to define the relative condition of the benthos (both at the B-IBI classification stage and in estimating segment-wide degradation) should produce more reliable results with fewer Type I errors (false alarms). On the other hand, if one quantitative benchmark is desirable to better track progress toward restoration, emphasizing the "severely degraded" category of current monitoring programs may be more informative than using a "meets goal" approach, and may add further resolution in the evaluation of patterns of degradation (Dauer and Llansó, this issue).

The B-IBI threshold for the "severely degraded" condition of current monitoring programs is, for most habitats in Chesapeake Bay, very close to the cut-off level used to define degraded conditions in the present study. Thus, Chesapeake Bay areas with high concentration of samples with severely degraded benthos (lower Potomac River, Maryland mainstem, Patapsco River, Figure 1) were identified as areas of concern in this analysis (Figure 2). However, the likelihood of observing benthos of intermediate quality was high for some areas that usually meet the restoration goals, indicating that a sizable portion of the samples in those areas had B-IBI values falling below best reference conditions. For example, the Virginia mainstem shows the least amount of degradation (Figure 1), but our analysis suggests that this region may benefit from further improvements in water quality. In deciding what approach to use, consensus among Chesapeake Bay partners must first be built if widespread applicability of the method is warranted.

One limitation of our analysis is small segment size. As more years of data are added, the temporal variability component of the estimate increases. Probability estimates are also average values for Bay segments and do not reflect local variation in benthic community condition. However, with sufficient sample size the condition of larger segments can be assessed with relatively high confidence, and further resolution with depth may help define the extent of degradation. For example, by comparing segments in Figure 3, the more extensive degradation affecting shallow waters in the Potomac River becomes evident. Regardless of the approach used to estimate benthic community degradation, our assessment agrees well with results of previous characterizations (e.g., Dauer *et al.*, 2000) which suggest dissolved oxygen stress as the more serious and widespread problem affecting benthic communities in the Bay. Our assessment of degradation by segment and water depth provides a greater resolution of relative condition than previously available. The study illustrates one application of the Chesapeake Bay index of biotic integrity that may be useful to management in their effort to identify areas most in need of restoration. Similar applications to other estuaries with other indices of benthic condition may be comparably useful.

Acknowledgments

We are grateful to the State of Maryland's Environmental Trust Fund which partially funded this work. We thank Ginny Mercurio for providing data analysis and Dr. Donald Strebel and Allison Brindley for expert GIS support.

References

Agresti, A. and Caffo, B.: 2000, 'Simple and Effective Confidence Intervals for Proportions and Differences of Proportions Result from Adding Two Successes and Two Failures', *The American Statistician* **54**, 280–288.

Alden, R. W., III: 1992, 'Uncertainty and Sediment Quality Assessments: Confidence Limits for the Triad', *Environmental Toxicology and Chemistry* **11**, 645–651.

Alden, R. W., III, Dauer, D. M., Ranasinghe, J. A., Scott, L. C. and Llansó, R. J.: 2002, 'Statistical Verification of the Chesapeake Bay Benthic Index of Biotic Integrity', Environmetrics, In Press.

Dauer, D. M., Ranasinghe, J. A. and Weisberg, S. B.: 2000, 'Relationships between Benthic Community Condition, Water Quality, Sediment Quality, Nutrient Loads, and Land Use Patterns in Chesapeake Bay', *Estuaries* **23**, 80–96.

Dauer, D. M., Rodi, A. J., Jr. and Ranasinghe, J. A.: 1992, 'Effects of Low Dissolved Oxygen Events on the Macrobenthos of the Lower Chesapeake Bay', *Estuaries* **15**, 384–391.

Hosmer, D. W. and Lemeshow, S.: 2000, Applied Logistic Regression, 2nd ed., John Wiley & Sons, New York, NY.

Llansó, R. J., Scott, L. C. and Kelley, F. S.: 2001, Chesapeake Bay Water Quality Monitoring Program, Long-term Benthic Monitoring and Assessment Component Level I Comprehensive Report, July 1984–December 2000, prepared for the Maryland Department of Natural Resources by Versar, Inc., Columbia, MD.

McGee, B. L., Fisher, D. J., Ashley, J. and Velinsky, D.: 2001, Using the Sediment Quality Triad to Characterize Toxic Conditions in the Chesapeake Bay (1999): An Assessment of Tidal River Segments in the Bohemia, Magothy, Patuxent, Potomac, James, and York Rivers, EPA 903-R-01-008, U.S. EPA Chesapeake Bay Program, Annapolis, MD.

Pearson, T. H. and Rosenberg, R.: 1978, 'Macrobenthic Succession in Relation to Organic Enrichment and Pollution of the Marine Environment', *Oceanography Marine Biology Annual Review* **16**, 229–311.

Ranasinghe, J. A., Weisberg, S. B., Gerritsen, J. and Dauer, D. M.: 1994, Assessment of Chesapeake Bay Benthic Macroinvertebrate Resource Condition in Relation to Water Quality and Watershed Stressors, prepared for the Governor's Council on Chesapeake Bay Research Fund and the Maryland Department of Natural Resources by Versar, Inc., Columbia, MD.

Ranasinghe, J. A., Weisberg, S. B., O'Connor, J. S. and Adams, D. A.: In Review, 'A Benthic Index of Biotic Integrity (B-IBI) for New York/New Jersey Harbor', Journal of Aquatic Ecosystem Stress and Recovery.

TMWA (Tidal Monitoring and Analysis Workgroup): 1999, Chesapeake Bay Program, Analytical Segmentation Scheme for the 1997 Re-evaluation and Beyond, prepared for the U.S. Environmental Protection Agency, Chesapeake Bay Program Office, by the Tidal Monitoring and Analysis Workgroup of the Chesapeake Bay Program Monitoring and Assessment Subcommittee, Annapolis, MD.

U.S. EPA: 1999, Targeting Toxics: A Characterization Report, A Tool for Directing Management and Monitoring Actions in the Chesapeake Bay's Tidal Rivers, EPA 903-R-99-010, U.S. EPA Chesapeake Bay Program, Annapolis, MD.

Van Dolah, R. F., Hyland, J. L., Holland, A. F., Rosen, J. S. and Snoots, T. S.: 1999, 'A Benthic Index of Biological Integrity for Assessing Habitat Quality in Estuaries of the Southeastern USA', *Marine Environmental Research* **48**, 269–283.

Weisberg, S. B., Ranasinghe, J.A., Dauer, D.M., Schaffner, L.C., Diaz, R.J. and Frithsen, J.B.: 1997, 'An Estuarine Benthic Index of Biotic Integrity (B-IBI) for the Chesapeake Bay', *Estuaries* **20**, 149–158.

SPATIAL SCALES AND PROBABILITY BASED SAMPLING IN DETERMINING LEVELS OF BENTHIC COMMUNITY DEGRADATION IN THE CHESAPEAKE BAY

DANIEL M. DAUER[1*] AND ROBERTO J. LLANSÓ[2]

[1]*Department of Biological Sciences, Old Dominion University, Norfolk, VA 23529;*
[2]*VERSAR Inc, Columbia, MD 21045*

[*]*(author for correspondence, email: ddauer@odu.edu)*

Abstract: The extent of degradation of benthic communities of the Chesapeake Bay was determined by applying a previously developed benthic index of biotic integrity at three spatial scales. Allocation of sampling was probability-based allowing areal estimates of degradation with known confidence intervals. The three spatial scales were: (1) the tidal Chesapeake Bay; (2) the Elizabeth River watershed; and (3) two small tidal creeks within the Southern Branch of the Elizabeth River that are part of a sediment contaminant remediation effort. The areas covered varied from 10^{-1} to 10^{4} km^{2} and all were sampled in 1999. The Chesapeake Bay was divided into ten strata, the Elizabeth River into five strata and each of the two tidal creeks was a single stratum. The determination of the number and size of strata was based upon consideration of both managerially useful units for restoration and limitations of funding. Within each stratum 25 random locations were sampled for benthic community condition. In 1999 the percent of the benthos with poor benthic community condition for the entire Chesapeake Bay was 47% and varied from 20% at the mouth of the Bay to 72% in the Potomac River. The estimated area of benthos with poor benthic community condition for the Elizabeth River was 64% and varied from 52–92%. Both small tidal creeks had estimates of 76% of poor benthic community condition. These kinds of estimates allow environmental managers to better direct restoration efforts and evaluate progress towards restoration. Patterns of benthic community condition at smaller spatial scales may not be correctly inferred from larger spatial scales. Comparisons of patterns in benthic community condition across spatial scales, and between combinations of strata, must be cautiously interpreted.

Keywords: spatial scales, benthic communities, macrobenthos, degradation, Benthic Index of Biotic Integrity, Chesapeake Bay, Elizabeth River, contamination, low dissolved oxygen

1. Introduction

A wide variety of sampling designs have been used in marine and estuarine environmental monitoring programs (e.g., see case studies reviewed recently in Kramer, 1994; Kennish, 1998; Livingston, 2001). Allocation of samples in space and time varies depending on the environmental problems and issues addressed (Kingsford and Battershill, 1998) and the type of variables measured (e.g., water chemistry, phytoplankton, zooplankton, benthos, nekton). In the Chesapeake Bay, the benthic monitoring program consists of both fixed-point stations and probability-based samples. The fixed-point stations are used primarily for the determination of long-term trends (e.g., Dauer and Alden, 1995; Dauer, 1997) and the probability-based samples for the determination of the areal extent of degraded benthic community condition (Llansó *et al.*, this issue). The probability-based sampling design consists of equal replication of random samples among strata and is, therefore, a stratified simple random design (Kingsford, 1998).

Environmental Monitoring and Assessment **81**: 175–186, 2003.

As part of the restoration of the Chesapeake Bay, the benthic monitoring program serves to evaluate and direct restoration efforts to reduce eutrophication and contaminant levels in the Bay (Dauer *et al.,* 1992; 1993; Dauer, 1993; Dauer and Alden, 1995; Dauer, 1997; Dauer, *et al.* 2000). Benthic community condition is assessed using the benthic index of biotic integrity (B-IBI) that evaluates the ecological condition of a sample by comparing values of key benthic community attributes to reference values expected under undegraded conditions in similar habitat types (Weisberg *et al*., 1997). In this paper we present an application of the B-IBI and probability-based sampling at multiple spatial scales to determine levels of benthic community degradation for areas ranging from 10^{-1} to 10^{4} km^{2}. The three spatial scales presented are the entire tidal Chesapeake Bay consisting of ten strata (total area of 11,607 km^{2}), the Elizabeth River watershed consisting of five strata (total area of 47 km^{2}) and an environmental restoration study of contaminated creeks consisting of two strata (both approximately 0.1 km^{2}). Within each stratum 25 random locations were sampled for benthic community condition. All data presented were collected in 1999 and sampling of the three spatial scales was separately designed and funded with independent monitoring objectives.

2. Materials and Methods

Within each stratum at each of the three spatial scales 25 random samples were collected in 1999 during the index period of the B-IBI (July 15 to September 30). All samples were collected using a Young grab with a surface area of 440 cm^{2}. Each sample was sieved on a 0.5 mm screen and preserved in the field. Samples were sorted, enumerated and identified to the lowest possible taxon. Ash-free dry weight biomass was determined for each taxon.

The tidal waters of the Chesapeake Bay were divided into ten strata (Figure 1). Probability-based sampling began in Maryland in 1994 and in Virginia in 1996 and has continued since. Here we present only the 1999 sampling of 250 random locations among the ten strata. The determination of the number and size of each stratum (Table 1) was based upon consideration of both managerially useful units for restoration and limitation of funding. The largest portion of the Bay, the Mainstem, was divided into three strata and five of the strata consisted of the major tributaries (James, York, Rappahannock, Potomac and Patuxent rivers). The final two strata included the remaining smaller tributaries on the Maryland western shore and Maryland eastern shore. The region labeled Deep Mainstem (Figure 1) is subjected to summer anoxia resulting in an azoic condition. The strata sampled represented the entire tidal region of the Bay from freshwater through polyhaline zones.

The Elizabeth River was divided into five strata (Figure 2). For the Lafayette River and Western Branch the downstream boundary was the north-south confluence with the Mainstem of the Elizabeth River. For the Eastern Branch and Southern Branch the downstream boundary was the confluence of both branches with the

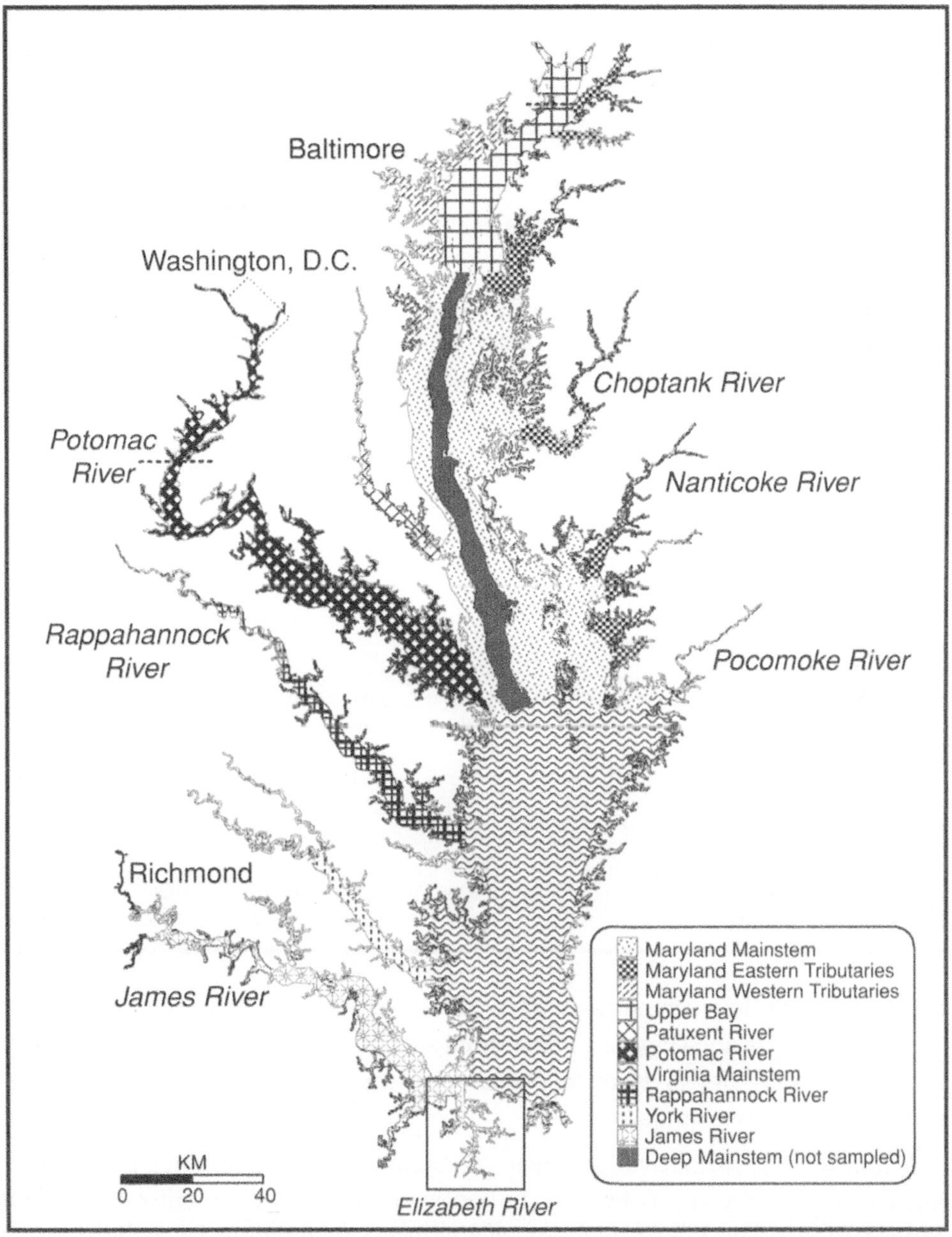

Figure 1. Location of ten strata of the Chesapeake Bay. Insert shows location of the Elizabeth River.

Table 1. Estimated tidal area in 1999 failing to meet the Chesapeake Bay Benthic Community restoration Goal. All columns except last two in km^2. A. Tidal Chesapeake Bay. B. Elizabeth River watershed. C. Sediment Contaminant Creeks. Abbreviations: SD = severely degraded, D = degraded, M = marginal, CI = 95% confidence interval in percentage. (See text for further information.)

	Total Area km^2	*SD*	*D*	*M*	*Total Failing*	*Percent Failing*	*CI*
A. Chesapeake Bay *(Bay Scale)*							
Combined Strata							
Chesapeake	11,607	3,121	1,648	681	5,450	47	7.7
Maryland Tidal Waters	6,244	2,423	1,137	374	3,935	63	9.3
Virginia Tidal Waters	5,362	698	510	306	1,515	28	12.6
Individual Strata							
Potomac River	1,276	663	153	102	918	72	18.0
Patuxent River	128	20	10	10	41	32	18.7
MD Western Tributaries	292	117	47	12	175	60	19.6
MD Eastern Tributaries	534	43	150	86	279	52	20.0
MD Upper Mainstem	785	188	63	63	314	40	19.6
MD Mid-Bay Mainstem	2,552	1,391	715	102	2,208	68	18.7
Virginia Mainstem	4,120	494	165	165	824	20	16.0
Rappahannock River	372	74	104	45	223	60	19.6
York River	187	75	22	15	112	60	19.6
James River	684	55	219	82	356	52	20.0
B. Elizabeth River Watershed *(River Scale)*							
Entire Elizabeth River	46.6	6.9	15.3	7.8	30.0	64	10.1
Individual Strata							
Mainstem	20.7	0.80	6.62	3.31	10.7	52	20.0
Lafayette River	5.7	0.68	2.05	0.91	3.6	64	19.2
Eastern Branch	5.7	0.68	2.28	0.68	3.6	64	19.2
Western Branch	6.9	1.38	2.21	1.38	5.0	72	18.0
Southern Branch	7.6	3.34	2.13	1.52	7.0	92	10.8
C. Sediment Contaminant Strata *(Creek Scale)*							
Scuffletown Creek	0.1	0.024	0.036	0.016	0.076	76	17.0
Jones-Gilligan Creeks	0.1	0.024	0.040	0.012	0.076	76	17.0

Mainstem. The downstream boundary of the Mainstem was set at the northern limit of Craney Island (Figure 2). The strata studied encompass salinities typically ranging from 19–24 psu. Here we present only the results from 125 random locations among the five strata.

Within the Elizabeth River, sediment contamination from heavy metals and polynuclear hydrocarbons has been identified as a major problem (Elizabeth River Project, 2000). The US Army Corps of Engineers is directing a feasibility study that includes a possible demonstration sediment clean-up effort in one small tidal creek (USACOE, 2001). As part of that study, two strata were sampled in 1999:

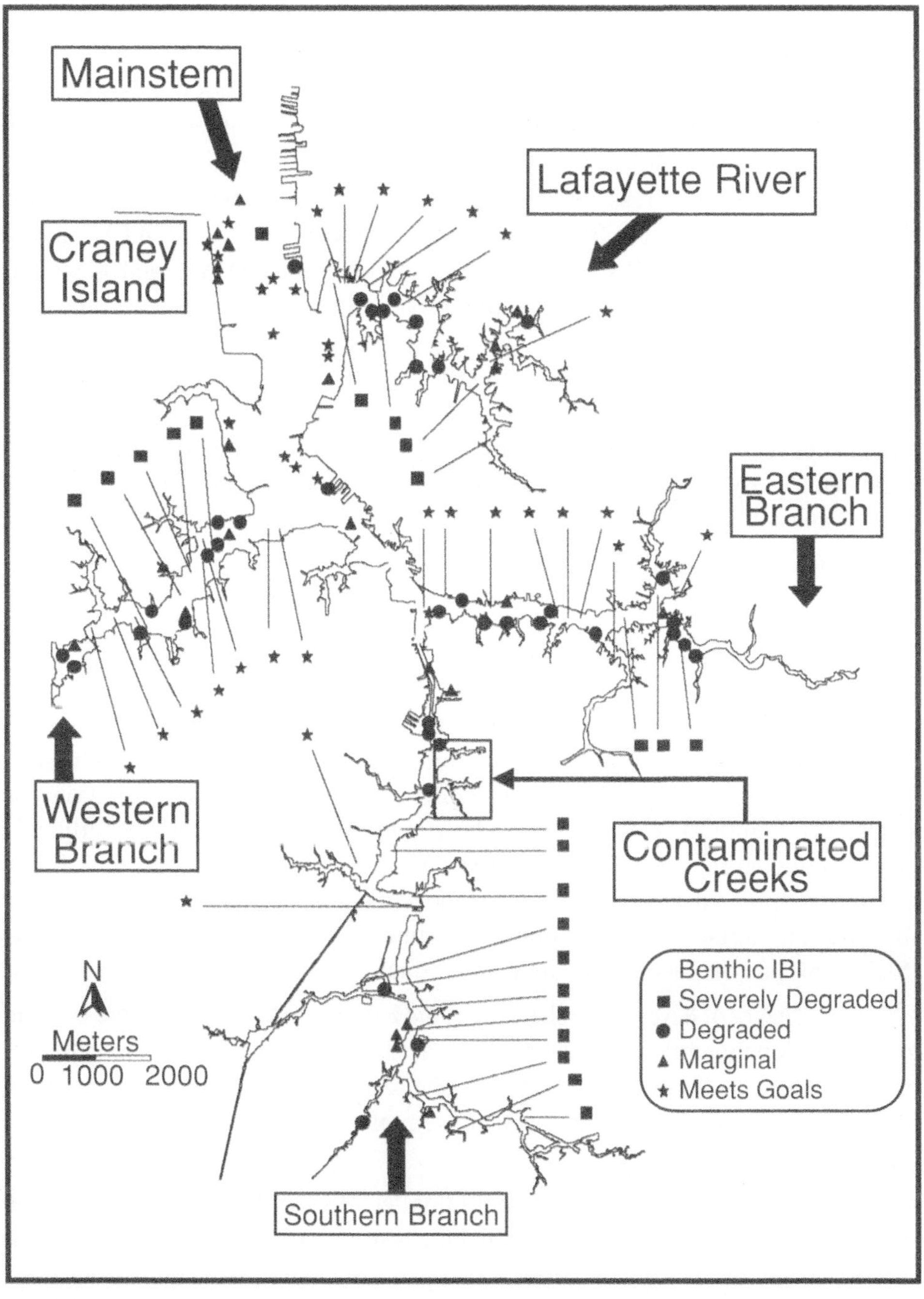

Figure 2. Elizabeth River watershed. Shown are five sampling strata and the location and condition of the benthic community at the 125 random locations sampled in 1999. For clarity, samples designated meets goal and severely degraded are shown outside of strata except for the Mainstem.

the demonstration project creek, Scuffletown Creek, and a similar creek system directly upstream, Jones and Gilligan creeks (see Contaminated Creeks of Figure 2). Fifty random locations were sampled among the two strata. Future sampling of these strata may occur if sediment remediation efforts occur.

The B-IBI was calculated for each of the 425 benthic samples of this study. The index was calculated by scoring values of benthic community metrics compared to values of the respective metrics at reference sites determined to be relatively free of anthropogenic stress. Selection of metrics and the values for scoring metrics were developed separately for each of seven benthic habitat types in Chesapeake Bay (Weisberg *et al.*, 1997). Metrics were scored 1, 3 or 5 and averaged. A score of 3.0 or greater indicated that the sample did not differ from reference conditions. The benthic community condition was classified as follows: meets goal (B-IBI $\geq$ 3.0), marginal (B-IBI = 2.7-2.9), degraded (B-IBI = 2.1-2.6) or severely degraded (B-IBI $\leq$ 2.0). The thresholds for scoring the benthic metrics can be considered benthic restoration goals for each habitat type. In that sense a B-IBI score of < 3.0 was characterized as failing the Chesapeake Bay benthic community restoration goal (Ranasinghe *et al.*, 1994).

Significant differences between strata were tested using the method of Schenker and Gentleman (2001). All reported differences are at $p = 0.5$.

3. Results

At the spatial scale of the tidal Chesapeake Bay, 47% of the bottom had benthic community condition that failed the restoration goal with 63% of Maryland tidal waters and 28% of Virginia tidal waters classified as failing the restoration goal (Table 1A, Figure 3). Among the ten strata, percentages of the benthos failing the restoration goal ranged from 20% to 72% with the best condition in the Virginia portion of the Mainstem of the Bay and the worst condition in the Potomac River. By absolute surface area, the middle stratum of the Mainstem (MD Mid Bay Mainstem) had the largest amount of poor quality benthic community condition.

At the Elizabeth River watershed scale, 64% of the bottom had benthic community condition that failed the restoration goal (Table 1B, Figure 2). Among the five strata, percentages of the benthos failing the restoration goal ranged from 52% to 92% with the best condition in the Mainstem of the river and the worst condition in the Southern Branch.

At the contaminated creek scale both tidal creeks had 76% of the bottom with benthic community condition that failed the restoration goal (Table 1C, Figure 4).

4. Discussion

Estuaries are dynamic and complex ecosystems with varying salinities, freshwater input, physical forces, bottom sediment types, etc. The spatial scale of potential an-

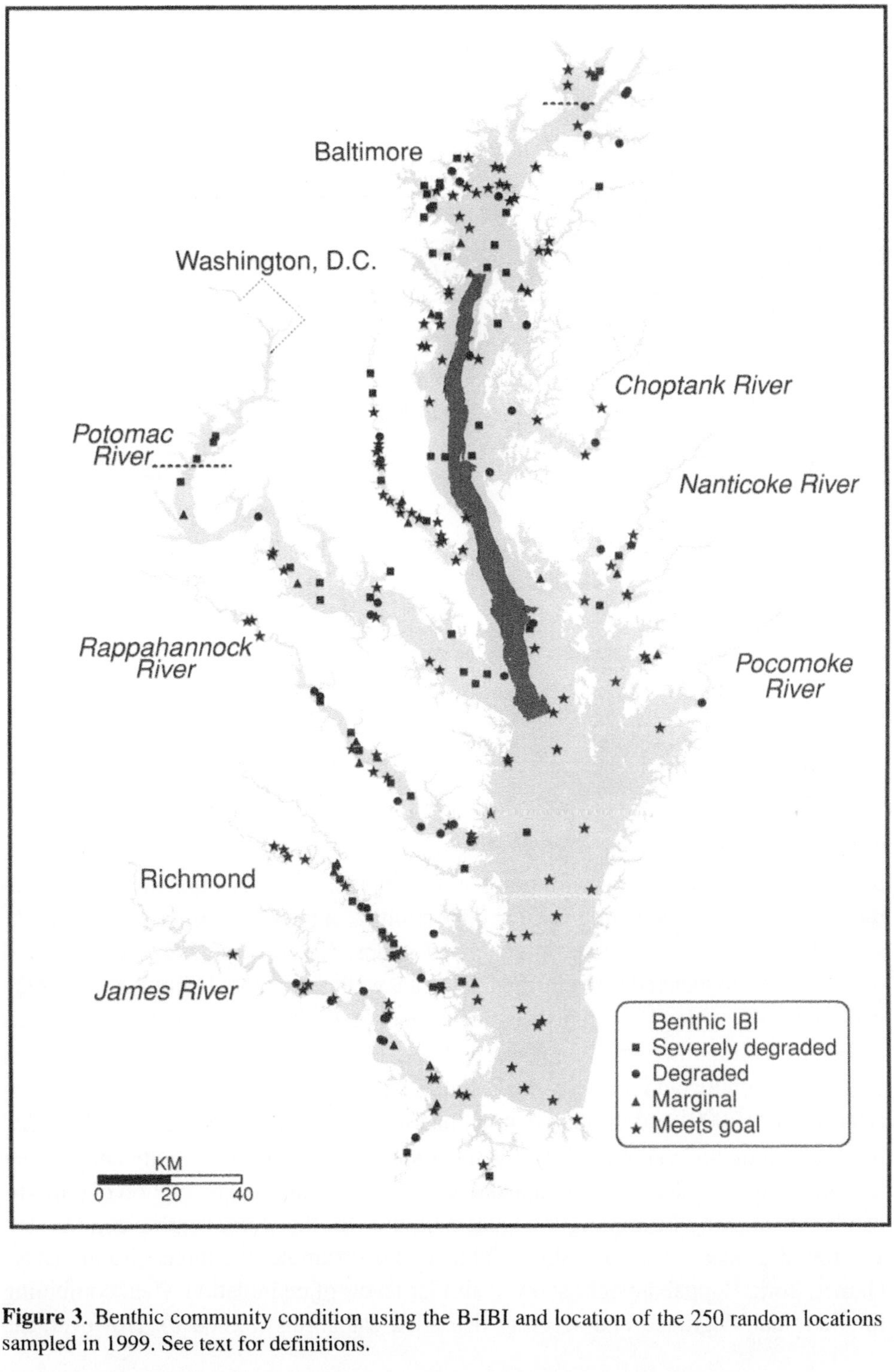

Figure 3. Benthic community condition using the B-IBI and location of the 250 random locations sampled in 1999. See text for definitions.

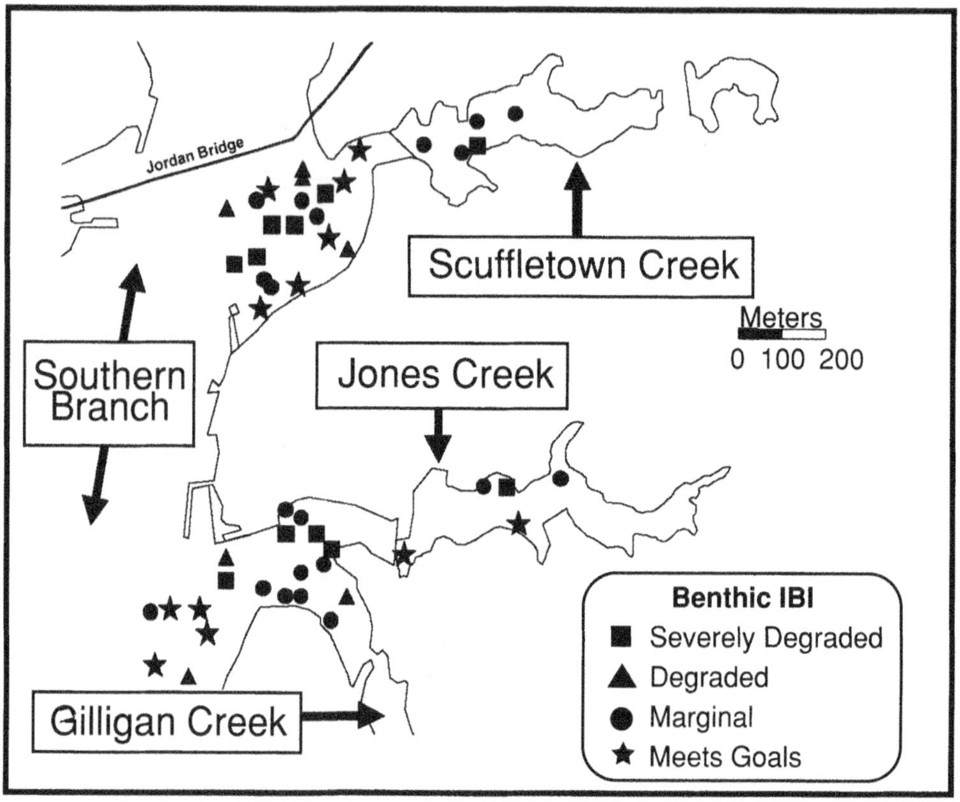

Figure 4. Scuffletown Creek, Jones Creek, and Gilligan Creek. Shown are the locations and condition of the benthic community at the 50 random locations sampled in 1999.

thropogenic stressors varies from large scale in the case of non-point inputs to potentially very small spatial scales for point-source inputs. Eutrophication is widely documented as adversely affecting benthic communities of the Chesapeake Bay (USEPA, 1999; Dauer *et al.*, 2000) over large spatial scales. At the Bay scale, low dissolved oxygen events associated with eutrophication are well documented (Smith *et al.*, 1992), geographically widespread in distribution (Dauer *et al.*, 2000) and have known effects on benthic community structure (Diaz and Rosenberg, 1995). Low dissolved oxygen events are common and severe in the Potomac River and the Maryland Mid Mainstem (Dauer *et al.*, 2000) and these two strata had the highest areal estimates of degraded benthic community condition, 72% and 68%, respectively and were significantly higher than the value for the entire Chesapeake Bay (47%). Additional resolution of patterns can be obtained by examining the areal estimates for the worst benthic condition—severely degraded areas (see SD in Table 1). For example, the three Virginia rivers (James, York, Rappahannock) showed similar levels of degradation when combining the marginal, degraded and severely degraded categories (52–60%). However, the percentages of severely degraded condition were 8% for the James River, 20% for the Rappahannock River and 40% for the York River. Bricker *et al.* (1999) characterized the eutrophic condition for the estuaries of the United States based upon a survey of over 300 experts on estuarine eutrophication. They listed the three Virginia rivers as follows: the James River as having a low eutrophic condition, the Rappahannock River as having a moderate eutrophic condition and the York River has having a high eutrophic condition. The areal extent of severely degraded benthic condition better conforms to the characterizations of Bricker *et al.* (1999) in comparison to the combined areas of marginal, degraded and severely degraded. Examining just the severely degraded condition, the Potomac River and Maryland Mid Mainstem still have the highest levels of severely degraded bottom with values of 52% and 55%, respectively. MD Eastern Tributaries have a total failing area of 50% but a value of 8% for the severely degraded condition. These tributaries have high agricultural land use, high nutrient load input, high chlorophyl values but low frequencies of low dissolved oxygen events compared to strata such as the Potomac River (Dauer *et al.*, 2000).

At the Bay scale, sediment contamination by toxic chemicals also adversely affects benthic communities of the Chesapeake Bay (Dauer, 1993; Dauer *et al.*, 1993; 2000) but serious problems are spatially concentrated in three regions, Baltimore Harbor in Maryland, the Anacostia River in Washington, D.C. and the Elizabeth River in Norfolk, Virginia (CBP, 1999; USEPA, 1999). The areal extent of degradation of benthic community condition of the Elizabeth River can not be inferred from sampling at the Bay scale. For example, in 1999 only two stations of the 25 stations allocated to the James River stratum were in the Elizabeth River watershed.

At the River scale, the Elizabeth River is characterized by poor water quality but bottom oxygen levels are not low enough to cause adverse effects on benthic community condition (Dauer *et al.*, 1999). The level of areal degradation of the Elizabeth River watershed (64%) was significantly higher than the Virginia value (28%) but not significantly different from the values of the three major rivers in Virginia (52-60%). Among the five strata at the River scale, the Mainstem had a significantly lower level of degradation (52%) compared to the other River strata and the Southern Branch has a significantly higher level of degradation (92%) compared to the remaining River strata. Examining just the severely degraded benthic condition, the values were 4% for the Mainstem, 12% for the Lafayette River and Eastern Branch, 20% for the Western Branch and 44% for the Southern Branch. The Western Branch and the Southern Branch are the sections of the river characterized by the highest levels of sediment contaminants (ERP, 2000). The Southern Branch is the only stratum characterized by severe problems due to levels of heavy metals and sediment organic compounds. In the Western Branch all stations summarized in ERP (2000) indicated a problem with heavy metals. In the Mainstem, Eastern Branch and Lafayette River measured levels of metals or organic compounds were not considered to be at problem levels.

The US Army Corps of Engineers, when considering a demonstration project for sediment contaminant remediation, restricted their consideration to small portions of the Elizabeth River that were less than 1 km^2 (USACOE, 2001). The areal extent of degradation of benthic community condition at the Creek scale can not be inferred from sampling at the River scale. For example, none of the 25 samples allocated to the Southern Branch were in either of the two creek systems.

In summary, a stressor such as low dissolved oxygen is generally more spatially widespread and larger sampling strata capture the resultant pattern of benthic community condition. Sediment contamination is spatially less extensive and smaller strata are necessary to capture the resultant pattern of benthic community condition. In our study the significant benthic community degradation due to sediment contamination of the Elizabeth River is not obvious from the Bay scale (11,407 km^3) nor at the spatial scale of the James River stratum (863 km^2). Even within the Elizabeth River strata of a spatial scale of 10^1 km^2 may still be too large to define local sediment contaminant problems. Comparisons of patterns in benthic community condition across spatial scales and between combinations of strata must be cautiously interpreted. For example, the Virginia Tidal Waters value of 28% failing is significantly lower then the Maryland Tidal Waters value of 63% (Table 1A). However, this pattern is driven by the very low failure level of the Virginia Mainstem which accounts for 77% of Virginia tidal waters. In fact the three major Virginia tributaries (James, York and Rappahannock) are not significantly different from the Potomac River when using total area failing. However when using the severely degraded condition the James River and Rappahanock River both have significantly lower levels of degradation compared to the Potomac River.

Benthic monitoring program designs must be flexible and modified according to the stressor(s) of concern. Further, they must provide quantitative ecological information to environmental managers including a characterization of the ecological condition of the system and the spatial extent of the problem addressed. A validated, unambiguous approach to characterizing the condition of the benthic community is essential. In the Chesapeake Bay Benthic Monitoring Program, we accomplish this by using the B-IBI and a stratified random sampling design. The B-IBI has been shown by Alden *et al.* (in press) to be sensitive, stable, robust, and statistically sound. Two additional advantages of the B-IBI are (1) several levels of degradation can be defined (e.g., severely degraded, degraded, marginal) and (2) the pattern of individual metrics of the index can be examined for insight into the reasons for the determined condition.

Acknowledgments

We acknowledge the support of the US Environmental Protection Agency, the US Army Corps of Engineers, the Maryland Department of Natural Resources, the Chesapeake Bay Program Office of the Virginia Department of Environmental

Quality, the Tidewater Regional Office of the Virginia Department of Environmental Quality and the Elizabeth River Project. The comments of three anonymous reviewers greatly improved the manuscript.

5. References

Alden, R.W. III, Dauer, D.M., Ranasinghe, J.A., Scott, L.C. and Llansó, R.J.: 2002, 'Statistical verification of the Chesapeake Bay Benthic Index of Biotic Integrity', *Environmetrics*. (In press).

Bricker, S.B., Clement, C.G., Pirhalla, D.E., Orlando, S.P. and Farrow, D.R.G.: 1999, 'National estuarine eutrophication assessment: effects of nutrient enrichment in the nation's estuaries', NOAA, National Ocean Service, Special Projects Office and the National Centers for Coastal Science. Silver Spring, Maryland, 71 pp.

CBP (Chesapeake Bay Program): 1999. 'Targeting toxics: A characterization report. A tool for directing management and monitoring actions in the Chesapeake Bay's tidal rivers', Chesapeake Bay Program, U.S. Environmental Protection Agency, Annapolis, Maryland, 49 pp.

Dauer, D.M.: 1993, 'Biological criteria, environmental health and estuarine macrobenthic community structure', *Marine Pollution Bulletin* **26**, 249–257.

Dauer, D.M.: 1997, 'Dynamics of an estuarine ecosystem: Long-term trends in the macrobenthic communities of the Chesapeake Bay, USA (1985–1993)', *Oceanologica Acta* **20**, 291–298.

Dauer, D.M.: 2000, 'Benthic Biological Monitoring Program of the Elizabeth River Watershed (1999)', Final Report to the Virginia Department of Environmental Quality, Chesapeake Bay Program, 73 pp.

Dauer, D.M. and Alden III, R.W.: 1995, 'Long-term trends in the macrobenthos and water quality of the lower Chesapeake Bay (1985–1991)', *Marine Pollution Bulletin* **30**, 840-850.

Dauer, D.M., Lane, M.F., Marshall, H.G., Carpenter, K.E. and Diaz, R.J.: 1999, 'Status and trends in water quality and living resources in the Virginia Chesapeake Bay: 1985–1998', Final report to the Virginia Department of Environmental Quality. 65 pp.

Dauer, D.M., Luckenbach, M.W. and Rodi, Jr., A.J.: 1993, 'Abundance biomass comparison (ABC method): effects of an estuarine gradient, anoxic/hypoxic events and contaminated sediments', *Marine Biology* **116**, 507–518.

Dauer, D.M., Ranasinghe, J.A. and Rodi, Jr., A.J.: 1992, 'Effects of low dissolved oxygen levels on the macrobenthos of the lower Chesapeake Bay', *Estuaries* **15**, 384–391.

Dauer, D.M., Ranasinghe, J.A. and Weisberg, S.B.: 2000, 'Relationships between benthic community condition, water quality, sediment quality, nutrient loads, and land use patterns in Chesapeake Bay', *Estuaries* **23**, 80–96.

Diaz, R.J. and Rosenberg, R.:1995. 'Marine benthic hypoxia: a review of its ecological effects and the behavioural responses of benthic macrofauna', *Oceanography and Marine Biology: An Annual Review* **33**, 245–305.

Elizabeth River Project: 2000, 'State of the River 2000,' Elizabeth River Project, Norfolk, Virginia, 24 pp.

Kennish, M.J.: 1998, *Pollution impacts on marine biotic communities*, CRC Press, Boca Raton, 310 pp.

Kingsford, M.J.: 1998, 'Analytical aspects of sampling design', in Kingsford, M. and Battershill, C. (eds), *Studying temperate marine environments. A handbook for ecologists*, CRC Press, Boca Raton, p. 49–83.

Kingsford, M.J. and Battershill, C. N.: 1998, 'Procedures for establishing a study', in Kingsford, M. and Battershill, C. (eds), *Studying temperate marine environments. A handbook for ecologists*, CRC Press, Boca Raton, p. 29–48.

Kramer, K.J.M.: 1994, *Biomonitoring of coastal waters and estuaries*, CRC Press, Boca Raton, 327 pp.

Livingston, R.J.: 2001, *Eutrophication processes in coastal systems*, CRC Press, Boca Raton, 327 pp.

Llansó, R.J., Dauer, D.M., Vølstad, J. H., and Scott, L.C.: 2001, ' Application of the Benthic Index of Biotic Integrity to environmental monitoring in Chesapeake Bay', *Environmental Monitoring and Assessment* **THIS ISSUE.**

Ranasinghe, J.A., Weisberg, S.B., Dauer, D.M., Schaffner, L.C., Diaz, R.J. and Frithsen, J.B.: 1994, 'Chesapeake Bay benthic community restoration goals', Report for the U.S. Environmental Protection Agency, Chesapeake Bay Office and the Maryland Department of Natural Resources. 49 pp.

Schenker, N. and Gentleman, J.F.: 2001, 'On judging the significance of differences by examining the overlap between confidence intervals', *The American Statistician* **55**, 182–186.

Smith, D.E, Leffler, M. and Mackiernan, G.: 1992, 'Oxygen dynamics in the Chesapeake Bay. A synthesis of recent research', Maryland Sea Grant College, College Park, Maryland, 234 pp.

USACOE (U.S. Army Corps of Engineers): 2001, Draft Feasibility Study and Draft Environmental Assessment, Main Report, Elizabeth River Basin, Environmental Restoration, U.S. Army Corps of Engineers, Norfolk District, Norfolk, Virginia, 188 pp.

USEPA (U.S. Environmental Protection Agency): 1999, The State of the Chesapeake Bay, EPA/903-R99-013, CBP/TRS222/108, U.S. Environmental Protection Agency, Annapolis, Maryland, 58 pp.

Weisberg, S.B., Ranasinghe, J.A., Dauer, D.M., Schaffner, L.C., Diaz. R.J. and Frithsen, J.B.: 1997, 'An estuarine benthic index of biotic integrity (B-IBI) for Chesapeake Bay', *Estuaries* **20**, 149–158.

AN APPROACH FOR IDENTIFYING THE CAUSES OF BENTHIC DEGRADATION IN CHESAPEAKE BAY

CORY S. CHRISTMAN AND DANIEL M. DAUER*

Department of Biological Sciences, Old Dominion University, Norfolk, VA 23529
**(author for correspondence, email: ddauer@odu.edu)*

Abstract: We developed an index to differentiate between low dissolved oxygen effects and sediment contamination effects for sites classified as degraded by the Chesapeake Bay Benthic Index of Biotic Integrity (B-IBI), using discriminant analysis. We tested 126 metrics for differences between sites with low dissolved oxygen and sites with contaminated sediments. A total of 16 benthic community metrics met the variable selection criteria and were used to develop a discriminant function that classified degraded sites into one of two stress groups. The resulting discriminant function correctly classified 77% of the low dissolved oxygen sites and 80% of the contaminated sites in the validation data.

Keywords: discriminant analysis, index of biotic integrity, Chesapeake Bay, sediment contamination, low dissolved oxygen, macrobenthos.

1. Introduction

Although two benthic indices are currently in use to assess benthic community condition in Chesapeake Bay, neither is designed to determine causes of degradation. The use of benthic macroinvertebrate communities as indicators of estuarine condition and trends has become a focus of the USEPA's Environmental Monitoring and Assessment Program (EMAP) (Summers *et al.*, 1991; Weisberg *et al.*, 1993; Engle *et al.*, 1994; Engle and Summers, 1998; 1999). Benthic invertebrates are used because they exhibit a wide range of consistent responses to multiple sources of stress (Pearson and Rosenberg, 1978; Bilyard, 1987; Dauer, 1993; Wilson and Jeffrey, 1994; Weisberg *et al.*, 1997). In Chesapeake Bay, the two indices were developed using different methods. EMAP developed an index for the Virginian Province, using a discriminant analysis approach (Paul *et al.*, 2001). The Chesapeake Bay Monitoring Program developed a Benthic Index of Biotic Integrity (B-IBI) (Weisberg *et al.*, 1997). Both indices perform well when used to assess whether sites were degraded or undegraded with overall correct classification percentages over 80% (Weisberg *et al.*, 1997; Paul *et al.*, 2001).

The two primary stressors on benthic communities in Chesapeake Bay are low dissolved oxygen (LDO) and sediment contamination (TOX). Although benthic indices assess both types as degraded, they may result in different community composition because the type, severity and frequency of a stress may result in the colonization and persistence of different groups of species (Connell, 1978; Gray, 1979; Sousa, 1979a,b; Menge and Sutherland, 1987).

LDO stress in the Chesapeake Bay is typically characterized by seasonal hypoxic events that occur suddenly and remain for a short period with community effects varying from partial to total elimination of the community (Diaz and

Rosenberg, 1995). Such areas are initially colonized by opportunistic species with extended periods of reproduction (Gray, 1979). These opportunists may then be replaced over time by other more competitive species as the community moves toward the pre-disturbance state.

TOX stress tends to occur persistently and have long lasting residual effects (Swartz and Lee, 1980). Chronic, persistent stresses alter community composition and favor more stress tolerant species (Gray, 1979). TOX affected benthic communities may be different than LDO affected communities, but these differences are likely small, particularly in estuaries where macrobenthic species are often characterized as eurytopic (Boesch, 1977).

Our objective was to determine if a function could be developed that would be able to differentiate between benthic communities stressed either by low dissolved oxygen levels or by sediment contaminants based on benthic monitoring information. To accomplish this we used discriminant analysis, a statistical technique that develops a mathematical function that maximizes the differences between two or more groups using multiple metrics (Klecka, 1980; Dillion and Goldstein, 1984). This function can then be used to predict into which of the a priori defined groups new sites are placed (McLachlan, 1992; Khattree and Naik, 2000). A successful function will have a high ability to correctly predict the classification of both LDO and TOX affected sites.

2. Methods

We conducted an analysis to determine the ability of discriminant analysis to successfully distinguish between benthic communities subjected to the two main causes of degradation in Chesapeake Bay. A five-step approach was used to develop a function that could distinguish between sites where benthic macroinvertebrates at degraded polyhaline mud sites in Chesapeake Bay were affected by low dissolved oxygen or by contaminated sediments. These steps were: (1) Site Selection; (2) Data Collection and Partitioning; (3) Candidate Metric Development; (4) Function Development; and (5) Function Validation.

2.1 Site Selection

We used data from four sites, two affected by low dissolved oxygen and two affected by sediment contaminants in Chesapeake Bay. Sites were chosen that were within the polyhaline region of Chesapeake Bay and with silt-clay content greater than 40% (the polyhaline mud habitat type of Weisberg *et al.*, 1997).

The two LDO sites were CB5.4 in the mainstem of Chesapeake Bay and LE3.2 at the mouth of Rappahannock River (Figure 1). Dauer *et al.* (1993) reported that 42.5% of semimonthly bottom-water dissolved oxygen measurements from June through September were less than 2 ppm at site CB5.4 and 45.9% at LE3.2. Dauer

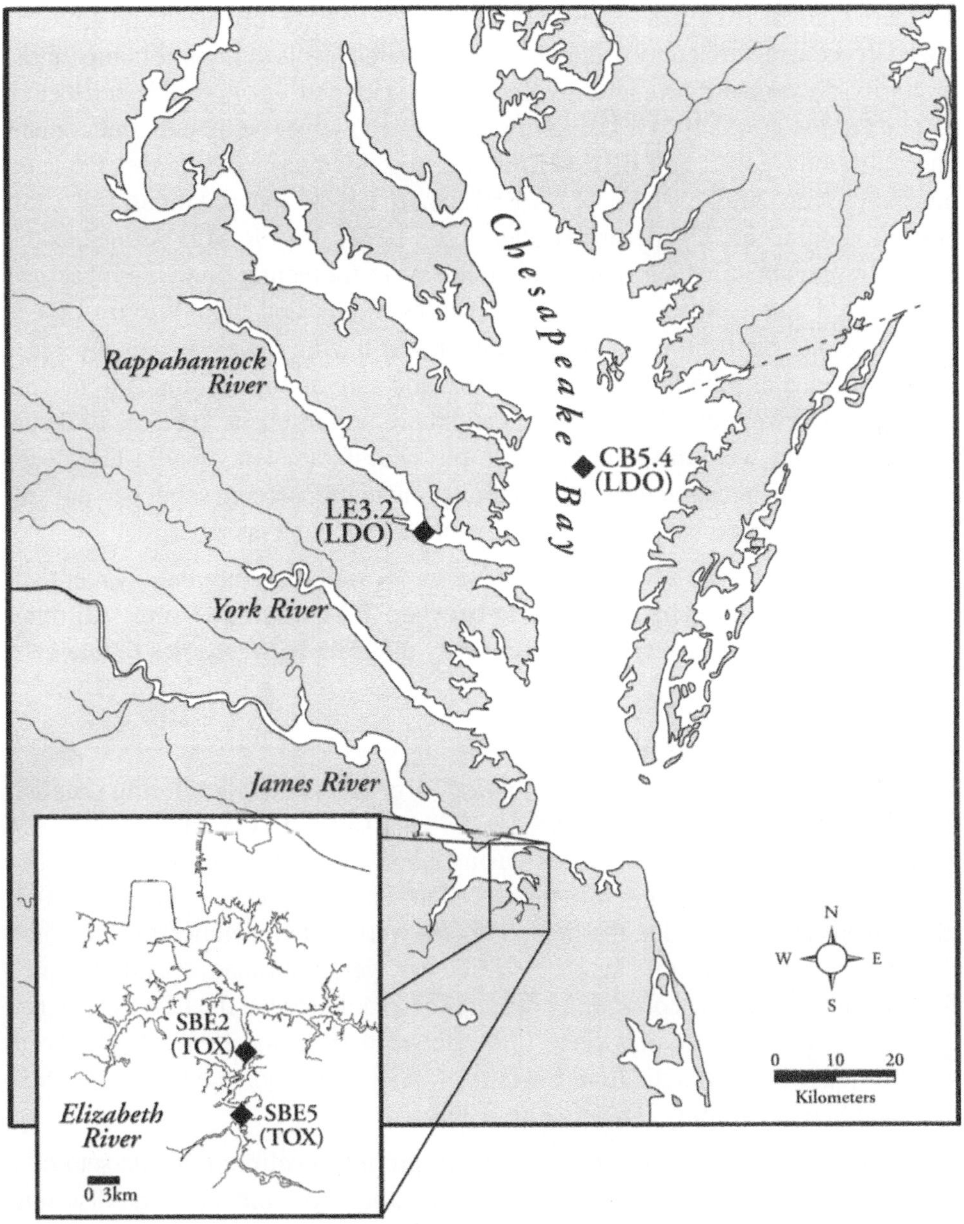

Figure 1. Map of Lower Chesapeake Bay showing the locations of the sediment contamination sites (TOX) SBE2 and SBE5, and the low dissolved oxygen sites (LDO) LE3.2 and CB5.4.

et al. (1992) used cluster analysis to classify these sites into a low dissolved oxygen affected site group distinct from all other polyhaline sites.

The two TOX sites (SBE2 and SBE5, Figure 1) are located in Elizabeth River's southern branch. The Elizabeth River watershed is heavily urbanized and industrialized. It is characterized by sediments with elevated levels of organic compounds, primarily polynuclear aromatic hydrocarbons (Huggett *et al.*, 1987) and heavy metals (Alden *et al.*, 1991). Using the Long *et al.* (1995) sediment contaminant concentrations, SBE2 had ER-M (Effects Range Median) exceedances for lead, mercury, zinc, and pyrene while SBE5 had ER-M exceedances for zinc, benzo(a)anthracene, and pyrene (Dauer *et al.*, 1992). An ER-M is defined as the concentration above which biological effects were frequently or always observed or predicted among species tested. Dauer (1993) indicated that all four of these sites were distinct from other polyhaline sites in having lower species diversity, less abundance of individuals, less biomass, and a community composition characterized by fewer equilibrium species and more opportunistic species. To determine if the sites were different based on physical parameters alone, t-tests were run on the physical characteristics of salinity, mean phi, percent sand, percent silt, percent clay, and percent silt-clay between the LDO and TOX groups.

The t-test run on the six physical measures showed that only one, percent silt was significantly different ($p < 0.001$) between TOX and LDO sites. All other parameters were not statistically significantly different between sites (Table 1).

2.2 Data Collection and Partitioning

We used September data from the on-going Virginia benthic monitoring program collected from 1985 to 1998 and 1989 to 1998 at LDO and TOX sites respectively. Three samples were collected at each using a box core with a surface area of 184 cm^2 and penetration of 25 cm into the sediment. Each sample was sieved on a 0.5-mm screen and preserved in the field. Samples were sorted, enumerated and identified to the lowest possible taxon. Ash-free dry weight biomass was determined for each taxon. Benthic community data has been collected at the LDO sites since 1985 and from the TOX sites since 1989. Between 1985 and 1996, all sites were sampled quarterly (March, June September, and December) and since 1996 in June and September only (Dauer *et al.*, 1998).

The data were divided at random into two groups, a calibration data set and a validation data set. The calibration data set for the development of the discriminant function consisted of five years, 1989, 1990, 1992, 1996, and 1997, of data randomly selected from the 10-year period 1989 to 1998 that was common to LDO and TOX sites. The remaining data were used for validation of function performance. By dividing the data set into two subsets, we are able to determine the accuracy of the classification of data by the function (Sharma, 1996).

Table 1. Summary of the physical measures. Mean and standard error are given for each metric for the LDO and TOX groups. The *p* values were calculated using a t-test.

Metric	LDO	TOX	*p*-value
Salinity	18.6 (0.58)	16.9 (0.95)	0.12
Mean phi	7.5 (0.15)	7.6 (0.32)	0.63
Percent sand	6.7 (1.66)	13.2 (4.07)	0.10
Percent silt	59.9 (1.95)	47.4 (2.34)	< 0.01
Percent clay	33.3 (1.77)	39.2 (2.69)	0.06
Percent silt-clay	93.2 (1.66)	86.7 (4.07)	0.10

2.3 CANDIDATE METRIC DEVELOPMENT

The three replicate samples taken at each site were combined to form the composite samples used in our analysis. Individual species were placed into functional and taxonomic groups (Table 2). For each of these groups all applicable candidate metrics from the development of the EMAP Virginian Province Index (Paul *et al.*, 2001) and the Chesapeake Bay B-IBI (Weisberg *et al.*, 1997) were calculated for evaluation by our function (Table 3). This resulted in 126 metrics being created for each composite site.

2.4 FUNCTION DVELOPMENT

Twenty sites, 10 from each group, were randomly placed into the calibration dataset. The use of only 20 sites in the discriminant analysis restricted the number of metrics that could be used in the analysis to 17 while still fulfilling the degree of freedom requirements (Klecka, 1980). To reduce the 126 candidate metrics to the required 17 or fewer needed, four variable selection techniques were employed: forward stepwise, backwards stepwise, combined stepwise, and t-test. The first three techniques employed a stepwise discriminant analysis using either a forward, backward or combined selection process. Variables were retained or eliminated if the partial R^2 changed by more than 0.05. All three methods were accomplished using the STEPDISC procedure in SAS/STAT (SAS, 1990).

The fourth method used a t-test to determine the candidate metrics to include in the function (Goldstein and Dillion, 1978; Sharma, 1996). The t-statistic was calculated for each of the candidate metrics between the two groups. Only those metrics found to be significantly different ($p < 0.05$) between the groups were retained for analysis. If the number of significant metrics was larger than the allowable degrees of freedom, the p-value was lowered in incremental steps until the allowable number of metrics was reached. A univariate t-test does not take into account the possible residual variation that may be present between two or more metrics. To accomplish this, a multivariate test such as a MANOVA would be

Table 2. Taxonomic and functional groups used in the generation of the discriminant function.

Taxonomic Groups	*Feeding Groups*
Oligochaetes	Carnivores
Polychaetes	Deposit feeders
Capitellid polychaetes	Suspension feeders
Spionid polychaetes	Interface feeders
Nereid polychaetes	
Molluscs	*Life History Groups*
Bivalves	Pollution-sensitive fauna[1]
Gastropods	Pollution-indicative fauna[1]
Amphipods	
Ampelisca amphipods	Entire composite

[1]Weisberg *et al.*, 1997

Table 3. Metrics calculated for the taxonomic and functional groups. Abundance and biomass are the total count for the composite sample. Relative abundance and biomass are the metric total divided by total for the composite sample. Community level metrics were calculated on abundance of the composite sample.

For Each Group in Table 1	*Additional Metrics for Entire Composite*
Abundance	Shannon's H'
Relative abundance	Pielou's evenness
Biomass	Simpson's D
Relative biomass	Gleason's D
Mean individual size	
Number of species	
Margalef's index	

needed. Unfortunately, with only 20 sites, a MANOVA could not be performed on all 126 metrics due to degree of freedom restrictions. For this reason if the function has a low classification rate, the use of the univariate t-test may be one of the causes.

The calibration data set and the validation data set combined were standardized to a mean of zero and standard deviation of one. The SAS procedure DISCRIM (SAS, 1990) was used to create a linear discriminant function for each of the sets of metrics obtained by the metric reduction techniques above.

2.5 Function Validation

The functions were tested to determine their ability to classify sites into the correct a priori determined groups with a two-step process. In the first step the calibration data used to create the function was used to test the classification efficiency of the function. If the classification efficiency was high (> 90 %), the second step tests the function with the validation data sub-set. The classification efficiency determined in this step gives a more unbiased estimation of the functions abilities. The classification performance for both TOX sites and LDO sites was used to compare each of the functions to determine which one was best.

3. Results

3.1 FUNCTION DEVELOPMENT

The forward, backward, combined, and t-test selection methods selected 16, 0, 15, and 16 metrics respectively. The metrics accounted for 99.9%, 0.0%, 99.9% and 98.7% of the variability in the calibration data. The forward and combined stepwise selection methods had 11 metrics in common. The t-test method had three and four metrics in common with the forward and combined stepwise selection methods, respectively.

3.2 CALIBRATION AND VALIDATION EFFICIENCIES

Each of the functions correctly classified the calibration data set with 100% efficiency. However, differences were seen in their ability to correctly classify the validation set. The forward and combined stepwise methods had identical classification efficiencies of 83% for the LDO sites (15/18) and 60% for the TOX sites (6/10) which gave them an overall efficiency of 75%. The t-test selection method was able to correctly classify 78% of the LDO sites (14 of 18) and 80% of the TOX sites (8 of 10) for an overall efficiency of 79%.

The t-test method has the best overall and within stress group classification efficiencies and, therefore, was chosen as the most appropriate data reduction method for differentiating the source of stress for new observtions. Table 4 presents the metrics for this function, the corresponding coefficients, and the mean value of the unstandardized data for the metrics by site group.

4. Discussion

Indices of benthic community condition previously developed for Chesapeake Bay are effective in distinguishing between degraded and undegraded conditions (Weisberg *et al.*, 1997; Paul *et al.*, 2001). For effective management of an estuary the ability to differentiate between communities affected by different causes of degradation is necessary. The discriminant function developed here for polyhaline mud habitats of Chesapeake Bay successfully classified two categories of stressors. We used a linear discriminant analysis and reduced the number of metrics using a t-test reduction method. The primary assumptions of a linear discriminant analysis are equality of covariance between stressor groups and multivariate nor-

Table 4. Discriminant function, selected metrics, metric coefficients (from standardized data) and unstandardized group means. Function takes the form of "site score = coefficient 1* metric 1 + ...+ coefficient 16* metric 16". If the site score is ≥ – 1.53 (cutoff value) then the site is designated as having a LDO problem and if the site score is < – 1.53 then the site is designated as having a TOX problem.

Coefficient	*Metric*	*LDO Mean*	*TOX Mean*
281.73	Number of isopod species	0.00	0.45
259.59	Margalef's index of spionid polychates	0.11	0.69
146.52	Mean individual size	0.00	$2.9E^{-4}$
134.45	Relative abundance of polychaetes	92.44	72.89
65.68	Relative biomass of nereid polychaetes	6.38	1.21
35.34	Relative abundance of carnivores	25.37	14.02
32.65	Relative biomass of depoist feeders	10.13	16.28
18.91	Biomass of oligochates	$1.4E^{-4}$	$1.9E^{-3}$
16.07	Number of interface feeding species	1.85	4.15
–19.15	Margalef's index of interface feeders	0.61	1.78
–57.98	Number of pollution indicative species	1.54	2.65
–83.32	Biomass of deposit feeders	$7.1E^{-3}$	$6.5E^{-3}$
–103.93	Relative abundance of oligochaetes	0.42	7.35
–159.66	Biomass of isopods	0.00	$9.5E^{-4}$
–212.56	Density of isopods	0.00	0.16
–235.16	Number of spionid polychaete species	1.17	2.10

mality of the metrics of each stressor group. Most environmental data violate these assumption (e.g., Alden *et al.*, 1982); however, discriminant analysis is robust to violations of assumptions (Lachenbruch, 1975; Everitt and Dunn, 1991; Sharma, 1996). When the goal is prediction, the percent correct classification of an independent validation data set is the best indicator of the effect of violations of assumptions. The function presented here has an overall error rate of 21%. This error rate appears to be an acceptable level within the context of ecological risk assessment (see e.g., Newman *et al.*, 2001) and any violations of assumptions of the discriminant analysis had minimal impact on the function.

Our approach in this study was to limit the analysis to two stressor groups within the same habitat type, the polyhaline mud habitat. By this approach we attempted to limit the range of variation of environmental variables that might confound the ability of biotic metrics to discriminate stressor groups. For estuarine macrobenthic communities, variation in salinity and sediment characteristics could increase greatly the error rates of the function. If our within-habitat study was unsuccessful, then across-habitat analyses would be less likely to succeed. We also limited our data sets to the index period of the Chesapeake Bay B-IBI—July 15 to September 30 (Weisberg *et al.*, 1997). Our approach was to limit the analysis to sites that were declared degraded by the B-IBI. This was essential to maximize the potential of identifying metrics that could distinguish between the two stressor groups. Future analytical efforts could evaluate the utility of this approach outside of the index period.

Many of the metrics of our function were correlated and further interpretation of the relative importance of any metric in the discriminant function is not possible. When multicolinearity exists no inference should be made about the importance of a metric (Sharma, 1996). However, metrics with a positive score contribute more toward a designation of a low dissolved oxygen effect while a negative coefficient contributes more toward a toxic contamination effect.

The results of our study indicate that a discriminant analysis approach can successfully separate sites with degraded benthic community condition due to different stressor categories. Future efforts in the development of a diagnostic discriminant approach should include expansion to different salinity zones and sediment types. We envision a two step process in which the present Chesapeake Bay B-IBI of Weisberg *et al.*, (1997) is used to define degraded sites followed by the application of a diagnostic discriminant analysis approach to assign the most probable cause of degradation. The combination of these methods will give managers information about what areas of the Bay need attention and insight into what should be done to correct problems. Understanding the cause and the effect will allow managers to better utilize their limited resources, by concentrating efforts where they will have the greatest impact.

Acknowledgments

The data used in this study were generated from support from the Virginia Department of Environmental Quality. We would also like to thank Anthony J. Rodi and Michael F. Lane and the anonymous reviewers for their assistance and insight.

References

Alden, R. W. III, Dauer, D. M. and Rule, H. H.: 1982, 'Environmental Studies at a Proposed Mid-Atlantic Dredged Material Disposal Site', in: Oceans '82, pp. 1034–1041.

Alden, R. W. III, Ewing, R. M., Helmstetter, M. F., Winfield, J. G., Burdige, D. J., Rule, J. H., Wong, G. T. F., Swift, D. J. P. and Muller, A.: 1991, 'Distribution and Management Implications of Contaminants in Elizabeth River Sediments', Technical Report, Applied Marine Research Laboratory Report No. 741-I, Old Dominion University, Norfolk, Virginia, 262 pp.

Bilyard, G. R.: 1987, 'The Value of Benthic Infauna in Marine Pollution Monitoring Studies', *Marine Pollution Bulletin* **18**, 581–585.

Boesch, D. F.: 1977, 'A New Look at the Zonation of Benthos Along the Estuarine Gradient' in: Coull, B. C. (ed), Ecology of Marine Benthos, Univ. of South Carolina Press, Columbia, SC, pp. 245–266.

VARIABILITY IN THE IDENTIFICATION AND ENUMERATION OF MARINE BENTHIC INVERTEBRATE SAMPLES AND ITS EFFECT ON BENTHIC ASSESSMENT MEASURES

J. ANANDA RANASINGHE[1*], DAVID E. MONTAGNE[2], STEPHEN B. WEISBERG[1]
MARY BERGEN[1,4] AND RONALD G. VELARDE[3]

[1]*Southern California Coastal Water Research Project, 7171 Fenwick Lane, Westminster, CA 92683-5218, USA;* [2]*Los Angeles County Sanitation Districts, P.O. Box 4998, Whittier, CA 90607, USA;* [3]*City of San Diego, 4918 N. Harbor Drive, Suite 101, San Diego, CA 92106, USA;* [4]*Present address: P.O. Box 729, Ojai, CA 93024-0729, USA.*

**(author for correspondence, email: anandar@sccwrp.org)*

Abstract: Studies designed to measure anthropogenic impacts on marine benthic communities depend on the ability of taxonomists to consistently discriminate, identify, and count benthic organisms. To quantify errors and discrepancies in identification and enumeration, 20 samples were completely reprocessed by another one of four participating laboratories. Errors were detected in 13.0% of the data records, affecting total abundance by 2.1%, numbers of taxa by 3.4%, and identification accuracy by 4.7%. Paired t-tests were used to test for differences in the Benthic Response Index (BRI), total abundance, numbers of taxa, and the Shannon-Wiener index between the original and the reanalysis data. Differences in the BRI were statistically insignificant. Although statistically significant differences were observed for numbers of taxa, total abundance, and the Shannon-Wiener index, the differences were small in comparison to the magnitude of differences typically observed between anthropogenically affected and reference sites.

Keywords: marine benthic invertebrates, identification, enumeration, quality assurance, quality control, inter-laboratory calibration, benthic assessment

1. Introduction

Many approaches to benthic community assessment depend on accurate species identification. Measures based on pollution tolerance ratings of the species in each sample, such as those proposed by Eaton *et al.* (2001) and Smith *et al.* (2001), are most dependent. Ordination and cluster analyses based on species abundance data also require accurate species separation. Many recently developed multi-metric benthic indices include species richness (numbers of species) or diversity (Engle *et al.*, 1994; Weisberg *et al.*, 1997; Engle and Summers, 1999; Van Dolah *et al.*, 1999; Paul *et al.*, 2001). Traditional methods such as the Abundance Biomass Comparison (ABC) method (Warwick *et al.*, 1987) and diversity indices (Washington, 1984) also depend on accurate species identification.

Several steps in benthic sample processing can introduce laboratory error into measurements of species richness and diversity. First, under-counts of organisms and species can result from failure to remove all organisms from the sediment during sorting. Second, misidentification of sorted organisms can cause underestimates of species richness if similar species are not distinguished, or over-counts if a single species is erroneously divided. Errors are also introduced if identified organisms are miscounted.

Environmental Monitoring and Assessment **81**: 199–206, 2003.

Most laboratories resort a subset of samples to quantify and minimize sorting error, but quality assurance practices to ensure taxonomic accuracy are typically less well developed although Ellis, (1988) identified the need. Many laboratories maintain collections of voucher specimens that are sent to outside experts for confirmation of identifications, but voucher specimens are typically limited to the best specimens for each species. There is also no guarantee that voucher materials accurately represent all specimens reported under a name. Inter-laboratory calibrations, like those of Ellis and Cross, (1981), may provide better quality assurance. Here we present the results of a calibration exercise in which four laboratories reprocessed samples to evaluate taxonomic and counting consistency and their effects on measures used in benthic assessments.

2. Methods

Samples were obtained from a regional survey of benthic infauna in the Southern California Bight (Bergen *et al.*, 2000). The samples were collected in August and September 1994 with a $0.1 m^2$ Van Veen grab, sieved through 1mm screens, relaxed for 30 min in $MgSO_4$ or propylene phenoxytol, fixed in sodium borate buffered 10% formalin, and transported to four laboratories for identification and enumeration analysis.

Taxonomic accuracy was assessed from twenty samples selected at random. Each sample was analyzed at one of the four laboratories and reanalyzed at another, selected at random. Taxonomists performing reanalysis had no access to original analysis results. When reanalysis was complete, we compared the original and reanalysis data and compiled a list of differences. These differences were classified as errors when they were caused by inaccurate identifications, incorrect counts, or specimens overlooked in the original analysis. They were classified as discrepancies, rather than errors, when they resulted from use of a junior synonym or other unconventional nomenclature, failure to note removal of specimens for vouchers, or differences in opinion about the taxonomic level to which an organism could be identified (e.g., *Polydora* sp. vs. *Polydora narica*). For each sample, error rates for total abundance and numbers of taxa were calculated as the ratio of the difference between the original and resolved values to the resolved value, expressed as a percentage; original values greater than resolved values resulted in negative rates. Error rates for identification accuracy were calculated as the ratio of misidentifications to resolved identifications, also expressed as a percentage. The resolved value represented "truth" by consensus agreement between the original and reanalysis taxonomists.

To assess the effects of differences in laboratory results on benthic assessments, the number of taxa per sample, total abundance, Shannon-Wiener diversity, and the Benthic Response Index (Smith *et al.*, 2001) were calculated for the original and the reanalysis data. Paired t-tests were then used to test for differences between the original and the re-analyzed data.

3. Results

Differences between original and reanalysis data were detected in 25.3% of the data records (Table 1), where a record consists of a taxon and its reported abundance. The differences were nearly equally divided between errors and discrepancies. Miscounts affected 4.8% of the data records and were the most common type of error; misidentifications (4.5%) and overlooked specimens (3.3%) were almost as common. The errors yielded total abundances, numbers of taxa and identifications that differed from "truth", on average, by 2.1%, 3.4% and 4.7%, respectively (Table 2).

Total abundance, numbers of taxa, and Shannon-Wiener index values were significantly different between the original and reanalysis data (Table 3). Mean differences between the original analysis and reanalysis were 4.75 per sample for total abundance, 2.25 per sample for number of taxa, and 0.037 for the Shannon-Wiener index. Differences in the Benthic Response Index were small and not statistically significant.

Table 1. Frequencies of differences in identification and enumeration for 20 samples. The data comprised 1,715 records each consisting of a taxon and its reported abundance. Negative values indicate that the net result of the error was an understatement of the true value; positive values indicate overstatement of the true value.

	Type of Difference	*Number*	*% of Differences*	*% of Records*
Errors	Miscount	83	19.1	4.8
	Mididentification	78	18.0	4.5
	Overlooked specimen (s)	57	13.1	−3.3
	Misapplication of identificationrules	5	1.2	0.3
	Total Errors	223	51.4	13.0
Discrepancies	Judgment differences	131	30.2	7.6
	Specimen loss or unrecorded voucher removal	57	13.1	3.3
	Unconventional nomenclature	23	5.3	1.3
	Total Discrepancies	211	48.6	12.3
	Total Differences	**434**		**25.3**

Table 2. Means (and ranges) of error rates for total abundance, numbers of taxa and identification accuracy.

		Mean Error Rate (%)		
Original Analysis Laboratory	*Number of Reanalyzed Samples*	*Total Abundance*	*Number of Taxa*	*Identification Accuracy*
A	6	3.1 (2.2 – 6.1)	4.8 (2.9 – 5.9)	6.9 (4.3 – 10.5)
B	2	1.0 (0.3 – 1.5)	1.8 (1.2 – 2.3)	3.6 (2.3 – 5.0)
C	6	2.2 (0 – 3.1)	4.5 (1.0 – 9.2)	3.0 (0 – 4.3)
D	6	1.5 (−1.2 – 4.9)	1.1 (0 – 2.0)	4.6 (2.0 – 11.7)
All	20	2.1 (−1.2 – 6.1)	3.4 (0 – 9.2)	4.7 (0 – 11.7)

Table 3. Mean values for the original and reanalysis data and, based on paired t-tests, the probability that the difference > 0. NS = Not significant.

Assessment Measure	*Mean Values*			
	Original Data	*Reanalysis Data*	*Difference*	*Probability*
Total Abundance (per sample)	334.25	339.00	4.75	0.040
Number of Taxa (per sample)	80.90	83.15	2.25	0.004
Shannon-Wiener Diversity Index	5.162	5.199	0.037	0.006
Benthic Response Index	14.242	14.238	0.004	0.979^{NS}

4. Discussion

The data presented in Tables 1 and 3 provide, for the first time, rates at which errors and discrepancies occur when different taxonomists process samples and their effect on commonly used assessment measures. While the percentage of records affected was high, the overall affect on conclusions that would be reached by application of the data for assessment purposes was not large. The observed differences of 4.75, 2.25 and 0.037 in abundance, numbers of taxa and the Shannon-Wiener index were small relative to the sample means of 334.25, 80.9 and 5.162, respectively. Moreover, the differences were much smaller than the differences typically observed between anthropogenically affected and reference sites (Weisberg *et al.*, 1997; Van Dolah *et al.*, 1999). The Benthic Response Index (BRI), which is the abundance-weighted pollution tolerance of the species in the sample (Smith *et al.*, 2001), was not affected by the observed differences. Apparently the BRI is robust to minor taxonomic errors, presumably because similar species have similar pollution tolerance values.

Two factors contributing to the small values of among-laboratory differences are the experience of the taxonomists and the extent of communication among them. Most of the taxonomists have more than two decades of experience working on the benthos of southern California. Moreover, recognizing the need for consistency, in 1982, taxonomists in southern California formed an organization, the Southern California Association of Marine Invertebrate Taxonomists (SCAMIT, http://www.scamit.org), that is dedicated to standardizing taxonomy in the region. SCAMIT publishes lists of accepted nomenclature, maintains reference collections, produces keys and other taxonomic tools, and fosters communication among its members through monthly workshops, newsletters, a web site and e-mail lists. Throughout the regional survey from which these samples were drawn, and independent of the quality assurance exercise presented here, descriptions and figures of unusual or unknown organisms were distributed using these communication mechanisms to keep other taxonomists abreast of developments as they occurred. Error rates in regions where this communication does not exist may be higher.

Another factor contributing to our low error rates was distinguishing between discrepancies and errors. One type of discrepancy was the use of unconventional nomenclature. We chose not to count this as an error because trained taxonomists easily recognize and correct synonymous names while merging data from different sources; however, this is an issue of concern because not all data users have the expertise to recognize synonyms when merging data sets. Although participating taxonomists made efforts to minimize these problems by relying on SCAMIT lists, discrepancies still occurred suggesting that larger numbers occur in areas where there is no authoritative standardization of nomenclature.

Of greater concern was the prevalence of instances where organisms were identified to different taxonomic levels in analysis and reanalysis. Some of these discrepancies resulted from differences of opinion whether condition of a specimen was sufficient for species-level identification while others reflected differences in experience of the taxonomists. We chose to classify these instances as discrepancies, rather than errors, because data can be lumped to the higher taxonomic level before analysis. However, lumping may affect measures of species richness and diversity (Wu, 1982; Wilson and Jeffrey, 1994) and this is of concern when richness and diversity data are compared to threshold values to infer condition of the benthos, as in the case of B-IBI measures (Weisberg *et al.*, 1997; Van Dolah *et al.*, 1999). Interpretations of assessment results are distorted if the level of taxonomy differs from the level used while developing thresholds.

This issue is of particular concern because taxonomy improves over time and, therefore, benchmarks developed from early data should be re-evaluated over time. Our regional surveys provide an example of improving taxonomy. In the survey providing data for these analyses, it was necessary to lump 43 taxa due to taxonomic uncertainty (Table 4). In a subsequent regional survey involving the same group of taxonomists it was only necessary to lump 16 taxa (Table 4). Our list does not include taxa that were recognized from the beginning as impractical to identify to species at the current state of knowledge; rather, it lists groups that the taxonomists thought they were identifying consistently and accurately but, after reanalysis and review of species lists and specimens, it became clear that they were not. The improvement between surveys was achieved in two ways. First, "failures" stimulated production of new keys and other identification tools by SCAMIT. Second, in the subsequent survey, all four laboratories referred a few groups (ceriantharian and edwardsiid anemones and euclymeninaen and lumbrinerid polychaetes) to single "specialty taxonomists" for identification and enumeration. Despite the efforts of SCAMIT, these groups still presented obstacles to consistent treatment unless one taxonomist identified all of them. The challenge is to ensure consistency between assessment tools and levels of taxonomy when the tools are applied.

Our study explored sources of variability and error often ignored when interpreting the results of benthic assessments (Ellis, 1988). Most importantly, our results provide a standard against which subsequent efforts may be judged. By integrating reanalysis by sample exchange among external taxonomists into quality assurance plans, multi-laboratory monitoring programs can greatly increase the

Table 4. Level of taxonomic identification assigned after re-identification of specimens and interlaboratory comparison of data. Bolded entries indicate taxonomic groups affected (all or in part) by reliance on a single taxonomist for identification in 1998.

Group	Name Adopted after Interlaboratory Comparison	Level	Number of Taxa Combined 1994	Number of Taxa Combined 1998
PHYLUM CNIDARIA, Class Anthozoa	**Ceriantharia**	**Order**	**4**	
	Actiniaria	**Order**	**5**	
Order Pennatulacea	*Acanthoptilum* spp.	Genus	3	3
PHYLUM NEMERTEA, Class Anopla	Anopla	Class	7	3
	Paleonemertea	Order	2	3
Class Enopla, Order Hoplonemertea	Hoplonemertea	Order	3	
	Lineidae	Family	12	9
	Amphiporus spp.	Genus	4	4
	Tetrastemma spp.	Genus	2	
PHYLUM MOLLUSCA, Class Aplacophora				
Order Aplacophora	Chaetodermatidae	Family	5	3
Class Gastropoda, Order	*Bittium* spp.	Genus	2	
Megagastropoda	*Lirobittium* spp.	Genus		3
	Asperiscala spp.	Genus	2	
	Nitidiscala spp.	Genus	2	
	Crepidula spp.	Genus	4	
Order Neogastropoda	*Ophiodermella* spp.	Genus	3	
Class Bivalvia, Order Veneroida	*Solen* spp.	Genus	2	
Order Myoida	*Corbula* spp.	Genus	2	
Order Septibranchida	*Cardiomya* spp.	Genus	2	
PHYLUM ANNELIDA, Class Polychaeta				
Order Orbiniida	*Levinsenia* spp.	Genus	3	
	Paradoneis spp.	Genus		4
Order Cossurida	*Cossura* spp.	Genus	2	
Order Spionida	*Boccardia* spp.	Genus		2
	Protocirrineris spp.	Genus	2	
	Monticellina spp.	Genus	5	
	Mediomastus spp.	Genus	3	3
Order Capitellida	***Clymenella* spp.**	**Genus**	**3**	
	Maldanidae	**Family**	**11**	
Order Opheliida	*Ophelina* spp.	Genus	2	
Order Phyllodocida	*Eusyllis* spp.	Genus		3
	Harmothoinae	Subfamily	15	
	Sthenelais spp.	Genus	3	
	Sphaerosyllis spp.	Genus	2	
Order Eunicida	***Lumbrineris* spp.**	**Genus**	**15**	
	Drilonereis spp.	Genus	3	
	Dorvillea (S.) spp.	Genus		3
	Arabella spp.	Genus		2
	Nothria spp.	Genus		2
Order Fauveliopsida	*Fauveliopsis* spp.	Genus	3	
Order Terebellida	*Terebellides* spp.	Genus	2	
Order Sabellida	*Demonax* spp.	Genus	2	
	Bispira spp.	Genus	2	2

continued on following page

Table 4. Continued

Group	Name Adopted after Interlaboratory Comparison	Level	Number of Taxa Combined 1994	1998
PHYLUM ARTHROPODA, Class Malacostraca				
Order Leptostraca	*Nebalia* spp.	Genus	3	
Order Isopoda	*Edotia* spp.	Genus	2	
	Synidotea spp.	Genus	2	
Order Amphipoda	*Aorides* spp.	Genus	6	
	Corophium spp.	Genus		3
	Photis spp.	Genus	4	
	Protomedeia spp.	Genus	2	
	Synchelidium spp.	Genus		3
PHYLUM ECHINODERMATA	Holothuroidea	Class	2	
PHYLUM CHORDATA	Ascidiacea	Class	4	

likelihood of producing results that are accurate and comparable. The levels of error measured in this study provide the first available data points about variability in identification and abundance measures during multi-laboratory taxonomic analysis. As additional data about these errors are accumulated they can be incorporated as targets and limits in quality assurance and quality control programs to ensure that laboratory data quality are maintained.

Acknowledgments

The authors gratefully acknowledge the contributions of Kelvin Barwick, Cheryl Brantley, Donald Cadien, Nancy Carder, Ann Dalkey, Douglas Diener, Thomas Gerlinger, Kathy Langen, Megan Lilly, John Ljubenkov, Lawrence Lovell, Ricardo Martinez-Lara, Thomas Parker, Dean Pasko, Charles Phillips, William Power, James Roney, Rick Rowe, Timothy Stebbins and Ronald Velarde, the participating taxonomists; their efforts and specialized expertise made this study possible. The Southern California Association of Marine Invertebrate Taxonomists (SCAMIT) provides a mechanism for standardizing the names of organisms on the mainland shelf of southern California and promotes communication among taxonomists; they were an integral part of this study. We appreciate the perseverance of the participants in the many meetings that culminated in the quality assurance plan and the many more that were required to carry it through. We also thank Virginia Engle, Dr. Daniel M. Dauer, Dr. Jack Word, and Dr. Derek Ellis who reviewed our manuscript. Finally, we acknowledge Dr. Derek Ellis, who has long recognized the importance and need for formal quality control and quality assessment of biological data that are used to assess anthropogenic impacts on marine invertebrate communities.

References

Bergen, M., Cadien, D., Dalkey, A., Montagne, D. E., Smith, R. W., Stull, J. K., Velarde, R. G. and Weisberg, S. B.: 2000, 'Assessment of benthic infaunal condition on the mainland shelf of southern California', *Environmental Monitoring and Assessment* **64**, 421–434.

Eaton, L.: 2001, 'Development and validation of biocriteria using benthic macroinvertebrates for North Carolina estuarine waters', *Marine Pollution Bulletin* **42**, 23–30.

Ellis, D. V.: 1988, 'Quality control of biological surveys', *Marine Pollution Bulletin* **19**, 506–512.

Ellis, D. V., and Cross, S. F.: 1981, 'A protocol for inter-laboratory calibrations of biological species identifications (Ring Tests)', *Water Research* **15**, 1107–1108.

Engle, V. D., Summers, J. K. and Gaston, G. R.: 1994, 'A benthic index of environmental condition of Gulf of Mexico estuaries', *Estuaries* **17**, 372–384.

Engle, V. D. and Summers, J. K.: 1999, 'Refinement, validation, and application of a benthic condition index for Northern Gulf of Mexico estuaries', *Estuaries* **22**, 624–635.

Paul, J. F., Scott, K. J., Campbell, D. E., Gentile, J. H., Strobel, C. S., Valente, R., Weisberg, S. B., Holland, A. F. and Ranasinghe, J. A.: 2001, 'Developing and applying a benthic index of estuarine condition for the Virginian Biogeographic Province', *Ecological Indicators* **1**, 83–99.

Smith, R. W., Bergen, M., Weisberg, S. B., Cadien, D., Dalkey, A., Montagne, D., Stull, J. K. and Velarde, R. G.: 2001, 'Benthic response index for assessing infaunal communities on the Southern California mainland shelf', *Ecological Applications* **11**, 1073–1087.

Van Dolah, R. F., Hyland, J. L., Holland, A. F., Rosen, J. S. and Snoots, T. R.: 1999, 'A benthic index of biological integrity for assessing habitat quality in estuaries of the southeastern USA', *Marine Environmental Research* **48**, 269–283.

Warwick, R. M., Pearson, T.H. and Ruswahyuni: 1987, 'Detection of pollution effects on marine macrobenthos: further evaluation of the species abundance/biomass method', *Marine Biology* **95**, 193–200.

Washington, H. G.: 1984, 'Diversity, biotic and similarity indices: a review with special relevance to aquatic ecosystems', *Water Resources* **19**, 653–694.

Weisberg, S. B., Ranasinghe, J. A., Dauer, D. M., Schaffner, L. C., Diaz, R. J. and Frithsen, J. B.: 1997, 'An estuarine benthic index of biotic integrity (B-IBI) for Chesapeake Bay', *Estuaries* **20**, 149–158.

Wilson, J. G. and Jeffrey, D. W.: 1994, 'Benthic biological pollution indices in estuaries', in: Kramer, K. J. M. (ed), Biomonitoring of coastal waters and estuaries, CRC Press, Boca Raton, FL, pp. 311–327.

Wu, R. S. S.: 1982, 'Effects of taxonomic uncertainty on species diversity indices', *Marine Environmental Research* **6**, 215–225.

PRODUCTION, RESPIRATION AND NET ECOSYSTEM METABOLISM IN U.S. ESTUARIES

JANE M. CAFFREY*

Center for Environmental Diagnostics and Bioremediation, University of West Florida, Pensacola, FL, USA
(author for correspondence, e-mail: jcaffrey@bellsouth.net)

Abstract. Primary production, respiration, and net ecosystem metabolism (NEM) are useful indicators of ecosystem level trophic conditions within estuaries. In this study, dissolved oxygen data collected every half hour between January 1996 to December 1998 by the National Estuarine Research Reserve System Wide Monitoring Program were used to calculate primary production, respiration, and net ecosystem metabolism. Data from two sites at each of 14 Reserves were analyzed. On average, three quarters of the data available could be used to calculate metabolic rates. Data from two of the Reserves were used to evaluate the assumption of homogeneity of water masses moving past the oxygen sensor. Temperature was the single most important factor controlling metabolic rates at individual sites, although salinity was also important at about half the sites. On an annual basis, respiration exceeded gross primary production demonstrating that all but 4 of the 28 sites were heterotrophic.

Keywords: estuary, production, respiration, net ecosystem metabolism, dissolved oxygen

1. Introduction

Primary production, respiration and the balance between the two, or net ecosystem metabolism (NEM) vary widely across ecosystems, from highly productive salt marshes to oligotrophic ocean waters. NEM is a useful indicator of the trophic condition within estuaries, whether autochthonous or allochthonous sources of organic matter dominate. If NEM is positive, the system is autotrophic suggesting that internal production of organic matter dominates, while if NEM is negative, the system is heterotrophic and reliant on external sources of organic matter.

In aquatic ecosystems, a variety of techniques have been used to measure production, respiration and NEM. H.T. Odum first developed the method of calculating metabolic rates from diel oxygen curve data in the 1950s (Odum, 1956; Odum and Hoskins, 1958). Since then it has been applied in a wide variety of systems, including many different estuaries (Nixon and Oviatt, 1972; Kemp and Boynton, 1980; Oviatt *et al.*, 1986; D'Avanzo *et al.*, 1996; Swaney *et al.*, 1999).

The National Estuarine Research Reserve (NERR) system currently includes 25 protected areas throughout the United States, including Puerto Rico. Each Reserve participates in the System Wide Monitoring Program (SWMP) to measure water quality parameters (temperature, salinity, dissolved oxygen, pH, turbidity, water depth) at a minimum of two sites. The goal of the NERR SWMP is to examine short-term variability and long-term changes in representative estuarine ecosystems (Wenner *et al.*, 2001).

This paper assesses the feasibility of using water quality monitoring data to calculate metabolic rates. Data were examined to determine whether advection

Environmental Monitoring and Assessment **81**: 207–219, 2003.

was important in controlling oxygen dynamics at different sites. The relationships among metabolic rates (gross photosynthesis, respiration, and NEM), salinity and temperature were also evaluated for each site.

2. Methods

I used existing dissolved oxygen data from the NERR SWMP from January 1996 to December 1998 for this analysis. All data had undergone extensive quality control and quality assurance as described in Wenner *et al.* (2001). Data from 2 sites at 14 Reserves were analyzed (Table 1). Sites were chosen based on several factors including hydrography, habitat types near deployment, geographic location, and availability of ancillary data such as nutrient and chlorophyll concentrations.

2.1 CALCULATIONS

Oxygen is produced as a by-product of photosynthesis and consumed by respiration. In aquatic environments, oxygen concentrations usually exhibit a characteristic diurnal pattern with concentrations increasing from morning into mid afternoon as photosynthesis outstrips respiration. Declining oxygen concentrations occur during the late afternoon or evening as photosynthetic rate declines and throughout the night when photosynthesis does not occur. In addition to these biological processes, physical processes can also affect oxygen concentrations. Diffusion of oxygen across the air-water interface can increase or decrease water column concentrations, with diffusion from the air into the water occurring when the water is under saturated and vice versa when the water is supersaturated.

The diffusion, or air-sea exchange, was estimated by equation (1) below

$$\text{Air-Sea Exchange} = \left(1 - \frac{DO_{\text{sat},\,t2} + DO_{\text{sat},\,t1}}{200}\right) * 0.5 * \text{dt}$$

where $(DO_{\text{sat},\,t2} + DO_{\text{sat},\,t1})$ are the oxygen concentrations (units -%) for t1 and t2 and dt is the time difference, in hours, between t2 and t1. The time interval for all data was 0.5 hours. A coefficient of 0.5 g O_2 m^{-2} hr^{-1} at zero O_2 was used to estimate the air-sea exchange (J. Hagy and W.R. Boynton, pers. comm.). The units for air-sea exchange are g O_2 m^{-2}. Thus, when the average oxygen concentration for the time interval $(DO_{\text{sat},\,t2} + DO_{\text{sat},\,t1})/200)$ is under saturated, air-sea exchange is positive and oxygen diffuses from the air into the water. If oxygen concentrations are supersaturated, air-sea exchange is negative and oxygen diffuses out of the water into the air.

This approach may underestimate exchange during periods of high winds and overestimate exchange during calm periods, since previous research has shown that the rate of diffusion is dependent on wind speed (Copeland and Duffer, 1964;

Table I. Location of NERR sites and dominant habitat near site. The total number of days of observations available, average length of deployment, and the percent of dates with negative gross production or respiration.

Site Code	*Reserve*	*Site*	*Dominant Habitat*	*Total Days*	*Length of Deployment*	*Negative Observations*
					(days)	*(%)*
A	ACE, SC	Big Basin	Salt marsh	619	20	21
a		St Pierre	Salt marsh	478	19	21
B	Apalachicola, FL	Surface	Open water	788	18	27
B		Bottom	Open water	506	18	34
C	Chesapeake Bay,	Jug Bay	Fresh marsh	168	12	24
c	MD	Patuxent Park	Fresh marsh	118	11	19
D	Chesapeake Bay VA	Goodwin Island	Eelgrass	413	12	5
d		Taskinas Creek	Brackish marsh	829	14	17
E	Elkhorn Slough, CA	Azevedo Pond	Uplands	946	27	3
e		South Marsh	Salt marsh	716	28	12
F	Great Bay, NH	Great Bay Buoy	Open water	631	17	31
f		Squamscott River	Open water	276	18	20
G	Hudson River, NY	Sawkill	Uplands	442	18	69
g		Tivoli South	Fresh marsh	486	19	24
H	Jobos Bay, PR	Station 9	Mangrove	289	14	3
h		Station 10	Mangrove	327	14	6
I	Narragansett Bay,	Potters Cove	Open water	733	30	26
i	RI	T-wharf	Open water	330	27	38
J	North Inlet-Winyah Bay, SC	Oyster Landing	Salt marsh	822	14	12
j		Thousand Acre Creek	Salt marsh	856	14	27
K	Padilla Bay, WA	Bay View	Eelgrass	714	29	29
k		Joe Leary Slough	Upland	840	16	43
L	Rookery Bay, FL	Upper Henderson	Mangrove	585	15	12
l		Blackwater River	Mangrove	175	14	21
M	Weeks Bay, AL	Fish River	Open water	302	14	23
m		Weeks Bay	Open water	320	14	25
N	Waquoit Bay, MA	Central Basin	Macroalgae	313	14	8

Hartman and Hammond, 1984; Marino and Howarth, 1993).

For each time interval, air-sea exchange was subtracted from the change in oxygen concentrations (DO) in g O_2 m^{-3} multiplied by water depth (m) to give oxygen flux (g O_2 m^{-2}) as described in equation (2) below.

$$\text{Oxygen flux} = (DO_{t2} - DO_{t1}) * \text{water depth} - \text{air-sea exchange} \qquad (2)$$

Oxygen fluxes during the daylight hours were summed to give net production and summed oxygen fluxes from night equaled night oxygen flux. Since respiration is defined as a positive quantity, night oxygen fluxes were multiplied by –1 to give a night respiration rate. Gross production and total (day + night) respiration rate were calculated using net production and night respiration values. Assuming a constant respiration during the day and night, night respiration divided by hours of night equals the hourly respiration rate (g O_2 m^{-2} h^{-1}). Total respiration equals the hourly respiration rate multiplied by 24 h (g O_2 m^{-2} d^{-1}). Gross production is the net production plus the respiration occurring during daylight hours and was calculated by adding net production to the hourly respiration multiplied by the daylight hours. Net ecosystem metabolism was calculated by subtracting total respiration from gross production, or more directly by net production minus night respiration.

A major assumption of the diel oxygen curve method is that water masses passing by the sensor are laterally and vertically homogenous, i.e. they have the same metabolic history. In areas where physical processes such as advection and diffusion dominate over biological processes, metabolic rates may be either underestimated or overestimated (Kemp and Boynton, 1980).

2.2 STATISTICAL ANALYSES

Mean, median, 25th and 75th percentiles were calculated for air-sea exchange values for all sites except Goodwin Island. All values were eliminated from the dataset on dates where gross production and total respiration values were less than zero. The percent of dates with negative gross production or total respiration observations was determined for each site. In addition, paired t-tests were used to examine whether stratification had a significant effect on metabolic rates from the surface and bottom deployments in Apalachicola Bay or whether there was a significant variation in sites 400 m apart in Waquoit Bay. These were the only two Reserves with deployments close enough to test the assumption of homogeneity of water masses.

Relationships among temperature and salinity and metabolic rates for each site were determined using correlation analysis. Daily rates of gross production and total respiration were compared using a paired t-test. If production and respiration rates were significantly different from one another, then net ecosystem metabolism was significantly different from zero. All results are reported as non-significant when p values were greater than 0.05.

3. Results and Discussion

The total number of days of data available for the metabolic calculations was variable between sites (Table 1). A variety of factors contributed to the gaps in the data record: from logistical and personnel constraints, to weather (winter ice and freezing temperatures) to problems with instrument calibrations. The average length of deployment varied between 11 to 29 days among the different sites (Table 1). In general, most Reserves in the Southeast and along the Gulf Coast deployed the instruments for 7 days during the summer to minimize biofouling and instrument drift (Wenner *et al.*, 2001). Data for the three-year period was available at most sites, except for 5 sites (Chesapeake Bay MD Jug Bay, Chesapeake Bay VA

Goodwin Island, Great Bay Squamscott River, Narragansett Bay T-wharf, Rookery Bay Upper Henderson) that had only two years of data and 5 sites (Chesapeake Bay MD Patuxent Park, Rookery Bay Blackwater River, Weeks Bay Fish River, Weeks Bay, Waquoit Bay Metoxit Point) that had a single year of data.

Exchange of oxygen across the air-water interface can have a significant effect on concentration in the water column, particularly when the water is under saturated or supersaturated. Air-sea exchange averaged 0.1 $gO_2\ m^{-2}$ or less at most sites and rarely exceeded 0.2 $gO_2\ m^{-2}$ (Table 2). These values were comparable with previous estimates of air-sea exchange (Odum, 1956; Odum and Hoskins, 1958; D'Avanzo *et al.*, 1996).

A fundamental assumption of this method is that water masses moving past the sensor are homogeneous. This assumption was examined in more detail using data from two of the Reserves where deployment sites were close. In Apalachicola Bay, one meter was deployed near the surface (0.6 m) and a second meter was deployed 1.2 m below the first and 0.4 m above the bottom. Gross production and respiration were not significantly different between the two depths (Figure 1), although NEM was significantly different ($p = 0.025$). Surface waters were significantly more heterotrophic than bottom waters, perhaps reflecting runoff from the adjacent bottomland hardwood swamp in the surface layer. In Waquoit Bay, two meters were deployed approximately 400 m apart for six weeks starting in November 1998 (Figure 2). Again there was no significant difference in gross production or respiration between the two deployment sites, while NEM was significantly different ($p = 0.001$). Previous research in Waquoit Bay has shown that mid-estuary deployments are representative of the estuary as a whole (D'Avanzo *et al.*, 1996).

In addition to heterogeneity of water masses, the signal to noise ratio can also affect the success of this method. When metabolic rates are very low, the oxygen fluxes will be small and difficult to detect. Thus, estimating rates can be difficult when temperatures are low or in oligotrophic environments.

Physical processes appeared to control oxygen dynamics at only two out of the 28 deployment locations. At Hudson River Sawkill, 69% of the dates had negative gross production or respiration values. At this site, the YSI data sonde was deployed just upstream of a dam in the creek. Physical processes such as advection of different water masses and enhanced exchange across the air-water interface probably control the oxygen dynamics rather than biological processes. Thus, when stream flow was greater than 0.4 m^3/s, gross production and respiration were usually negative values. The other site with a relatively low percentage of useable observations was Padilla Bay Joe Leary Slough where 43% of the dates had negative gross production or respiration values. This small, intermittently flushed Slough drains agricultural fields and pastures. A dam and one-way tide gates restrict water flow between the Slough and Padilla Bay. High salinity and high DO water from Padilla Bay eelgrass beds seeped through the tide gates at high tides increasing

DO concentrations within the Slough. If high tide occurred at night, DO concentrations often increased which led to negative estimates of respiration.

The percentage of dates with negative values of gross production and respiration was also variable from 3% to 69%, although only 5 sites had percentages greater than 30%. On average, 23% of the data was eliminated (Table 1). Removing all data on dates with either negative gross production or respiration led to higher average annual gross production and respiration than if all data was used, although this had a minimal effect on net ecosystem metabolism. Several reasons justify removing data on dates with negative production and respiration. Respiration cannot be negative—there is no biological process that leads to the production of oxygen at night. Close scrutiny of the results from Hudson River Sawkill and Padilla Bay Joe Leary Slough as described above clearly demonstrate how heterogeneity of water masses led to impossible numbers. These same sort of physical processes occur at other sites. Some of these negative values were associated with

Table 2. Air-sea exchange in $gO_2\ m^{-2}$ using constant diffusion coefficient = 0.5 and calculated from equation 1. Number of observations used to estimate mean, median, 25th and 75th percentiles.

Site Code	*Site*	*Number of Observations*	*Mean*	*Median*	*25th*	*75th*
A	Big Basin	37,919	0.09	0.07	0.02	0.18
a	St Pierre	27,369	0.08	0.05	0.02	0.13
B	Surface	26,960	0.03	0.02	-0.02	0.08
b	Bottom	35,578	0.02	0.02	-0.01	0.05
C	Jug Bay	8,016	0.14	0.15	0.09	0.21
c	Patuxent Park	5,445	0.05	0.07	0.02	0.10
d	Taskinas Creek	40,060	0.04	0.04	0.01	0.08
E	Azevedo Pond	44,874	0.05	0.07	-0.01	0.14
e	South Marsh	34,116	0.04	0.04	0.01	0.06
F	Great Bay Buoy	29,466	0.00	0.00	-0.02	0.02
f	Squamscott River	13,106	0.02	0.02	0.00	0.03
G	Sawkill	20,920	-0.00	-0.00	-0.02	0.00
g	Tivoli South	23,221	0.03	0.03	0.01	0.05
H	Station 9	14,603	0.08	0.08	0.02	0.15
h	Station 10	17,024	0.03	0.03	-0.01	0.08
I	Potters Cove	35,360	0.01	0.00	-0.02	0.03
i	T-wharf	842	0.02	0.02	0.01	0.03
J	Oyster Landing	41,592	0.04	0.03	0.00	0.08
j	Thousand Acre Creek	40,252	0.06	0.05	0.02	0.10
K	Bay View	40,940	0.01	0.02	-0.02	0.04
k	Joe Leary Slough	39,464	0.12	0.12	0.06	0.17
L	Upper Henderson	9,807	0.15	0.16	0.12	0.19
l	Blackwater River	27,349	0.12	0.13	0.08	0.18
M	Fish River	14,346	0.02	0.04	0.00	0.06
m	Weeks Bay	15,024	0.01	0.00	-0.05	0.06
N	Central Basin	15,035	-0.02	-0.01	-0.05	0.02

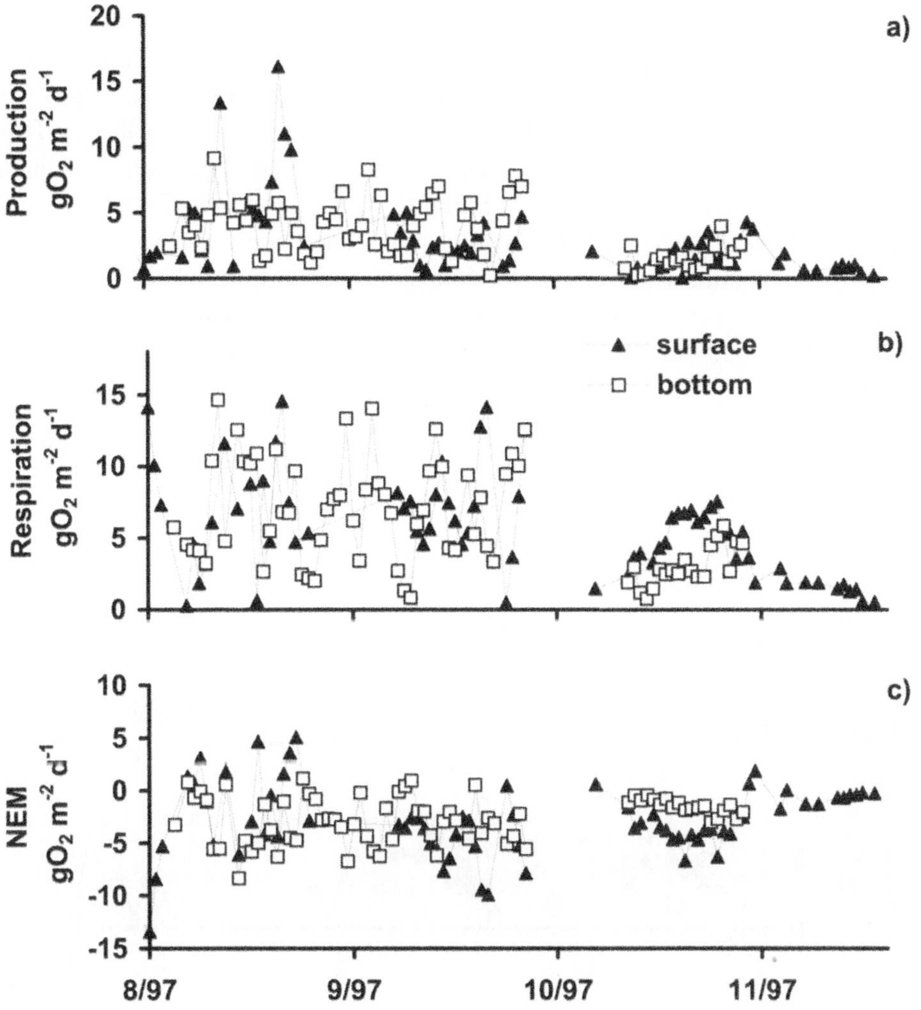

Figure 1. Comparison of a) gross production b) total respiration and c) net ecosystem metabolism (g O_2 m^{-2} d^{-1}) between Apalachicola Bay surface and bottom deployments.

a rapid change in salinity, suggesting the movement of a different water mass past the sensor. These physical processes probably occur intermittently perhaps associated with fronts or strong currents. A more complete analysis of the data is needed to determine whether heterogeneity of water masses is responsible for negative values at other sites.

Average annual gross production and total respiration rates were calculated at each site. The lowest average annual gross production and total respiration rates occurred at the Hudson River Sawkill site, a small freshwater creek. Rates were also generally low at Hudson River Tivoli South, North Inlet-Winyah Bay Thousand Acre Creek and Weeks Bay Fish River (Figure 3). The highest rates occurred at the Elkhorn Slough Azevedo Pond, Padilla Bay Bayview and ACE Basin St. Pierre, which were nearly double the rates at the other sites. Across all sites average gross production and respiration were significantly correlated (Figure 3, $r = 0.82$, $p < 0.01$). Average annual respiration was significantly greater than average annual gross production, except at 4 of the sites where gross production was sig-

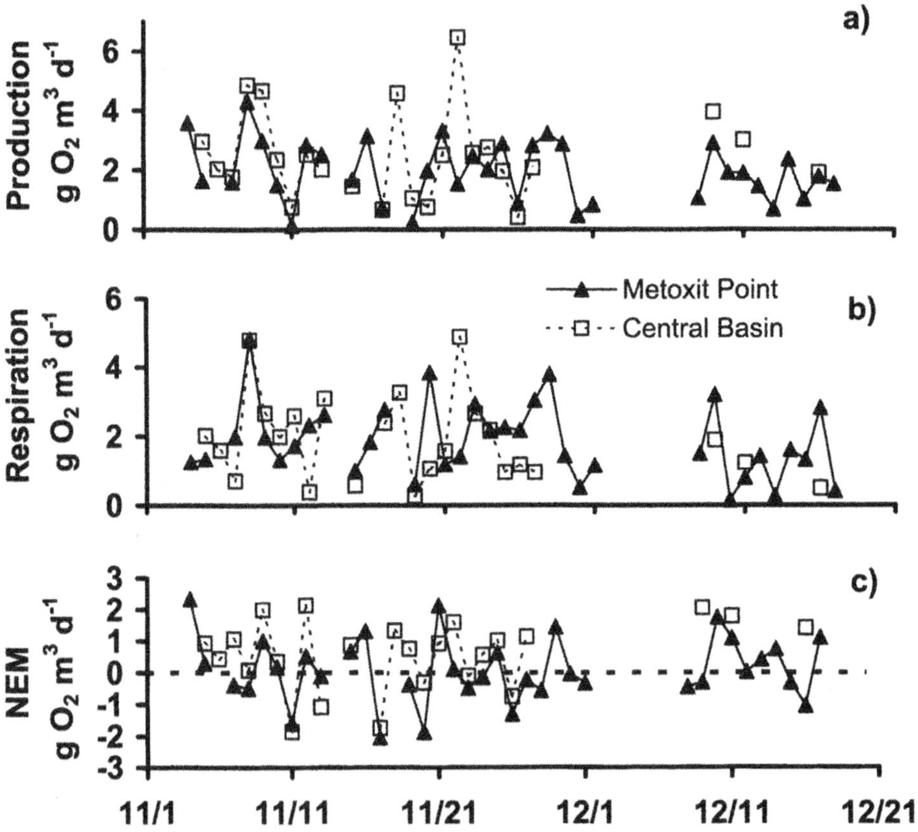

Figure 2. Comparison of a) gross production b) total respiration and c) net ecosystem metabolism (g O_2 m^{-2} d^{-1}) between Central Basin and Metoxit Point stations in Waquoit Bay in 1998.

nificantly greater than respiration (Chesapeake Bay VA Goodwin Island) or they were equal (Waquoit Bay Central Basin, Padilla Bay Bayview and Hudson River Sawkill) (Figure 3). When daily gross production and respiration rates were compared using a paired t-test, respiration was significantly greater than gross production at all sites except for Padilla Bay Bayview, Waquoit Bay Central Basin, and Hudson River Sawkill where there was no significant difference between rates. At Chesapeake Bay VA Goodwin Island, gross production was significantly greater than respiration. At three of these sites, eelgrass (Chesapeake Bay VA Goodwin Island, Padilla Bay Bayview) or macro algae (Waquoit Bay Central Basin) dominated. It was surprising that the Padilla Bay Bayview site was not net autotrophic given the extensive intertidal beds of eelgrass in the Bay. However, the instrument was deployed in a deep channel adjacent to the grass beds rather than in the grass beds proper. Another site that was probably influenced by nearby eelgrass beds was Great Bay Buoy, which was only slightly heterotrophic. Sites from Chesapeake Bay MD, Rookery Bay and ACE Basin were very heterotrophic, probably because of high organic inputs from adjacent marshes and mangroves.

Temperature and metabolic rates were significantly correlated at most of the 27 YSI deployment sites (Waquoit Bay Metoxit Point was not included due to the small amount of data available) (Table 3). Gross production and temperature were correlated at 23 sites; respiration and temperature were correlated at 26 sites, while

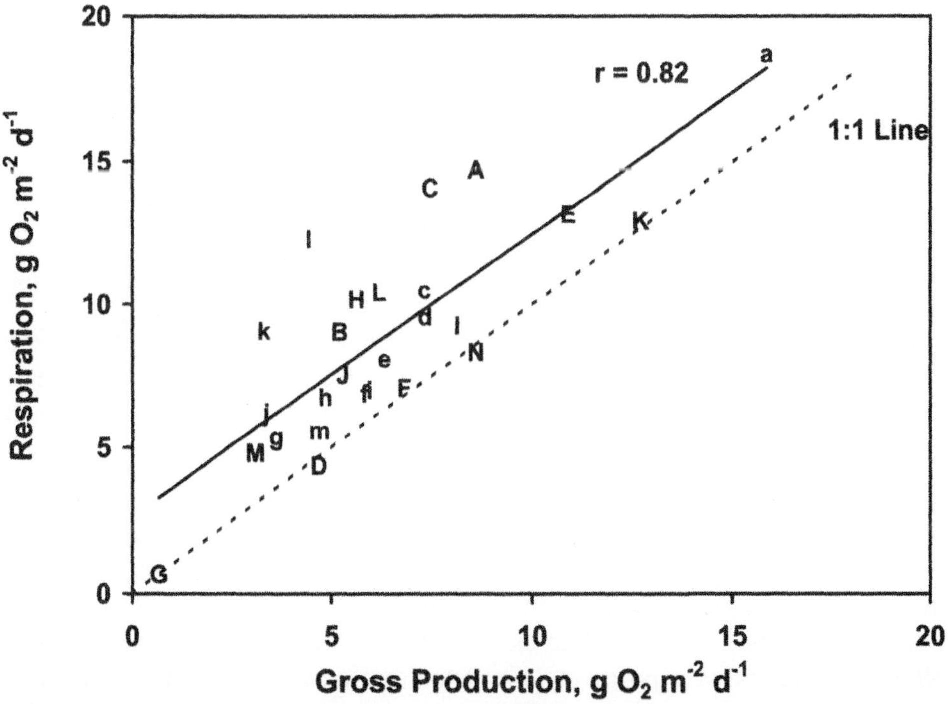

Figure 3. Average annual rate of production and respiration in g O_2 m^{-2} d^{-1} among NERR sites. Site codes in Table 1.

Table 3. Correlation coefficients for temperature, salinity, Gross Production (Pg), Total Respiration (R) and net ecosystem metabolism (NEM)($p < 0.05$ at NERR sites). ns – non significant ($p > 0.05$). # is number of observations

Site Code	Site	#	Temperature effects			Salinity effects		
			Pg	*R*	*NEM*	*Pg*	*R*	*NEM*
A	Big Basin	489	0.33	0.35	–0.16	0.13	ns	0.23
a	St Pierre	378	0.34	0.48	–0.18	ns	ns	ns
B	Surface	575	0.36	0.27	–0.08	0.17	ns	ns
b	Bottom	334	0.27	0.23	ns	0.20	0.29	–0.22
C	Jug Bay	128	ns	0.23	–0.38	ns	ns	–0.34
c	Patuxent Park	96	0.58	0.43	0.21	ns	ns	ns
D	Goodwin Island	392	0.64	0.67	–0.68	–0.16	–0.19	ns
d	Taskinas Creek	688	0.52	0.65	–0.38	0.13	0.16	–0.10
E	Azevedo Pond	918	0.48	0.52	0.09	0.39	0.46	ns
e	South Marsh	630	0.35	0.38	–0.11	0.23	0.22	ns
F	Great Bay Buoy	435	0.44	0.49	–0.06	0.29	0.25	0.14
f	Squamscott River	221	0.39	0.40	ns	ns	ns	ns
G	Sawkill	137	–0.34	0.19	–0.51	ns	ns	ns
g	Tivoli South	369	0.50	0.61	–0.45	ns	ns	ns
H	Station 9	280	0.30	0.41	–0.23	0.14	0.31	–0.25
h	Station 10	307	0.22	0.21	ns	ns	–0.20	0.21
I	Potters Cove	542	0.47	0.55	–0.33	ns	ns	ns
i	T–wharf	205	0.19	0.19	ns	–0.18	–0.21	ns
J	Oyster Landing	723	0.47	0.71	–0.53	0.16	0.16	ns
j	Thousand Acre Creek	625	0.44	0.56	–0.34	0.29	0.37	–0.24
K	Bay View	507	0.30	0.30	ns	–0.11	ns	–0.18
k	Joe Leary Slough	479	–0.18	0.30	0.46	–0.09	0.27	0.33
L	Upper Henderson	515	ns	0.30	–0.47	ns	ns	ns
l	Blackwater River	138	ns	0.36	–0.51	0.24	ns	0.48
M	Fish River	233	ns	ns	ns	0.37	ns	0.38
m	Weeks Bay	240	0.68	0.64	ns	ns	ns	–0.33
N	Central Basin	288	0.47	0.46	ns	ns	ns	ns

net ecosystem metabolism and temperature were correlated at 19 sites. This is consistent with the idea that biological processes are controlling metabolic rates at these sites, with warmer temperatures leading to higher metabolic rates. At most sites, higher temperatures generally resulted in more heterotrophic conditions.

In contrast, salinity and metabolic rates were significantly correlated at about half of the sites (Table 3). At most sites, gross production and respiration were low when salinity was low. Low salinities likely reflect greater freshwater inputs that may reduce the water residence time, leading to a greater flushing of the planktonic communities out of tidal creeks or estuaries. In contrast, at half the sites, NEM and salinity were positively correlated (more autotrophic at higher salinities) and half the sites were negatively correlated (more heterotrophic at higher salinities). This may reflect differing inputs of nutrients or organic matter associated with freshwater runoff at individual sites. Nutrient rich freshwater runoff could enhance production over respiration leading to more autotrophic conditions, while organic rich freshwater runoff could lead to more heterotrophic conditions.

Average salinity ranged between 0 and 39 at the Reserve deployment sites with freshwater conditions occurring at Chesapeake Bay Maryland and Hudson River sites throughout the three-year period. Jobos Bay sites were consistently hypersaline. The range in salinity within each site could also be large, up to 30 PSU or more at several sites (ACE St Pierre, Apalachicola East Bay Surface and Bottom, Elkhorn Slough Azevedo Pond, Great Bay Squamscott River, and North Inlet-Winyah Bay Oyster Landing). Higher gross production generally occurred at sites with salinity between 25–30 that had a moderate salinity range less than 10 PSU (Padilla Bay Bayview, Waquoit Bay Central Basin, Narragansett Bay Potters Cove), although production was also high at the freshwater sites in Chesapeake Bay Maryland. However, there was no trend among respiration, net metabolism, salinity, or salinity range.

4. Conclusions

Dissolved oxygen data collected as part of a routine water quality monitoring program can be used to estimate metabolic rates. However, it is important to select sites where biological processes dominate over physical processes. Despite the fact that NERR SWMP monitoring stations were not chosen with the intention of measuring metabolic rates, most of the data collected at these sites provided useful estimates of metabolic rates. Only two sites, Hudson River Sawkill Creek and Padilla Bay Joe Leary Slough, were the exception where physical processes consistently dominated oxygen dynamics.

Metabolic parameters are useful indicators of ecosystem condition, although their interpretation may not be as clear-cut as other indices such as degree of hypoxia or benthic diversity. These data can be used to answer such questions as: 1) how productive are estuaries relative to one another; 2) what is the balance between production and consumption of organic matter within these systems; and 3) how are these systems changing over time?

Most of the NERR sites (23 out of 27) were significantly heterotrophic with respiration exceeding production on an annual basis. Some of the sites were very heterotrophic suggesting high inputs of organic matter from adjacent marshes and mangroves. Three of the sites that were not heterotrophic were dominated by either eelgrass or macro algae.

Temperature and salinity were important variables explaining some of the daily variations in metabolic rates. However, other factors such as nutrient and chlorophyll concentrations, solar irradiance, nutrient loading, organic loading and residence time may be important in controlling metabolic rates. Nutrient loading (D'Avanzo *et al.*, 1996), organic loading (Smith and Hollibaugh, 1997) and the ratio of nutrient to organic loading (Kemp *et al.*, 1997) have been hypothesized to explain variations in rates of net ecosystem metabolism in different estuaries. Future manuscripts will examine how these factors explain variations in metabolic rates both within and between NERR sites.

Acknowledgements

This analysis would not have been possible without the dedication of the National Estuarine Research Reserve Research Coordinators and technicians. I thank Betty Wenner, Mike Arendt, and others at SC DNR, along with the Central Data Management Office, Belle Baruch, USC for critically evaluating the data used for this analysis. Chuck Nieder made available stream flow data from Sawkill Creek. He, Lee Edmiston and Doug Bulthius also provided useful insights about their NERR sites. I thank Fred Holland, Walter Boynton, Jim Hagy, and Mike Murrell for many helpful discussions and Jim Kremer, Jonathan Pennock, Erik Smith, and Virginia Engle for their critical reviews of this manuscript. The Estuarine Reserve Division of NOAA provided funding for this project.

References

Copeland, B.J. and Duffer, W.R.: 1964, 'Use of a Clear Plastic Dome to Measure Gaseous Diffusion Rates in Natural Waters', *Limnology and Oceanography*, **9**, 494–499.

D'Avanzo, C., Kremer, J.N. and Wainright, S.C.: 1996, 'Ecosystem Production And Respiration In Response To Eutrophication In Shallow Temperate Estuaries', *Marine Ecology Progress Series*, **141**, 263–274.

Hartman, B. and Hammond, D.E.: 1984, 'Gas Exchange Rates Across the Sediment Water and Air-Water Interfaces in South San Francisco Bay', *Journal of Geophysical Research*, **89**, 3593–3603.

Kemp, W.M. and Boynton, W.R: 1980, 'Influence of Biological and Physical Processes on Dissolved Oxygen Dynamics in an Estuarine System: Implications for Measurement of Community Metabolism', *Estuarine, Coastal and Marine Science*, **11**, 407–431.

Kemp, W.M., Smith, E.M., Marvin-Dipasquale, M., and Boynton, W.R.: 1997, 'Organic Carbon Balance and Net Ecosystem Metabolism in Chesapeake Bay', *Marine Ecology Progress Series*, **150**, 229–248.

Marino, R. and Howarth, R.W.: 1993, 'Atmospheric Oxygen Exchange in the Hudson River: Dome Measurements and Comparison with Other Natural Waters', *Estuaries*, **16**, 433–445.

Nixon, S.W. and Oviatt, C.A.: 1972, 'Preliminary Measurements of Midsummer Metabolism in Beds of Eelgrass, Zostera marina', *Ecology*, **53**, 150–153.

Odum, H.T.: 1956, 'Primary Production in Flowing Waters', *Limnology And Oceanography*, **1**, 102–117.

Odum, H.T. and Hoskins, C.M.: 1958, 'Comparative Studies on the Metabolism of Marine Waters', *Publications of the Institute of Marine Science, Texas* **5**, 16–46.

Oviatt, C.A., Keller, A.A., Sampou, P.A., and Beatty, L.L.: 1986, 'Patterns of Productivity During Eutrophication: a Mesocosm Experiment', *Marine Ecology Progress Series*, **28**, 69–80.

Smith, S.V. And Hollibaugh, J.T.: 1997, 'Annual Cycle and Interannual Variability of Ecosystem Metabolism in a Temperate Climate Embayment', *Ecological Monographs,* **67,** 509–533.

Swaney, D.P., Howarth, R.W. and Butler, T.J.: 1999, 'A Novel Approach for Estimating Ecosystem Production and Respiration in Estuaries: Application to the Oligohaline and Mesohaline Hudson River', *Limnology and Oceanography,* **44,** 1509–1521.

Wenner, E.L., Holland, A.F., Arendt, M.D., Edwards, D. And Caffrey, J.M.: 2001, 'A Synthesis of Water Quality Data from the National Estuarine Research Reserve System-Wide Monitoring Program', Report to NOAA Estuarine Reserve Division.

FORAMINIFERA AS BIOINDICATORS IN CORAL REEF ASSESSMENT AND MONITORING: THE FORAM INDEX

PAMELA HALLOCK[1*], BARBARA H. LIDZ[2]
ELIZABETH M. COCKEY-BURKHARD[3], AND KELLY B. DONNELLY[4]

[1]*College of Marine Science, Univ. of South Florida, 140 Seventh Ave. S., St. Petersburg, FL 33701, USA;* [2]*U.S. Geological Survey Center for Coastal and Regional Marine Studies, 600 Fourth St. S., St. Petersburg, FL 33701, USA;* [3]*Minerals Management Service, 381 Elden Street, Herndon, VA 20170, USA;* [4]*Florida Fish and Wildlife Conservation Commission, 100 Eighth St. SE, St. Petersburg, FL 33701, USA*

*(author for correspondence, e-mail: pmuller@marine.usf.edu)

Abstract. Coral reef communities are threatened worldwide. Resource managers urgently need indicators of the biological condition of reef environments that can relate data acquired through remote-sensing, water-quality and benthic-community monitoring to stress responses in reef organisms. The "FORAM" (Foraminifera in Reef Assessment and Monitoring) Index (FI) is based on 30 years of research on reef sediments and reef-dwelling larger foraminifers. These shelled protists are ideal indicator organisms because:

- Foraminifers are widely used as environmental and paleoenvironmental indicators in many contexts;
- Reef-building, zooxanthellate corals and foraminifers with algal symbionts have similar water-quality requirements;
- The relatively short life spans of foraminifers as compared with long-lived colonial corals facilitate differentiation between long-term water-quality decline and episodic stress events;
- Foraminifers are relatively small and abundant, permitting statistically significant sample sizes to be collected quickly and relatively inexpensively, ideally as a component of comprehensive monitoring programs; and
- Collection of foraminifers has minimal impact on reef resources.

USEPA guidelines for ecological indicators are used to evaluate the FI. Data required are foraminiferal assemblages from surface sediments of reef-associated environments. The FI provides resource managers with a simple procedure for determining the suitability of benthic environments for communities dominated by algal symbiotic organisms. The FI can be applied independently, or incorporated into existing or planned monitoring efforts. The simple calculations require limited computer capabilities and therefore can be applied readily to reef-associated environments worldwide. In addition, the foraminiferal shells collected can be subjected to morphometric and geochemical analyses in areas of suspected heavy-metal pollution, and the data sets for the index can be used with other monitoring data in detailed multidimensional assessments.

Keywords: bioindicators, coral reefs, Foraminifera, zooxanthellae, algal symbiosis

1. Introduction

Human activities are changing environmental conditions on a global scale. Roughly half the Earth's land area has been transformed or degraded (Vitousek *et al.*, 1997). Human activities have effectively doubled the annual transfer of nitrogen from the atmospheric pool of N_2 to biologically available fixed nitrogen (Schnoor *et al.*, 1995). Much of this fixed nitrogen, along with nitrous oxide gases from burning of fossil fuels (Prinn *et al.*, 1990), is washed into aquatic systems by rain. As a result of strato-

Environmental Monitoring and Assessment **81**: 221–238, 2003.

spheric ozone depletion, the intensity of biologically damaging ultraviolet radiation (UV_b) at 20° N latitude between April and August now exceeds the June 1969 (summer solstice) maximum (Shick *et al.*, 1996). Carbon dioxide concentration in the atmosphere has increased by nearly 30% since the beginning of the Industrial Revolution (Shimel *et al.*, 1995), with impacts ranging from global climate change to changes in ocean chemistry that inhibit calcification.

The U.S. Environmental Protection Agency's (EPA) Environmental Monitoring and Assessment Program (EMAP) was created to develop and monitor indicators of pollution exposure and habitat condition. Objectives are to determine the magnitude and geographic distribution of resources that are adversely impacted by pollution and other environmental stresses (Messer *et al.*, 1991). Coral reefs are among the ecosystems most threatened by human activities. Bryant *et al.* (1998) estimated that nearly 60% of the Earth's coral reefs are threatened by relatively local impacts including nutrients and other chemical pollutants, sedimentation, destructive fishing practices, and shipping. An assessment of the status of the world's coral reefs in 2000 concluded that more than one quarter have been lost (Wilkinson, 2000), about 15% to mortality following 1997–98 mass bleaching events. Risk (1999) noted that resource managers have monitored water quality and reef conditions for decades but they lack bioindicators that can link those measures to meaningful efforts to preserve remaining reef resources. Attempts to interpret water-quality data continue to be confounded by the reef community's ability to sequester nutrients, making them unavailable to monitoring yet readily effective in inducing community change, an enigma recognized by Laws and Redalje (1979) but widely misunderstood (Risk, 1999).

This paper utilizes EPA "Evaluation Guidelines for Ecological Indicators" (Jackson *et al.*, 2000) to present a simple index, the FORAM (Foraminifera in Reef Assessment and Monitoring) Index (FI), based on foraminiferal assemblages from surface sediments. The FI is intended to provide resource managers with a measure, which is independent of coral populations, to determine whether water quality in the environment is sufficient to support reef growth or recovery. A major advantage of foraminifers in reef assessment is their short life span, as compared with long-lived reef-building corals. Foraminiferal assemblages can potentially facilitate differentiation between long-term reef decline associated with declining water quality and temporary reef decline associated with episodic mortality events (Cockey *et al.*, 1996).

2. Conceptual Relevance

2.1 RELEVANCE TO ASSESSMENT

The FI applies historic observations that healthy coral reefs had abundant mixotrophic larger foraminifers (e.g., McKee *et al.*, 1956; Hallock 1981a, 1988), which were important sediment constituents. On coral reefs that are subject to significant nutrification (i.e., increase in nutrient flux that results in change in biological community structure), with or without increased terrigenous sedimentation, populations of smaller heterotrophic foraminifers proliferate and their shell numbers overwhelm those of declining larger foraminifers (Hirshfield *et al.*, 1968; Cockey *et al.*, 1996).

Unfortunately, over the past 25 years, reef-building corals have declined nearly worldwide in response to a variety of factors unrelated to nutrification, including new diseases, bleaching in response to temperature stress, and physical impacts such as hurricanes and ship groundings. Coral reef communities are subject to a myriad of stresses and are declining from most of them. Thus, it is critical to have an indicator of water-quality conditions that will support reef development, even in the absence of healthy coral populations following mass mortality events. Cockey *et al.* (1996) argued that larger foraminiferal populations, which are immune to coral-specific diseases and recover much more quickly from physical impacts than long-lived coral populations, are sensitive indicators of water-quality conditions that support reef development.

According to Engle (2000, p. 3-1), "An ideal indicator of the response of benthic organisms to perturbations in the environment would not only quantify their present condition in ecosystems but would also integrate the effects of anthropogenic and natural stressors on the organisms over time (Boesch and Rosenberg 1981; Messer *et al.* 1991)." This information is precisely what foraminiferal tests in the sediments provide.

2.2 RELEVANCE TO ECOLOGICAL FUNCTION

Environmental perturbations of critical concern to coral reefs fall into three major categories (e.g., Hallock 2000; 2001): local impacts, new diseases of regional extent, and global change. The FORAM Index can be used to address local impacts and to assist in differentiating between local impacts that affect water quality and impacts that result from regional- to global-change issues.

The principal physiological analogy between reef-building corals and larger foraminifers is the dependence of both groups on algal symbionts to enhance growth and calcification (e.g., Lee and Anderson, 1991). This analogy and examples from both groups were used to develop a model to predict the energetic benefits of algal symbiosis (Hallock, 1981b), with the conclusion that this mode of life was energetically most advantageous when dissolved nutrients (i.e., NH_4^+, NO_2^-, NO_3^-, PO_4^{3-}) and particulate food resources were scarce. Physiological studies of corals (e.g., Falkowski *et al.*, 1993; Steven and Broadbent, 1997) and of larger foraminifers (Lee, 1998) have since demonstrated that fixed nitrogen limitation is crucial to maintenance of the host-symbiont relation.

Birkeland (1977; 1988) recognized that nutrient availability on the local or regional scale is a major control on benthic-community structure, especially in subtropical and tropical seas (see also Hallock and Schlager, 1986; Hallock *et al.*, 1993). Coral reefs thrive in the most nutrient-depleted oceanic waters where mixotrophic nutrition, i.e., the recycling of nutrients between host and algal symbionts, is most advantageous. As the nutrient supply increases, reef-building coral domination of the benthos gradually gives way to macroalgal domination as algal symbiosis becomes less advantageous. Slightly higher nutrient flux promotes phytoplankton blooms in the water column, limiting light penetration to the benthos. Light attenuation promotes the dominance of the benthos by non-symbiotic filter-feeding animals such as sponges and ascidians and of detritus-feeding echinoderms and crustaceans that do not directly require sunlight for survival (as observed, e.g., in Kaneohe Bay, Hawaii by Smith *et al.*, 1981).

Benthic foraminiferal assemblages respond similarly to nutrient flux (Hallock, 1987, 1988). In very low-nutrient marine environments, such as those found around most Pacific atolls, larger foraminifers that host algal endosymbionts (Table 1) completely dominate sand-size sediments in reef systems (e.g., McKee *et al.*, 1956; Hallock, 1981a). As nutrient supplies increase, bioeroded coral fragments, calcareous algae, molluscan debris, and smaller herbivorous and detritivorous foraminifers (Table 1) become more common as sediment constituents (Hirschfield *et al.*, 1968; Hallock, 1988; Cockey *et al.*, 1996).

As the environment becomes unsuitable for the survival of foraminifers with algal symbionts, their dead tests become rare in the sediments, and remnants become increasingly corroded (e.g., Cottey and Hallock, 1988). These changes occur with nutrification, which does not result in a measurable increase in dissolved nutrients in the water column because the planktic and benthic communities are able to incorporate and utilize all available nutrients (e.g., Laws and Redalje, 1979). True "eutrophication," which is nutrification to the degree that organic carbon buildup occurs in bottom waters and sediments (e.g., Cockey *et al.*, 1996), results in further change in benthic-community structure including domination by opportunistic taxa (Table 1) that can tolerate episodic anoxia (e.g., Alve, 1995).

A critical application of the FI is differentiation between nutrification-induced decline in coral dominance in a reef environment and decline in response to episodic stress or mortality events (e.g., temperature extremes or hurricanes) that are independent of water quality. While the immediate cause of coral population decline is often a mortality event, if chronic nutrification has also occurred, coral populations continue to decline rather than recover from that event.

3. Methods

3.1 SAMPLE COLLECTION AND PROCESSING

Several sample collection procedures have been successfully used by numerous researchers who have assessed foraminiferal assemblages in surface sediments. For example, samples can be collected by SCUBA divers using a scoop and plas-

Table 1. Functional Groups of Foraminifers Used in Coral Reef Assessments

Functional Group	*Order*	*Family*	*Genus*	*Distribution*
Symbiont-Bearing	Rotaliida	Amphisteginidae	*Amphistegina*	Circumtropical
		Calcarinidae	5 genera	Indo-Pacific
		Nummulitidae	*Heterostegina*	Circumtropical
			3 other genera	Indo-Pacific
	Miliolida	Alveolinidae	*Alveolinella*	Indo-Pacific
			Borelis	Circumtropical
		Peneroplidae	Several genera	Circumtropical
		Soritidae	*Sorites*	Circumtropical
			Amphisorus	Circumtropical
			3 genera	Caribbean
			Marginopora	Indo-Pacific
Opportunistic*	Trochamminida	Trochamminidae	Several genera	Cosmopolitan
	Textulariida	Lituolidae	Several genera	Cosmopolitan
	Buliminida	Bolivinidae	Several genera	Cosmopolitan
		Buliminidae	Several genera	Cosmopolitan
	Rotaliida	Rotaliidae	*Ammonia*	Cosmopolitan
		Elphidiidae	*Elphidium*	Cosmopolitan
Other Small Taxa	Miliolida	Most except larger taxa noted above		Cosmopolitan
	Rotaliida	Most except those noted above		Cosmopolitan
	Textulariida	Most		Cosmopolitan
	Other	Most		Cosmopolitan

*Full range of opportunistic genera under local conditions is not well known.

tic bags (e.g., Donnelly, 1993) or minicore (e.g., Cockey *et al.*, 1996). When grab samples or box cores are routinely taken for invertebrates and/or sediment analyses in larger monitoring efforts (e.g., McRae *et al.*, 1998; Engle, 2000), a small surface subsample (upper 2 cm) for foraminiferal analysis can simply be processed separately.

The ideal sample size is 10 grams dry weight, which is a volume of approximately 10–20 cm^3. A sample of this size can be split in half, with one half saved as backup or archived. The other half is divided into an approximately 1 gram portion and a 4 gram portion. The 4 gram portion is used for routine sediment grain-size analysis (Folk, 1974). If such analysis is being performed as part of the larger monitoring effort, then it need not be repeated on the foraminiferal sample.

The 1-gram portion of the sample is washed with fresh water over a 63-μm mesh sieve to remove mud-size sediments, then dried on filter paper at 40–50°C until the sample is thoroughly dry. The dried sample is gently disaggregated, thoroughly mixed, and poured into a mound on a clean, smooth surface. With a fine spatula, a small scoop of the sample (approximately 0.1 g) is removed from the center of the mound and weighed to the nearest milligram. The weighed subsample is sprinkled over a small, gridded tray and examined using a conventional stereomicroscope or a video-imaging system. A very fine artist's brush (tip size 3/0 to 5/0), moistened with water, is used to remove foraminiferal specimens from the sediment (heavily worn and reworked specimens are excluded). Each specimen is placed onto a cardboard micropaleontological faunal slide, which is lightly coated with water-soluble glue. Then a preliminary count is made of each individual. If the number of foraminiferal specimens is approximately 150–200, the subsample is sufficient. If fewer specimens were present in the first portion, then a second portion is removed from the mound, weighed, and sorted. This procedure is repeated until 150–200 specimens are obtained or until the entire gram of sample is processed.

3.2 DATA ANALYSIS AND INFORMATION MANAGEMENT

Once an adequate subsample is obtained, the foraminifers are sorted by genus and counted. Generic-level identification is recommended because it is historically well established and an excellent basic reference is available (Loeblich and Tappan, 1987); species-level identifications tend to be inconsistent across investigators. If deformed specimens are observed, the proportions of deformed specimens of abundant taxa should also be noted. Deformed foraminifers are well-known indicators of heavy-metal pollution (Alve, 1995; Yanko *et al.*, 1998).

Raw counts are entered onto a spreadsheet, with appropriate sample location and identification codes, environmental data, and the weight of the sample picked. Basic data from the counts include relative abundance (proportions of the subsample) and absolute abundance (# specimens/gram of sediment) of each genus identified. Information management should be standardized to that of the larger monitoring program (e.g., McRae *et al.*, 1998; Engle, 2000). As resources permit, data analyses such as multivariate and multidimensional analyses can be performed consistent with those applied to other data sets from the monitoring program.

Procedures for calculating the FI are presented in Table 2. Foraminiferal relative-abundance data are summed into functional groups, which include taxa of larger foraminifers that host algal symbionts, pollution-tolerant opportunistic foraminifers that dominate high-stress environments, and small taxa that proliferate in response to nutrification. The basic premise upon which this index is based is that environments suitable for proliferation of symbiont-bearing organisms have sediments in which at least 25–30% of the foraminiferal tests were produced by taxa that hosted algal symbionts. A sample that contains 25% larger foraminiferal tests and 75% tests of other small taxa has a FI = 4. Environments with sediments devoid of larger foraminiferal tests by definition have a FI $\leq$ 2. FI values between 2 and 4 in sediments from areas with existing coral reefs indicate that conditions are marginal to unsuitable for recovery of coral communities after a mortality event. Several of the data sets upon which these interpretations are based are presented and discussed in Section 4.

Table 2. Calculating the FORAM Index (FI)

Step 1. From each subsample examined, sort all foraminiferal specimens by genus, count, and record in a spreadsheet, with genera arranged by functional group. (See Table 1.)

Step 2. Calculate the proportion (P) of specimens for each functional group by summing the specimens of each genus of that group (N) and dividing by the total number of specimens counted (T).

a) $P_s = N_s/T$, where subscript "s" represents symbiont-bearing foraminifers

b) $P_o = N_o/T$, where subscript "o" represents opportunistic foraminifers

c) $P_h = N_h/T$, where subscript "h" represents other small, heterotrophic foraminifers

Step 3. Weight proportions to calculate the FORAM Index (FI):

$FI = (10 \times P_s) + (P_o) + (2 \times P_h)$

Step 4. Interpretation:

$FI > 4$ indicates environment conducive to reef growth

FI varying between 3 and 5 indicates environmental change (Coefficient of Variation > 0.1)

$2 < FI < 4$ indicates environment marginal for reef growth and unsuitable for recovery

$FI < 2$ indicates stressed conditions unsuitable for reef growth

3.3 QUALITY ASSURANCE

Foraminifers can be used as benthic indicators in new or existing monitoring programs that include routine sampling for analysis of sediment parameters (e.g., texture, chemical pollutants, nutrients) and/or benthic invertebrates. Basic Quality Assurance and Quality Control (QA/QC) procedures developed for field sampling, laboratory processing and data analysis for benthic data under EMAP guidelines are ideal (e.g., Engle, 2000). Foraminiferal assemblage analysis is relatively insensitive to the method of sediment collection employed, although it is important to collect the sediment surface layer, and it is important that the surface layer be undistrubed until it is sampled. Whatever collection method is employed, the sampling team should be instructed in the technique and should use it consistently.

To ensure comparability of taxonomic identification, the scientist or technician performing the identifications should be trained and supervised by a specialist in benthic foraminiferal identification, and QA/QC procedures should be employed (e.g., Engle, 2000). For monitoring efforts in remote areas or developing countries where resources are severely limited, technicians can be trained and selected samples can be sent to a specialist at a local university or even in another country to check for consistency in identifications.

3.4 MONETARY COSTS

The monetary costs of collecting foraminifers in sediment samples depends upon the type of monitoring program into which this procedure is being incorporated.

If grab samples or box cores are already being collected for benthic invertebrates and sediment characterization, the per-sample cost to include foraminiferal samples is minimal. Supplies required are small plastic bags or vials and labels. A maximum of 5-10 minutes/sample is needed for label preparation, sample collection, sample sealing, and sample recording. The major cost of including foraminiferal assemblages in monitoring studies is the cost of a technician to analyze samples. The procedure described above provides not only data needed for the FI, but also assemblage data that can be incorporated in more extensive multidimensional analyses. Using this procedure, one dedicated technician can analyze up to three samples per day, depending upon other responsibilities. If screening of large numbers of samples is a priority for the project, an experienced technician can directly count the foraminifers into the three functional groups described above without separating specimens onto slides, roughly tripling the number of samples that can be processed per day. These samples can be economically stored for later detailed analyses as time and resources permit.

Preferred minimum qualifications for the foraminiferal analyst are a Bachelor's degree in a biological or geological science with some experience in taxonomic identification and statistical analysis; Master's-level education with specialized training in foraminiferal ecology is ideal. Basic laboratory needs include a stereomicroscope, a computer to routinely enter data, taxonomic resources (i.e., Loeblich and Tappan, 1987, and local references), and assorted supplies. Recurring costs include supplies and computer and software upgrades.

The FI can be economically adopted in developing countries. Slides and holders can be made at minimal cost from glue, paint, recycled cardboard and aluminum beverage cans (Bayu Ludvianto, Personal Communication). Samples can be collected from a canoe or raft, if necessary, using a small grab, which can be locally fabricated, or by a proficient snorkel or SCUBA diver. Because data required for the FI are simple counts, basic calculations and graphics can be done by hand or with an inexpensive hand calculator. A computer is required only to archive and analyze data further. A researcher at a field laboratory could easily perform sample collection, microscopic analysis, data recording and FI calculations. Data sheets can then be sent to an in-country research center or university or to a project coordinator in another country for more detailed analyses of data sets. Also, because unprocessed samples are relatively small and do not require refrigeration, they can be shipped relatively inexpensively from a field location to a laboratory for processing and analysis.

4. Response Variability

4.1 SAMPLE SIZE

Historically, most foraminiferal researchers have counted 300 specimens per sediment sample to obtain data for community analyses. To determine if smaller sample sizes provide consistent data, Dix (2001) used the basic procedure outlined previously to pick three subsamples of approximately 100 foraminifers per subsample from each of 12 samples. Each subsample was picked, the foraminifers identified to genus, and the relative abundances of each genus calculated for that subsample. The raw data from each set of three subsamples were pooled and used as the "expected" distribution for relative abundances. The Kolmogorov-Smirnov Goodness of Fit test was used to compare all pairs of subsamples, as well as all individual subsamples against their respective pooled "expected" distribution. Out of the 72 resultant comparisons, only one pair of subsamples was significantly different at the 0.05 probability level. Dix concluded that, for generic identifications, a sample size of 150-200 provided a useful compromise between the added precision of larger samples (i.e., 300) and added cost of processing (i.e., double the time of picking and identification).

4.2. SPATIAL AND TEMPORAL VARIABILITY

Two major data sets illustrate spatial variability in the FI and its ability to reflect decadal-scale decline of environmental conditions in the Florida Keys. The changes in dominant taxa reflect changes in the suitability of Keys environments for the symbiotic organisms essential to coral-reef habitats (Figure 1). In 1961, sediment samples were taken along two traverses off Key Largo, Florida. Published reports on these samples (Rose and Lidz, 1977; Lidz and Rose, 1989) provided baseline data for comparison with a limited set of samples taken from the same traverses in 1982, and more extensive sets collected in 1991 and 1992 (Cockey *et al.*, 1996). Mean FI values for samples collected from thriving reef environments in 1961 were 7, with a coefficient of variation (CV) of 0.04 (Table 3). In 1982, the mean FI had plummeted to 4, while the CV had increased to 0.15. By the early 1990s, the mean FI had stabilized at approximately 3, and the CV had returned to < 0.01.

The temporal differences between index values in 1991 and 1992 reflect seasonal differences more than interannual differences. The 1991 samples were taken in September at the end of a climatologically quiet summer and represent peak accumulation of smaller tests of heterotrophic foraminifers. The 1992 samples were taken in May and reflect the effects of higher energy conditions during the winter months. Sites more exposed to storm waves show higher values that reflect those seasonal effects. The 1982 samples are particularly interesting because they reflect a reef ecosystem under stress (e.g., Dustan and Halas, 1987), but not yet in the collapse observed in the 1990s (Jaap *et al.*, 2001). In 1982, shells of smaller foraminifers dominated locally, but in locations more exposed to wave energy, the long-term accumulation of larger foraminiferal shells had not yet been overwhelmed by the more recent production of smaller, more easily sorted shells.

4.3. DISCRIMINATORY ABILITY

Another set of FI values (Figure 2) was calculated from samples collected in 1985 along onshore-offshore transects at La Parguera, Puerto Rico (Donnelly, 1993).

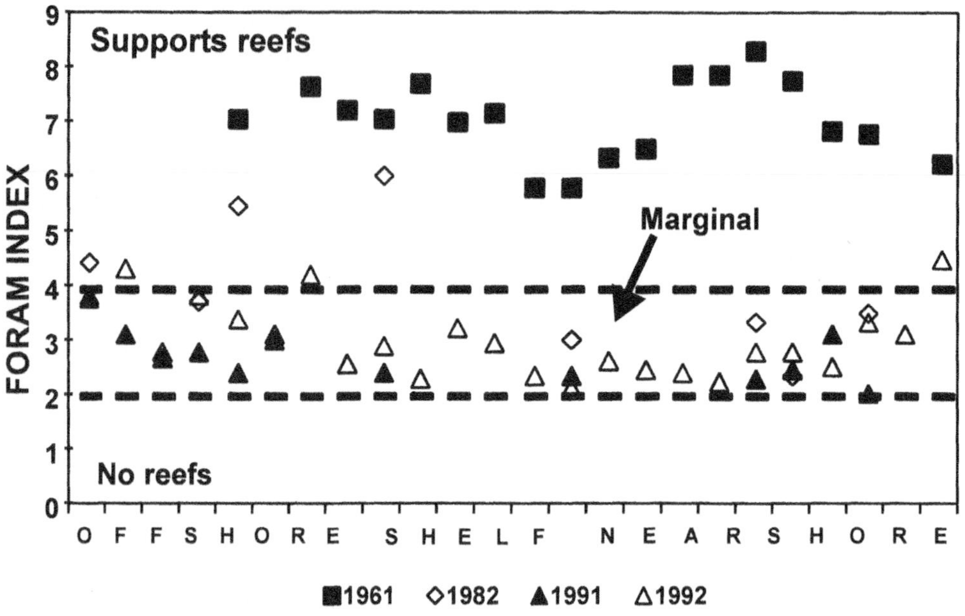

Figure 1. Spatial and temporal variability and interdecadal changes in the FORAM Index for samples from traverses off Key Largo, Florida.

Table 3. Intersample variability in FI values presented in Figures 1 and 2.

Location	*Comparison*	*# Samples*	*Mean FI*	*CV**
Key Largo	1961	18	7.09	0.039
Key Largo	1982	8	4.02	0.145
Key Largo	1991	12	2.75	0.038
Key Largo	1992	24	3.07	0.028
Puerto Rico	Inshore	20	2.69	0.026
Puerto Rico	Midshelf	19	3.74	0.197

*Coefficient of Variation

Inshore reefs were in collapse from coastal nutrification and mid-shelf reefs were exhibiting stress (Hallock, 1988). FI values from mid-shelf samples are higher and more variable than nearshore FI values. The shells of smaller foraminifers locally overwhelmed the larger foraminiferal shells that had accumulated over decades of reef growth, although the larger taxa were still common at midshelf sites. On the inner reefs the shells of smaller, fast-growing foraminifers overwhelmingly dominated in the sediments.

An early concern with the FI was that the index might be unduly influenced by sediment grain size. However, plotting indices against median grain size indicates that, particularly in the most common sediment-size ranges of medium to coarse sands, the FI seems to be relatively unaffected (Figure 3). Intuitively, it is obvious that the shells of smaller foraminifers, as well as juveniles and fragments of larger taxa, should be concentrated in fine sands and muds (phi > 3). However, it is also well known that coral reefs thrive where there is significant water motion, so the presence of muddy sediments can indicate a relatively unfavorable environment, particularly if muddy sediments cover slightly older reef sands and hardbottoms. Hallock (1988) noted that nutrification promotes the production and accumulation of carbonate muds in reef environments for three reasons: (1) Limited nutrification may increase growth and production by calcareous green algae, which break down to mud-size sediments; (2) Nutrification increases rates of bioerosion of reef substrate by endolithic sponges, which etch out mud-size fragments; and (3) Nutrification can promote the growth of algal and bacterial biofilms and mats that baffle and bind finer sediments, allowing those sediments to accumulate. During sampling of inner reefs at La Parguera, Puerto Rico, in 1985 and at Tennessee Reef, Florida Keys, throughout the 1990s, several centimeters of gelatinous carbonate muds overlying sand, rubble and dead-reef substrate in depressions among diseased and heavily bioeroded corals was commonly observed (Pamela Hallock, Personal Communication). The very presence of abundant gelatinous muds in a reef environment indicates a declining environment and will be reflected by a low FI.

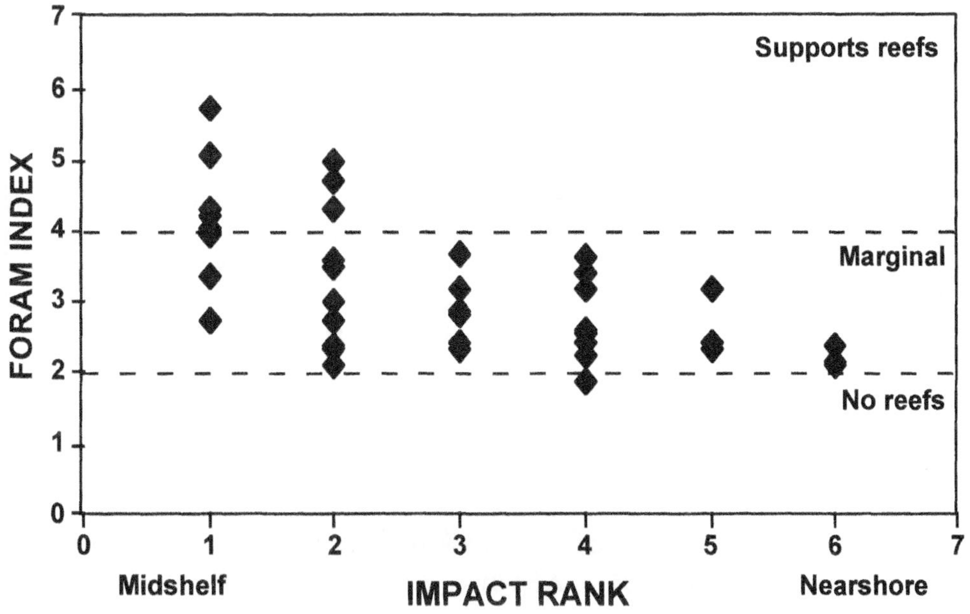

Figure 2. FORAM Index for samples collected from a declining reef system off La Parguera, Puerto Rico.

those in the western Atlantic region (i.e., including the Caribbean and Gulf of Mexico) have dramatically declined over the past 30 years as a result of three major problems: new diseases, coral bleaching, and declining water quality. How interrelated these problems are, i.e., coral bleaching and disease, or disease and water quality, is still not fully known. Certainly coral bleaching is related to anomalously high sea-surface temperatures over the past 25 years, implicating global warming. Synthesis of global climate models and coral-bleaching studies indicates that coral reefs may cease to exist as ecosystems over the next 20 to 50 years (Hoegh-Guldberg, 1999). Ozone depletion and resultant increased intensities of biologically damaging ultraviolet radiation (UV_b) have also increased over the past 25 years and very likely contribute to both temperature stress and susceptibility to disease.

Thus, the major hope for the future of western Atlantic coral reefs as ecosystems is that surviving corals can adapt to temperature stress and disease, and that their descendants can repopulate reef environments. New studies of coral bleaching indicate that surviving corals may exchange lost symbionts for more heat- or light-tolerant strains (e.g., Rowan, 2000; Baker, 2001). In addition, the possibility exists that scientists can genetically engineer heat- and disease-resistant strains of corals and zooxanthellae that can be reintroduced to reef environments. But for either of those optimistic scenarios to be possible, water quality of coastal environments either must be maintained or, where currently contributing to the decline of mixotroph-based communities, must be improved.

Thus, a major question remains unanswered in areas such as the Florida Keys reefs, where the vast majority of coral populations have been lost over the past 25 years and where the reefs will presumably lack significant coral populations for the foreseeable future. How can resource managers recognize whether environments are suitable for coral-reef regrowth if wild populations or genetically engineered corals can adapt to global change? That is one basic question that the FI can address. Larger foraminifers are not dependent upon corals per se, but on the high water quality commonly associated with coral reefs.

Many Indo-Pacific reefs were severely damaged by the 1997–98 mass bleaching event (Hoegh-Guldberg, 1999; Wilkinson, 2000). Nevertheless, there are still significant surviving coral populations and, of course, much higher diversities within coral communities (e.g., Veron, 1995). With very high oceanic water quality offshore from reef resources, for example, in Hawaii, Guam, the Mariana Islands, and American Samoa, the prognosis for survival of Pacific reef resources is much better than in the Florida Keys. Thus, the FI can be applied more conventionally to provide an independent assessment of the suitability of local environments for continued reef growth or recovery in the event of a mass mortality event. Further targeted testing of the index in Pacific reefs is needed to confirm and refine it where regional (oceanic) conditions still support reef growth.

7. Summary

Coral reefs are among the most threatened ecosystems worldwide. The FORAM Index provides a metric for determining whether water quality is suitable for mixotroph-based (i.e., algal-symbiotic-dominant) communities. In the western Atlantic region, where disease and bleaching have profoundly damaged coral communities, the FI can be used to assess whether water quality is sufficient to support reef recovery even in the absence of significant coral populations. In the Indo-Pacific, where significant coral communities still exist, the FI can provide a local, independent indicator of environmental suitability for the continuation of reef growth following a mass mortality event resulting from bleaching, a typhoon, or ship grounding.

Acknowledgments

Development and assessment of foraminifers as bioindicators was funded by the U.S. Environmental Protection Agency ORD-STAR-GAD-R825869. Funding contributing to collection of the foraminiferal data sets includes the National Science Foundation EAR-8407781 and OCE-9203278; U.S. Geological Survey 1434-94-A-1185 and 99HQAG0004; and the National Oceanic and Atmospheric Administration National Undersea Research Program Subcontracts No. 9120, 9204.4, 9221, 9322, 9515, 9609, 9703.66 and 9922. Sampling in the Florida Keys was carried out under Florida Department of Natural Resources Permit Numbers 92S-0031, 93S-0156, and 94S-0034; Florida Department of Environmental Protection Permit Numbers 95S-00063, 96S-016, and 00S-016; and Florida Keys National Marine Sanctuary Permit Numbers KLNMS-23-92, FKNMS 09-93, -(UR)-09-94, 07-95, -(UR)-29-96, -006-97, and -2000-011. We thank Drs. D. Griffin, B. Melzian, L. Robbins, T. Smith, and four unidentified reviewers for their helpful comments on the manuscript.

References

Alve, E.: 1995, 'Benthic foraminiferal responses to estuarine pollution: a review', *Journal of Foraminiferal Research* **25**, 190–203.

Baker, A.C.: 2001, 'Reef corals bleach to survive change', *Nature* **411**, 765–766.

Birkeland, C.: 1977, 'The importance of the rate of biomass accumulation in early successional stages of benthic communities to the survival of coral recruits', Proceedings, 3rd International Coral Reef Symposium, Miami, Florida 1, 16–21.

Birkeland, C.: 1988, 'Geographic comparisons of coral-reef community processes', Proceedings, 6th International Coral Reef Symposium, Townsville, Australia, 1, 211–220.

Boesch, D.F. and Rosenberg, R.: 1981, 'Response to stress in marine benthic communities', in: Barrett, W.G. and Rosenberg, R (eds), Stress Effects on Natural Ecosystems, Wiley-Interscience, New York, pp. 179–200.

Bryant, D., Burke, L., McManus, J. and Spalding, M.: 1998, 'Reefs at Risk: A map-based indicator of potential threats to the world's coral reefs', Washington, DC, World Resources Institute, (web site: www.wri.org/indictrs/reefrisk.htm).

Cockey, E.M., Hallock, P. and Lidz, B.H.: 1996, 'Decadal-scale changes in benthic foraminiferal assemblages off Key Largo, Florida', *Coral Reefs* **15**, 237–248.

Cottey, T.L. and Hallock, P.: 1988, 'Test surface degradation in *Archaias angulatus*', *Journal of Foraminiferal Research* **18**, 187–202.

Dix, T.L.: 2001, The Distribution and Ecology of Benthic Foraminifera of Tampa Bay, Florida, Ph.D. Dissertation, University of South Florida, Tampa, Florida, 197 pp.

Donnelly, K.B.: 1993, Characterization of the Foraminiferal Assemblage Off the Southwest Coast of Puerto Rico at La Parguera, M. Sc. Thesis, University of South Florida, Tampa, Florida, 272 pp.

Dustan, P. and Halas, J.C.: 1987, 'Changes in the reef-coral community of Carysfort Reef, Key Largo, Florida: 1974–1982', *Coral Reefs* **6**, 91–106.

Engle, V.D.: 2000, Chapter Three: Application of the indicator evaluation guidelines to an Index of Benthic Condition for Gulf of Mexico estuaries', in: Jackson, L.E., Kurtz, J.C., and Fisher, W.S. (eds.), Evaluation Guidelines for Ecological Indicators, EPA/620/R-99/005. U.S. Environmental Protection Agency, Research Triangle Park, NC, pp. 3–1 to 3–29.

Environmental Protection Agency: 1998, 'Florida Keys National Marine Sanctuary Water Quality Protection Program, Coral Reef and Hardbottom Monitoring Project', 1997 Annual and 4th Quarterly Report, 39 pp.

Falkowski, P.G., Dubinsky, Z., Muscatine, L. and McCloskey, L.: 1993, 'Population control in symbiotic corals', *BioScience* **43**, 606–611.

Folk, R.L.: 1974, Petrology of Sedimentary Rocks, Hemphill, Austin, TX, 159 pp.

Hallock, P.: 1981a, 'Production of carbonate sediments by selected foraminifera on two Pacific coral reefs', *Journal of Sedimentary Petrology* **51**, 467–474.

Hallock, P.: 1981b, 'Algal symbiosis: a mathematical analysis', *Marine Biology* **62**, 249–255.

Hallock, P.: 1987, 'Fluctuations in the trophic resource continuum: A factor in global diversity cycles?', *Paleoceanography* **2** 457–471.

Hallock, P.: 1988, 'The role of nutrient availability in bioerosion: consequences to carbonate buildups', *Palaeogeography, Palaeoclimatology, Palaeoecology* **63**, 275–291.

Hallock, P.: 2000, 'Symbiont-bearing foraminifera: harbingers of global change', *Micropaleontology* **46** (Suppl. 1), 95–104.

Hallock, P.: 2001, 'Coral Reefs in the 21st Century: Is the past the Key to the Future?' Proceedings, 10th Symposium on the Geology of the Bahamas and other Carbonate Regions, Gerace Research Center, San Salvador, Bahamas, pp. viii-xxiv.

Hallock, P. and Schlager, W.: 1986, 'Nutrient excess and the demise of coral reefs and carbonate platforms', *Palaios* **1**, 389–398.

Hallock, P., Müller Karger, F.E. and Halas, J.C.: 1993, 'Coral reef decline—Anthropogenic nutrients and the degradation of western Atlantic and Caribbean coral reefs', *Research and Exploration* **9**, 358–378.

Hirshfield, H.I., Charmatz, R. and Henson, L.: 1968, 'Foraminifera in samples taken from Eniwetok Atoll in 1956', *Journal of Protozoology* **15**, 497–502.

Hoegh-Guldberg, O.: 1999, 'Climate change, coral bleaching and the future of the world's coral reefs', *Marine and Freshwater Research* **50**, 839–866.

Hughes, T.: 1994, 'Catastrophes, phase shifts, and large-scale degradation of a Caribbean coral reef', *Science* **265**, 1547–1551.

Jaap, W.C., Porter, J.W., Wheaton, J., Hackett, K., Lybolt, M., Callahan, M., Tsokos, C., Yanev, G. and Dustan, P.: 2001, 'Coral Reef Monitoring Project Executive Summary', EPA Science Advisory Panel, Key Colony Beach, Dec. 5–6, 2000, website: ftp://ftp.fmri.usf.edu/pub/uploads/CRMP/

Jackson, L.E., Kurtz, J.C. and Fisher, W.S. (eds): 2000. Evaluation Guidelines for Ecological Indicators, EPA/620/R-99/005. U.S. Environmental Protection Agency, Office of Research and Development, Research Triangle Park, NC, 107 pp.

Laws, E. A. and Redalje, D.G.: 1979, 'Effect of sewage enrichment on the phytoplankton populations of a tropical estuary', *Pacific Science* **33**, 129–144.

Lee, J.J.: 1998, 'Living sands: Larger Foraminifera and their endosymbiotic algae', *Symbiosis* **25**, 75–100.

Lee, J.J. and Anderson, O. R.: 1991, 'Symbiosis in Foraminifera', in: Lee, J.J. and Anderson, O.R. (eds.), Biology of Foraminifera, Academic Press, New York, pp. 157–220.

Lidz, B.H. and Rose, P.R.: 1989, 'Diagnostic foraminiferal assemblages of Florida Bay and adjacent shallow waters - a comparison', *Bulletin of Marine Science* **44** (1), 399–418.

Loeblich, A.R. Jr and Tappan, H.: 1987, Foraminiferal Genera and Their Classification, Van Nostrand Reinhold Company, New York, 2 vol., 1182 pp.

McKee, E.D., Chronic, J. and Leopold, E.B.: 1956, 'Sedimentary belts in the lagoon of Kapingimarangi Atoll', *American Association of Petroleum Geologists Bulletin* **43**, 501–562.

McRae, G., Camp, D.K., Lyons, W.G. and Dix, T.L.: 1998, 'Relating benthic infaunal community structure to environmental variables in estuaries using nonmetric multidimensional scaling and similarity analysis', *Environmental Monitoring and Assessment* **51**, 233–246.

Messer, J.J., Linthurst, R.A. and Overton, W.S.: 1991. 'An EPA program for monitoring ecological status and trends', *Environmental Monitoring and Assessment* **17**, 67–78.

Prinn, R., Cunnold, D., Rasmussen, R., Simmonds, P., Alyca, F., Crawford, A., Fraser, P. and Rosen, R.: 1990, 'Atmospheric emissions and trends of nitrous oxide deduced from 10 years of ALE-GAGE data', *Journal of Geophysical Research* **95**, 369–85.

Risk, M.J.: 1999, 'Paradise lost: how marine science failed the world's coral reefs', *Marine and Freshwater Research* **50**, 831–837.

Rose, P.R. and Lidz, B.H.: 1977, 'Diagnostic foraminiferal assemblages of shallow-water modern environments: south Florida and the Bahamas', Sedimentia VI, pp. 1–56.

Rowan, R.: 2000, 'Evolutionary and ecological dynamics of coral-zooxanthella symbioses', American Zoologist 40(6): 1193–1193.

Schnoor, J.L., Galloway, J.N. and Schlesinger, W.H.: 1995, 'Nitrogen fixation: Anthropogenic enhancement-environment response', *Global Biogeochemical Cycles* **9**, 235–252.

Shick, J.M., Lesser, M.P. and Jokiel, P.L.: 1996, 'Effects of ultraviolet radiation on corals and other coral reef organisms', *Global Change Biology* **2**, 527–545.

Shimel, D., Enting, I.G., Heimann, M., Wigley, T.M.L., Raynaud, D., Alves, D. and Siegenthaler, U.: 1995, 'CO_2 and the carbon cycle', in: Houton, L.G., Meira Filho, L.G., Bruce, J., Lee, H., Callander, B.A., Haites, E., Harris, N. and Maskell, K. (eds.), Climate Change 1994, Cambridge University Press, Cambridge, pp. 35–71.

Smith S.V., Kimmerer, W.J. Laws, E.A. Brock, R.E. and Walsh, T.W.: 1981, 'Kaneohe Bay sewage diversion experiment: Perspectives on ecosystem responses to nutrient perturbation', *Pacific Science* **35**, 279–402.

Steven, A.P.L. and Broadbent, A.D.: 1997, 'Growth and metabolic responses of Acropora palifera to long-term nutrient enrichment', Proceedings, 8th International Coral Reef Symposium, Panama 1, 867–872.

Summers, J.K., Paul, J.F. and Robertson, A.: 1995, 'Monitoring the ecological condition of estuaries in the United States', *Toxicological and Environmental Chemistry* **49**, 93–108.

Veron, J.E.N.: 1995, Corals in Space and Time, Comstock/Cornell, Ithaca and London, 321 pp.

Vitousek, P.M., Mooney, H.A., Lubshenco, J. and Melillo, J.M.: 1997, 'Human domination of Earth's ecosystems', *Science* **277**, 494–499.

Wilkinson, C.: 2000, 'Status of Coral Reefs of the World: 2000', Global Coral Reef Monitoring Network, Australian Institute of Marine Science, Townsville. http://www.aims.gov.au/pages/research/coral-bleaching/scr2000/scr-00.html

Yanko, V., Ahmad, M. and Kaminski, M.: 1998, 'Morphological deformities of benthic foraminiferal tests in response to pollution by heavy metals: implications for pollution monitoring', *Journal of Foraminiferal Research* **28**, 177–200.

MONITORING NEKTON AS A BIOINDICATOR IN SHALLOW ESTUARINE HABITATS

KENNETH B. RAPOSA[1*], CHARLES T. ROMAN[2], AND JAMES F. HELTSHE[3]

[1]*Graduate School of Oceanography, University of Rhode Island, Narragansett, RI, U.S.A.;*
[2]*USGS Patuxent Wildlife Research Center, University of Rhode Island, Narragansett, RI, U.S.A.;*
[3]*Department of Computer Science and Statistics, University of Rhode Island, Kingston, RI, U.S.A.*
[*]*(current address for correspondence author: Narragansett Bay National Estuarine Research Reserve, Prudence Island, RI, U.S.A; e-mail: kenny@gso.uri.edu)*

Abstract. Long-term monitoring of estuarine nekton has many practical and ecological benefits but efforts are hampered by a lack of standardized sampling procedures. This study provides a rationale for monitoring nekton in shallow (< 1 m), temperate, estuarine habitats and addresses some important issues that arise when developing monitoring protocols. Sampling in seagrass and salt marsh habitats is emphasized due to the susceptibility of each habitat to anthropogenic stress and to the abundant and rich nekton assemblages that each habitat supports. Extensive sampling with quantitative enclosure traps that estimate nekton density is suggested. These gears have a high capture efficiency in most habitats and are small enough (e.g., 1 m^2) to permit sampling in specific microhabitats. Other aspects of nekton monitoring are discussed, including spatial and temporal sampling considerations, station selection, sample size estimation, and data collection and analysis. Developing and initiating long-term nekton monitoring programs will help evaluate natural and human-induced changes in estuarine nekton over time and advance our understanding of the interactions between nekton and the dynamic estuarine environment.

Keywords: nekton, monitoring, throw trap, salt marsh, eelgrass, habitat value

1. Introduction

Estuarine ecosystems are constantly changing in response to numerous activities and processes, including nutrient enrichment, habitat loss, over fishing, climate change, sea level rise, shoreline alteration, and other factors. Comprehensive long-term monitoring programs are needed to detect estuarine ecosystem changes and to understand the factors that contribute to these changes. The purpose of this paper is to provide a compelling rationale for including nekton sampling in long-term estuarine monitoring programs and then to highlight some of the fundamental components that should be addressed when designing a nekton monitoring protocol in temperate estuaries. Identification of monitoring questions, selection of sampling gear, and comments on spatial and temporal sampling frequency and data analysis techniques are included in this paper. Our discussion is limited to shallow, subtidal (< 1 m depth) estuarine habitats that retain water throughout the tidal cycle, such as marsh creeks, marsh pools, seagrass beds, and other nearshore habitats. Much of the work presented in this paper is based on quantitative data that we have collected from previous studies and sampling programs in five southern New England estuaries (Figure 1).

Environmental Monitoring and Assessment **81**: 239–255, 2003.

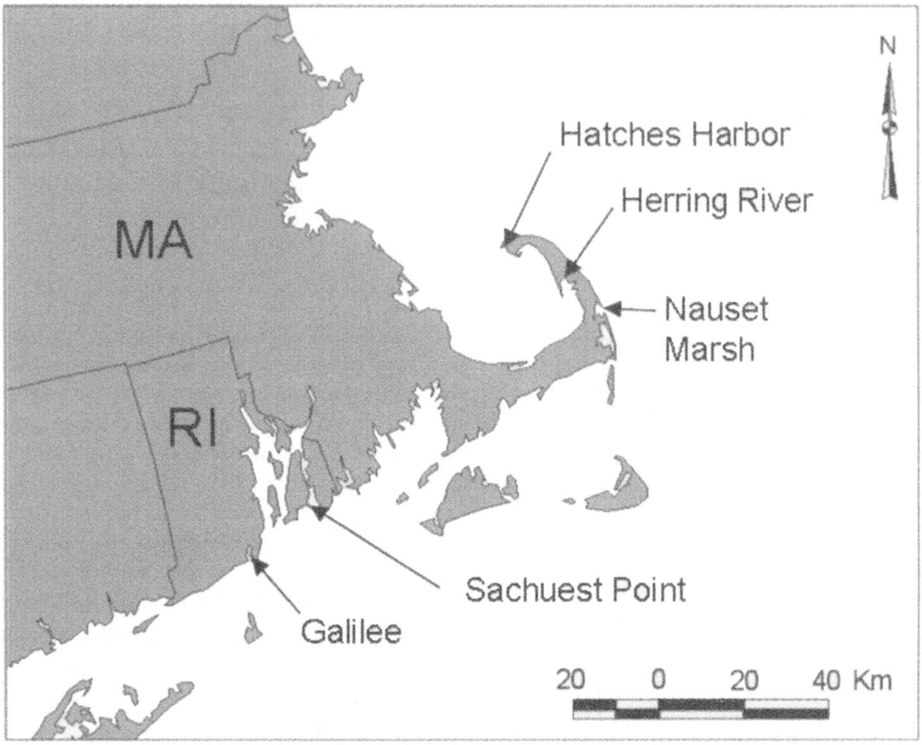

Figure 1. Locations of the five study sites in southern New England where nekton data were collected with a 1 m^2 throw trap from 1997–1999.

2. Why Monitor Nekton in Shallow Estuaries?

Nekton, defined here as an assemblage of free-swimming fishes and decapod crustaceans, is an abundant estuarine fauna with unique responses to environmental change that make them desirable for inclusion in a coastal monitoring program. Development of the Index of Biotic Integrity (Karr, 1981) and the Estuarine Index of Biotic Integrity (Deegan *et al.*, 1997) attests to the value of monitoring nekton to document ecosystem level responses to anthropogenic stress. The foundation of these indices lies in the notion that fishes incorporate and reflect multiple ecosystem processes, and therefore indicate overall ecosystem integrity (Karr, 1981; Fausch *et al.*, 1984; Deegan *et al.*, 1997).

Nekton responds to ecosystem changes resulting from anthropogenic impacts. For example, fish abundance, species richness, and growth rates of mummichogs (*Fundulus heteroclitus*) increased in response to enhanced nitrogen loading (LaBrecque *et al.*, 1996; Tober *et al.*, 1996). Matheson *et al.* (1999) documented a shift in nekton community structure resulting from changes in seagrass distribution and standing crop in Florida.

Estuarine nekton is an integral link between primary producers, consumers, and top predators and is likely to respond to either top-down or bottom-up estuarine perturbations (Kneib, 1997). For example, nutrient enrichment (a bottom-up perturbation) could affect nekton by altering submersed vegetative habitats (Valiela *et al.*, 1992; Harlin, 1995). Conversely, removal of predatory fishes through overfishing (top-down) could induce responses in the forage or prey nekton guild (Carpenter and Kitchell, 1985). Nekton also represents a significant portion of the diets of many piscivorous birds, economically valuable fishes, and, when in estuaries, marine mammals (Friedland *et al.*, 1988; Sekiguchi, 1995; Smith, 1997). In addition, the public generally values nekton and public perception of resource value is often an important consideration for managers making decisions.

There are many factors that make nekton a potentially useful and informative monitoring tool in estuaries. Figure 2 identifies many of the linkages between human-induced and natural environmental stressors (e.g. altered hydrology, nutrient enrichment, storms), associated changes in estuarine habitat structure, and responses of the nekton community.

3. Some Monitoring Questions

Long-term monitoring of nekton will be especially valuable for addressing questions related to long-term/large-scale ecosystem changes or processes. In addition, the issue of habitat restoration is especially prevalent in the northeastern US and elsewhere, and nekton monitoring can be an important component to restoration programs. Some specific monitoring questions that may pertain to evaluating the response of nekton to long-term or large-scale perturbations and processes are:

1. How do nekton respond to long-term human-induced or natural changes in the structure and distribution of estuarine habitats?
2. How do nekton respond to regional or large-scale processes such as global climate fluctuations, sea level rise, ocean temperature changes, or watershed development and nutrient enrichment?
3. To what degree do nekton attributes vary inter-annually and how can natural variability be isolated from human-induced variability?
4. Are invasive species present in the nekton community, are new invasive species being introduced, are they changing in abundance, and are they affecting the structure and function of the estuarine nekton community?

Regarding habitat restoration, recent studies document the rapid responses of nekton to restoration of tidal wetlands, both in New England and elsewhere (Rey *et al.*, 1990; Vose and Bell, 1994; Taylor *et al.*, 1998; Able *et al.*, 2000; Raposa, 2000). However, the complete effects of restoration on nekton are generally attained over several years, and therefore require long-term monitoring (Vose and Bell, 1994; Raposa, 2000). Long-term monitoring of nekton will help address the

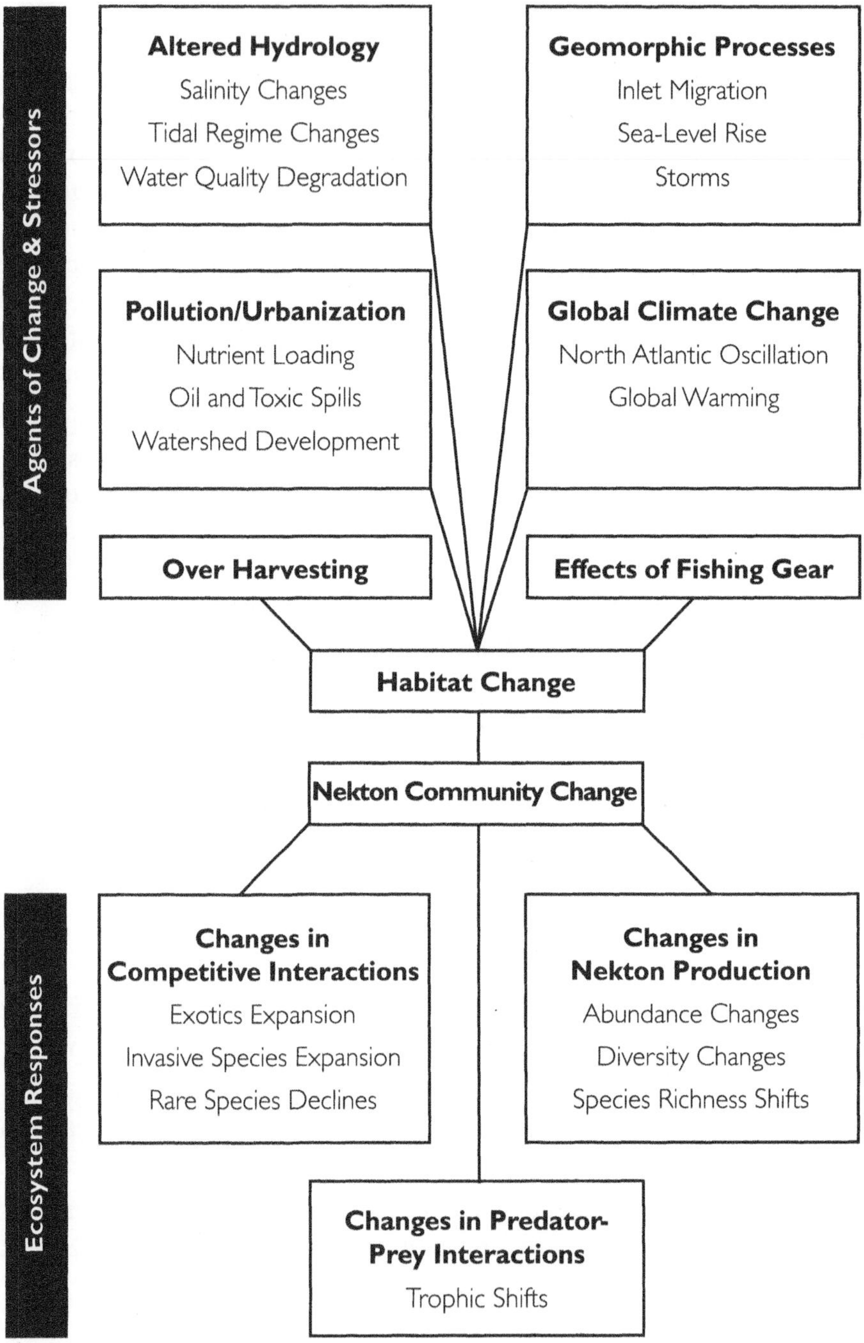

Figure 2. Linkages among environmental stressors and nekton responses in shallow estuarine environments.

following questions as they pertain to restoration. These questions are specific to salt marshes, but they could apply to the restoration of other estuarine habitats, such as seagrass beds:

1. How do nekton communities in impacted salt marshes differ from those in reference marshes?
2. What are responses of nekton to restoration of impacted salt marshes?
3. What is the time frame for nekton communities in restoring salt marshes to achieve functional equivalency when compared to reference marshes?
4. How are changes in nekton related to changes in other ecosystem components such as vegetation, benthos, birds, and water quality during salt marsh restoration?
5. Can the response of nekton to restoration practices be predicted prior to implementation of restoration management?

4. Factors to Consider When Designing a Nekton Monitoring Protocol

4.1 SAMPLING GEAR

Many sampling gears are used to collect nekton in shallow (< 1 m) estuarine habitats. The large body of work devoted to gear comparisons and describing gear characteristics illustrates the importance of sampling gear selection (see review in Rozas and Minello, 1997). The goals of individual projects will ultimately dictate gear selection, but pull nets (e.g. seines) and enclosure traps (e.g. throw traps, pop nets) are two of the more common gear types for sampling nekton in shallow water.

The capture efficiency of pull nets is generally low and is variable among different habitats (Rozas and Minello, 1997). There is evidence that some pull nets preferentially capture water column fishes and under-represent benthic nekton (Zedler, 1990). In contrast, the capture efficiency of enclosure traps is generally high and consistent among most habitat types (Jordan *et al.*, 1997; Kneib, 1997; Rozas and Minello, 1997). Small enclosure traps may preferentially sample smaller nekton, while larger, faster, or less abundant species may be underrepresented in samples (Kushlan, 1981). No gear can effectively sample the entire nekton assemblage in all habitats, but the high and consistent capture efficiency is a primary advantage of enclosure traps over pull nets. Higher capture efficiencies may also lower sample variance, and thus, sample size during monitoring (Peterson and Rabeni, 1995). For these reasons, we concur with Rozas and Minello (1997) and suggest using enclosure traps (e.g. throw traps) for monitoring nekton in shallow (< 1m) estuarine habitats. However, if a study is intended to examine larger, faster nekton or if sampling is to occur in deeper water, trawls or seines may be more appropriate than enclosure traps.

4.2 SPATIAL AND TEMPORAL VARIABILITY IN NEKTON

Spatial variability in nekton abundance is much higher than temporal variability in freshwater systems due to habitat heterogeneity (Peterson and Rabeni, 1995). These authors found that collecting a larger number of samples on fewer dates would optimize sampling efforts, as opposed to taking a smaller number of samples spread out over multiple dates. To our knowledge, a similar detailed analysis of spatio-temporal variability does not exist for estuarine nekton. However, an analysis using nekton densities in tidal creeks from three southern New England salt marshes suggests that variability patterns may be similar for estuarine nekton (Table 1). Temporal variability in density among sampling dates was on average 21 times smaller than spatial variability (i.e., variability among samples taken on the same sampling date). Because of this, we adopt the sampling strategy suggested by Peterson and Rabeni (1995) and suggest that a larger number of samples be collected on fewer dates to address spatial variability and improve sampling precision.

4.3 SPATIAL SAMPLING FREQUENCY

There are at least two approaches for selecting nekton sampling stations in seagrass. First, random samples can be collected solely from within seagrass beds on each sampling date. The extent and distribution of seagrass changes temporally, so station locations must be flexible among sampling dates so that each $1 m^2$ sample is located in seagrass. This method was used in studies in New Jersey (Sogard and Able, 1991) and Florida Bay (Matheson *et al.*, 1999). A second approach is to randomly establish permanent locations in an area that supports seagrass. With this method sample locations may occur where seagrass is absent due to patchi-

Table 1. Spatial and temporal variability in nekton density in three New England salt marshes. All values are variance component estimates of temporal and spatial variability calculated using untransformed density data in the SAS variance component estimation procedure (PROC VARCOMP; SAS Institute, Inc., 1997). Nekton was collected in tidal creeks with throw traps between June and October 1997.

	Hatches Harbor		*Galilee*		*Sachuest Point*		*Average*	
	Spatial	*Temporal*	*Spatial*	*Temporal*	*Spatial*	*Temporal*	*Spatial*	*Temporal*
Total nekton	1977.0	95.0	1735.0	187.6	1048.1	47.1	1586.7	109.9
Total fish	1850.8	72.4	528.6	110.9	930.6	29.7	1103.3	71.0
Total decapods	252.6	0.0	1128.5	39.9	18.5	0.0	466.5	13.3
Fundulus heteroclitus	1814.7	66.8	170.3	34.2	890.4	0.0	958.5	33.7
Carcinus maenas	3.2	0.1	1.0	0.0	0.0	0.0	2.1	0.1
Fundulus majalis	1.5	0.1	17.2	1.9	0.0	0.0	9.3	1.0
Palaemonetes pugio	0.0	0.0	1123.2	42.0	18.6	0.0	570.9	21.0
Menidia menidia	0.8	0.0	48.9	3.2	2.2	0.9	17.3	1.4

ness in cover. In addition, even if all stations were originally established in seagrass, this might not be the case on future sampling dates due to changes in seagrass cover and distribution. The first approach is appropriate if the goal of monitoring is to assess changes in seagrass-associated nekton assemblages. However, if the goal is to document overall changes in estuarine nekton over time in response to changes in seagrass habitat, including seagrass expansion, die-off, or replacement with macroalgae, then the second approach is more appropriate.

Selecting sampling stations in salt marshes is more involved because of the number of distinct sub-habitats present in a marsh. Before monitoring is initiated a choice must be made between sampling in one habitat type that may be of special interest (e.g., creeks) or in all habitats that are available to nekton (e.g. creeks, pools, ditches, marsh surface, etc.). Unless there is a single marsh microhabitat that is of special interest, or if human impacts will clearly affect nekton in only one habitat type, samples should ideally be collected from as many available habitats as possible when monitoring nekton in salt marshes. A stratified random sampling approach, where habitat types are identified and the number of stations is based on the area of each habitat or the level of nekton variability in each habitat, could then be used to select stations within each habitat type (Krebs, 1989).

4.4 SAMPLE SIZE

As previously noted, densities of estuarine nekton are highly variable, especially over spatial scales (Table 1). One way to address this variability and improve the ability to detect biological differences (e.g., species richness, density) among treatments is to increase sample size. However, determining the appropriate sample size depends on a number of factors, such as the desired level of precision or if statistical comparisons are to be made, the desired difference among treatments one wishes to detect (Sokal and Rohlf, 1981; Krebs, 1989). Sample size also varies among different nekton species and depends on different attributes of the nekton community that are under consideration (e.g., density, richness). A simple formula is available to estimate the required sample size to reach a desired level of precision (Snedecor and Cochran, 1980):

$$N = (t^2\ CV^2)/L^2$$

In this formula, N is the required number of samples, t is a constant that varies with the desired confidence level, CV is the coefficient of variation (CV = standard deviation/mean), and L is the desired level of precision.

We calculated the number of samples required to reach 20% precision around the mean (i.e., SE $\leq$ 0.2 times the mean), at the 95% confidence level (t = 1.96) for densities of total nekton and for common species in eelgrass, marsh edge, creek, and pool habitats (Figure 3). The 20% level has been used in other nekton sampling studies (Pihl Baden and Pihl, 1984; Peterson and Rabeni, 1995). Sample size clearly depends on the habitat and the level of community organization that is of interest (e.g., individual species vs. total nekton). When considering total nekton, the number of samples required in eelgrass beds and along the edge of fringing marsh or marsh-lined embayments was determined to be relatively low. In contrast, sample size was substantially greater in tidal creek and pool habitats. If there is interest in evaluating long-term trends in the density of individual, common species (e.g., *Fundulus heteroclitus, Palaemonetes pugio*), then an even larger sample size would be necessary to attain 20% precision (Figure 3). Although not calculated here, it is expected that sample sizes for uncommon or rare species would be significantly higher.

In addition to using a classic sample size formula to establish minimum sample size, we also conducted a power analysis. The objective of a power analysis is to determine the minimum number of sample replicates that are necessary to detect changes between nekton data sets. Power is a function of the differences between two populations, sample size, alpha level of the test (probability of a Type I Error),

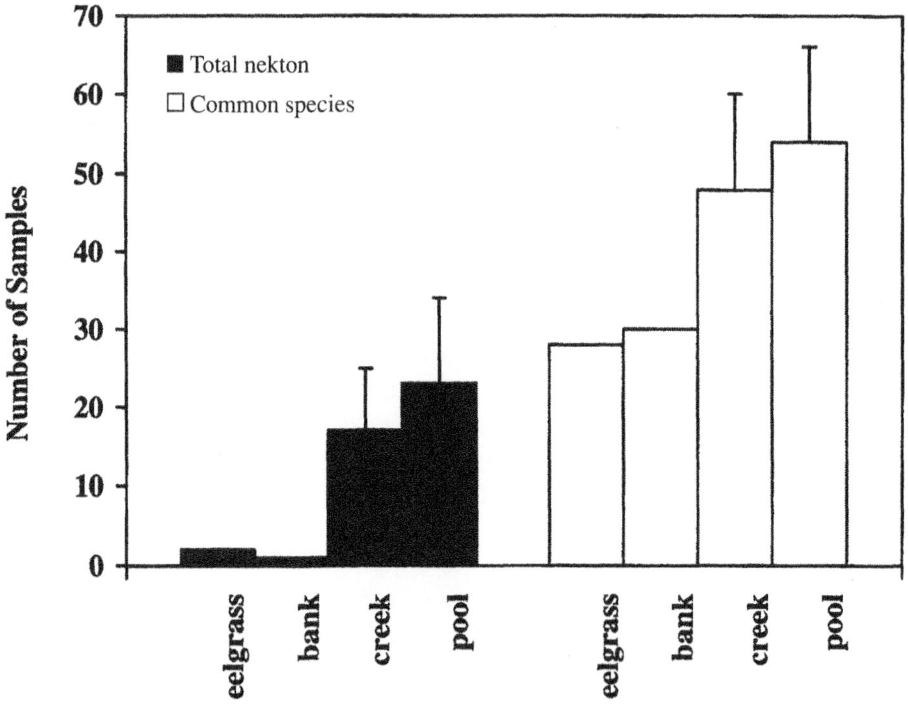

Figure 3. Sample size estimates in four shallow estuarine habitats. Eelgrass and marsh edge sample size estimates were made using Nauset Marsh data; creek and pool estimates were made for Nauset Marsh, Galilee, and Hatches Harbor and then averaged across sites (± SE). In each habitat, sample size estimates were made for the total nekton density and for individual species that are commonly abundant (i.e., any species that was collected in (50% of the samples from that habitat) and then averaged across all species. Data were log (x+1) transformed prior to analysis.

and variability of the measured response. For this analysis, nekton community data (species composition and abundance) from several southern New England marshes (Herring River, Hatches Harbor, Nauset Marsh, MA; Galilee salt marsh and Sachuest Point salt marsh, RI), collected with a throw trap from marsh creeks during the summer and fall seasons, were used. Pairs of data sets that exhibited a range from similar nekton composition (e.g., Herring River in fall vs. summer) to quite different (e.g., Herring River in summer vs. Galilee in summer) were identified. A Euclidean distance similarity index was used to evaluate similarity between data sets (Krebs, 1989). From Figure 4 we can estimate the statistical power of detecting a difference between two nekton community data sets. As noted, with an n = 5 there is a low power to detect differences, even for many cases where the differences between the two data sets are great. Increasing the sample size to n = 10 or n = 15 dramatically increases the power to differentiate two marsh nekton data sets, even between data sets that are quite similar. With a power above 0.9, there is a > 90% chance of detecting a difference between data sets when a difference actually exists. With a low power there is an increased probability of not detecting a difference when the data sets are actually different (i.e., Type II Error). From the power analysis and associated power curve, an investigator could determine that if detecting subtle differences between nekton density data sets was of interest (e.g., comparing nekton density in Marsh A over two consecutive sample years), then it may be appropriate to have a large number of replicates. If dramatic

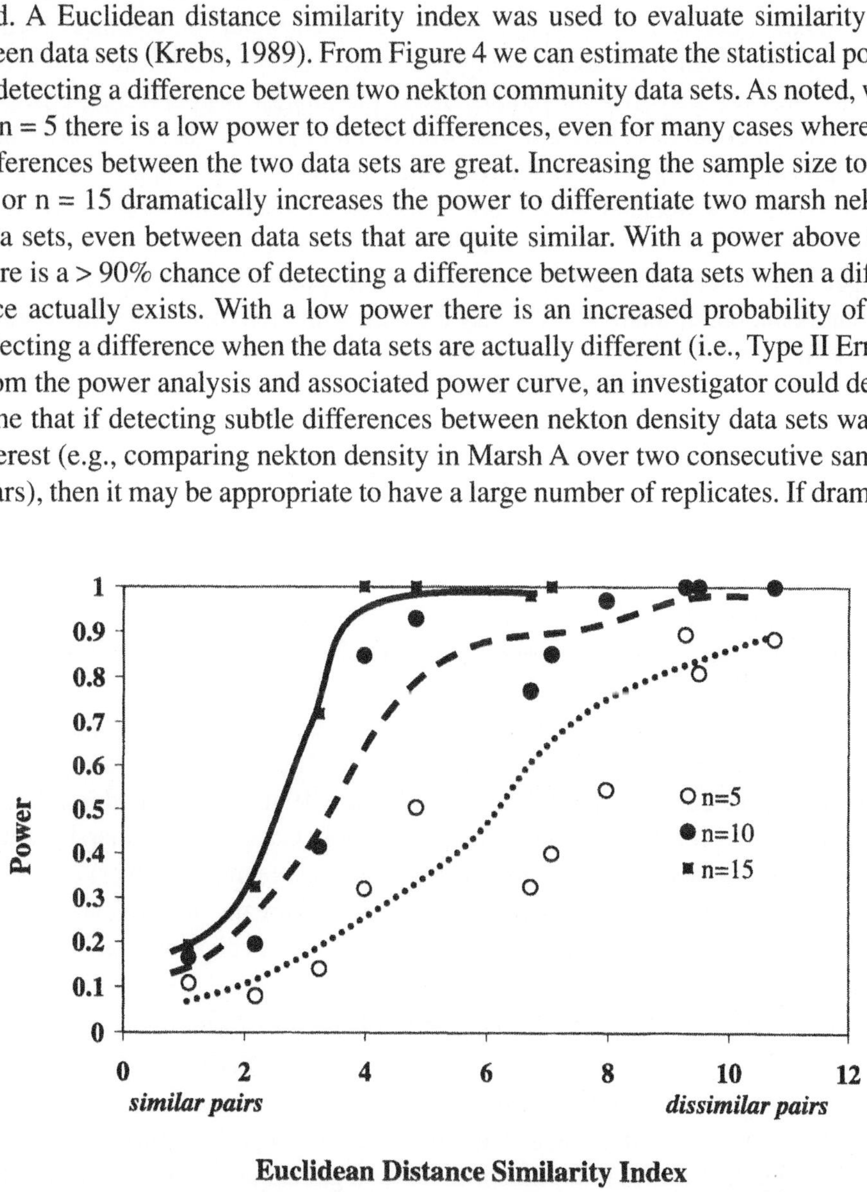

Figure 4. Power curves for sample sizes of 5, 10, and 15 with an alpha level of 0.05. Nekton density data from pairs of data sets that range in similarity from similar to dissimilar are compared.

changes were to be detected (e.g., comparing pristine Marsh A with highly impacted Marsh B), then perhaps a smaller number of replicates would suffice.

To summarize the discussion on sample size, we have presented several estimates for determining an appropriate sample size. If based solely on the sample size formula and if interested in assessing changes in total nekton density, it is suggested that an appropriate number of throw trap samples may vary from n = 5 for eelgrass beds to n = 20 – 25 for marsh pools and creeks (Figure 3). However, if there is an interest in understanding trends in the density of individual species that are common, then the sample size would increase substantially, based on the sample size formula. The power analysis provides guidance on sample size if the interest is in detecting change in nekton communities (species composition and abundance), and the results are remarkably similar to the sample size estimates for density; suggesting an appropriate sample size of n = 15. A power analysis was not performed for just density or species richness data. It should be noted, however, that these sample size estimates were calculated using data from southern New England; in other geographic regions this work offers a starting point, but different regions may encounter different degrees of variability in their shallow water nekton.

4.5 TEMPORAL SAMPLING FREQUENCY

Nekton abundance and richness were highest in either summer or fall in Hatches Harbor, Nauset Marsh, Herring River, and Galilee (Table 2). Similar peaks during warm temperatures are common in other temperate estuarine habitats (Pearcy and Richards, 1962; Recksiek and McCleave, 1973; Adams, 1976; Cain and Dean, 1976; Hoff and Ibara, 1977; Heck and Orth, 1980; Orth and Heck, 1980; Pihl and Rosenberg, 1982; Pihl Baden and Pihl, 1984; Ayvazian *et al.*, 1992; Rountree and Able, 1992; Able *et al.*, 1996; Lazzari *et al.*, 1999).

In some cases the exact timing of nekton peaks depends on latitude and/or habitat type. For example, nekton abundance in eelgrass beds peaked in June in Chesapeake Bay (Heck and Orth, 1980; Orth and Heck, 1980), but peaked in late summer and fall in Nauset Marsh (Heck *et al.*, 1989). In Cape Cod salt marshes, abundance peaked in landward habitats (marsh pools, upstream tidal river) later in the year than in seaward habitats (marsh creeks, downstream tidal river) (Table 2), probably as a result of autumnal movements of some species into landward overwintering habitats (Fritz *et al.*, 1975; Smith and Able, 1994). A similar pattern was also observed in New Jersey salt marshes (Able *et al.*, 1996).

Despite the variability in the timing of abundance and richness peaks, both parameters are generally highest between June and October in temperate estuaries regardless of habitat and latitude. Therefore, monitoring efforts should be concentrated during this time to maximize information gained per sampling effort. However, there are species and processes unique to every season (e.g., anadromous fish immigrations in spring) and seasonal sampling should reflect the goals of individual monitoring programs.

Table 2. Seasonal patterns of nekton density (animals m^{-2}) and species richness in four New England estuaries. Spring = April to mid-June; Summer = mid-June to mid-September; Fall = mid-September to mid-December; Winter = mid-December to March.

		Density				*Number of Species*			
Habitat	*Site*	*Spr.*	*Sum.*	*Fall*	*Win.*	*Spr.*	*Sum.*	*Fall*	*Win.*
Tidal Creeks	Hatches Harbor	2.34	27.95	12.26	1.38	8	10	8	5
	Nauset Marsh	0.40	22.60	6.40	0.00	3	6	4	0
	Galilee	8.74	89.90	25.53	1.04	7	10	12	4
Marsh Pools	Hatches Harbor	14.64	13.03	70.32	13.54	7	7	6	5
	Nauset Marsh	3.66	16.32	16.30	0.52	8	9	8	3
	Galilee	3.67	225.67	49.87	8.40	6	9	9	4
Tidal River									
Downstream	Herring River	40.00	78.50	38.35	3.15	10	13	7	3
Upstream	Herring River	24.95	18.48	81.85	1.93	9	15	11	4
Eelgrass	Nauset Marsh	66.28	335.56	94.68	1.43	14	18	16	2

To summarize, we suggest collecting samples during at least two periods: once during early summer (June–July) and once during late summer-early fall (August–October). The two sample times are supported by work in the Hudson River estuary where nekton assemblages collected in early summer were different from those collected in late summer (Able *et al.*, 1998). In Galilee and Hatches Harbor creeks, nekton communities in June or July differed from communities collected in August, September, or October at the same site 83% of the time (Analysis of Similarity, ANOSIM; $p < 0.05$ in 15 out of 18 comparisons). Sampling in both early summer and late summer-early fall will more comprehensively document nekton use of the study area. In addition, since abundance and richness vary among sites and habitats this sampling scheme will improve the probability of sampling during peaks in nekton use.

Each sampling period should extend over multiple days (e.g., 3–5 days). Nekton parameters can vary considerably over consecutive day periods in salt marshes (Varnell *et al.*, 1995). These authors showed that sampling on only one day would often produce inaccurate results, depending on the parameter in question. Sampling over multiple days will not only provide more accurate results, but also will allow for a larger sample size and increased sampling precision.

5. Some Specifics on Data Collection and Analysis

Species composition, richness, and density are obtained simply by identifying and enumerating all captured animals in each trap. Accurate length estimates can be obtained by measuring as few as five individuals per throw trap sample (Table 3), but we suggest a more conservative approach by randomly measuring at least 15

Table 3. Mean lengths (mm) of common nekton species obtained by measuring between 5 and 30 individuals captured per throw trap sample at Galilee (n = 26 throw trap samples). *Fundulus heteroclitus* were collected in July 1997, *Menidia menidia* were collected in August 1999, and *Palaemonetes pugio* were collected in August 1997. For each species, comparisons among treatments were made using one-way ANOVA, and the resultant p value is presented.

		Number of Individuals Measured per Sample					
Species	*p*	*5*	*10*	*15*	*20*	*25*	*30*
Fundulus heteroclitus	0.99	30.04	29.53	30.05	30.05	29.71	29.61
Menidia menidia	0.66	40.91	42.21	43.52	44.62	44.82	44.48
Palaemonetes pugio	0.87	24.89	24.94	24.71	24.75	24.55	24.54

individuals of each species. By measuring nekton lengths researchers gain information on habitat use by different life history stages. For example, by measuring mummichog (*Fundulus heteroclitus*) sizes from Hatches Harbor throw trap samples, we were able to demonstrate changes in the size distributions of this species throughout the year, emphasizing the influx of young-of-the-year in summer (Figure 5).

In addition to nekton sampling, companion environmental data should be collected, including salinity, water temperature, water depth, dissolved oxygen, and a description of the sampled habitat. Secchi depth or some other measure of water clarity should also be collected since it is possible that the efficiency of throw traps is influenced by turbidity. These site-specific data, coupled with associated monitoring data sets on habitat change throughout the entire estuarine system, land use, water quality, and others, can be invaluable when attempting to understand why nekton communities are changing. There are innumerable techniques for analyzing nekton monitoring data. Analyses of univariate measures (density, length) can be compared among treatments (e.g., over time, among habitats, among sites) using analysis of variance (ANOVA), followed by post-hoc multiple comparison procedures to elucidate specific differences. Density data are generally log (x+1) transformed to meet the assumptions of normality and equal variance that are associated with ANOVA. If the data fail to meet the parametric assumptions even after transformation, non-parametric tests can be used (e.g., Kruskal-Wallis rank test, log-linear contingency tables).

Figure 6 provides an example of how nekton density data have been successfully used to monitor the effects of restoring tidal flow to a previously tide-restricted salt marsh in Rhode Island. As noted, in 1997, prior to tidal restoration, nekton density was significantly lower than in an adjacent tide-unrestricted or reference marsh. With restoration of tidal flow in 1998 and similarly after two years of tidal restoration in 1999, nekton density in the now restoring marsh was equivalent to the reference marsh.

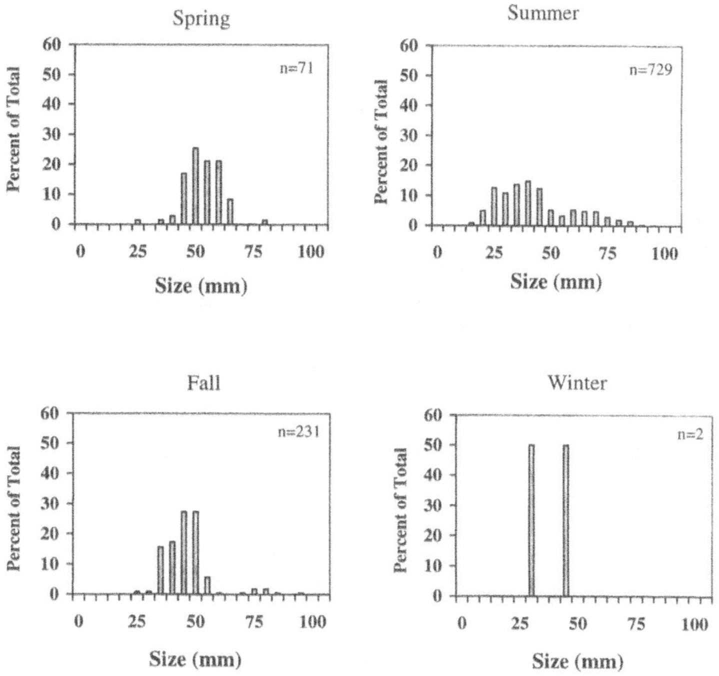

Figure 5. Length-frequency histograms for the mummichog (*Fundulus heteroclitus*) during four seasons at Hatches Harbor salt marsh. All fish were captured in tidal creeks with 1 m^2 throw traps.

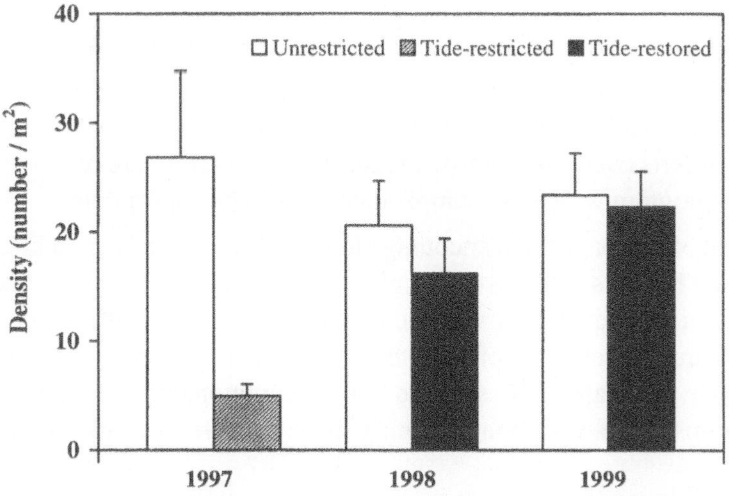

Figure 6. Density of nekton (number of individuals m^{-2}) at Sachuest Point from shallow water habitats (marsh creeks and pools) for the unrestricted and restricted/restoring marshes during pre-restoration (1997) and restoring conditions (1998 and 1999). Data were collected using a 1 m^2 throw trap from August to October of each year. Analysis by 2-way ANOVA with significant comparisons by SNK multiple range test.

Community composition can be analyzed with analysis of similarity (ANOSIM), as well as several other multivariate classification and ordination techniques. ANOSIM is a non-parametric test, similar to multivariate analysis of variance (MANOVA) but without the generally unattainable assumptions (Clarke and Warwick, 1994; Carr, 1997). A nice feature of ANOSIM is that significant differences among treatments can be followed with a similarity percentages (SIMPER) procedure that identifies individual species that contribute to any observed differences.

Species richness in a habitat or system can be estimated using the procedure developed by Heltshe and Forrester (1983). Analysis between two treatments is compared by Student's t-test with Bonferroni adjusted alpha if pairwise comparisons are considered. This species richness jackknife procedure is more desirable than simply tallying the number of species collected since it takes into account sample size as well as the number of samples collected.

6. Conclusions

This paper describes some of the basic factors that should be considered when designing an estuarine nekton sampling protocol for shallow water habitats. Much of our presentation is centered around the minimum sampling effort necessary to detect significant changes in the density, richness, and nekton community composition over time and in response to environmental and habitat changes. If additional time, personnel, or funds are available, supplementary sampling can be initiated; for example, additional sampling in spring, concurrent sampling on the marsh surface, or measurements of nekton biomass. We acknowledge some limitations associated with our discussion. For example, sacrificing more sampling dates in favor of a large sample size on fewer dates increases the possibility of missing short-term pulses of migrating species or newly hatching young-of-the-year. It would also not be possible to estimate growth rates by tracking modal lengths of cohorts over time. If growth rates (or production) are of interest, then a monitoring program with more sample dates would be appropriate.

The intent of this paper is to encourage long-term monitoring of nekton in shallow estuarine habitats. Moreover, we want to stress the necessity for identifying specific monitoring questions and encourage the collection of quantitative data that will insure the detection of long-term trends with a degree of statistical certainty. Finally, this paper may serve to inspire commonality among nekton sampling programs, thereby enabling meaningful geographic comparisons to be made.

Acknowledgements

This research was supported, in part, by the US Geological Survey in association with the design and testing of monitoring protocols for the Long-term Coastal Ecosystem Monitoring Program at Cape Cod National Seashore. Support was also provided by the NOAA-National Marine Fisheries Service, Restoration Center, with funds administered through the NOAA Cooperative Marine Education Research Program at the University of Rhode Island.

References

Able, K. W., McBride, R. S., Rountree, R. A. and Smith, K. J.: 1996, 'Fishes of polyhaline estuarine shores in Great Bay-Little Egg Harbor, New Jersey: a case study of seasonal and habitat influences', in: Nordstrom, K. F. and Roman, C. T. (eds.), Estuarine Shores: Evolution, Environments and Human Alterations, John Wiley & Sons, Ltd., West Sussex, England, pp. 335–353.

Able, K. W., Manderson, J. P. and Studholme, A. L.: 1998, 'The distribution of shallow water juvenile fishes in an urban estuary: the effects of manmade structures in the lower Hudson River', *Estuaries* **21**, 731–744.

Able, K. W., Nemerson, D. M., Light, P. R. and Bush, R. O.: 2000, 'Initial response of fishes to marsh restoration at a former salt hay farm bordering Delaware Bay', in: Weinstein, M.P. and Kreeger, D.A. (eds.), Concepts and Controversies in Tidal Marsh Ecology, Kluwer Academic Publishers, The Netherlands, pp. 749–773.

Adams, S. M.: 1976, 'The ecology of eelgrass, *Zostera marina* (L.), fish communities. I. Structural analysis', *Journal of Experimental Marine Biology and Ecology* **22**, 269–291.

Ayvazian, S. G., Deegan, L. A. and Finn, J. T.: 1992, 'Comparison of habitat use by estuarine fish assemblages in the Acadian and Virginian zoogeographic provinces', *Estuaries* 15, 368–383.

Cain, R. L. and Dean, J. M.: 1976, 'Annual occurrence, abundance and diversity of fish in a South Carolina intertidal creek', *Marine Biology* **36**, 369–379.

Carpenter, S. R. and Kitchell, J. F.: 1985, 'Cascading trophic interactions and lake productivity', *Bioscience* **35**, 634–639.

Carr, M. R.: 1997, Primer User Manual: Plymouth Routines in Multivariate Ecological Research, Plymouth Marine Laboratory, Plymouth, England, 42 pp.

Clarke, K. R. and Warwick, R. M.: 1994, Change in marine communities: an approach to statistical analysis and interpretation, Plymouth Marine Laboratory, Plymouth, England, 144 pp.

Deegan, L. A., Finn, J. T., Ayvazian, S. G., Ryder-Kieffer, C. A. and Buonaccorsi, J.: 1997, 'Development and validation of an estuarine biotic integrity index', *Estuaries* **20**, 601–617.

Fausch, K. D., Karr, J. R. and Yant, P. R.: 1984, 'Regional application of an index of biotic integrity based on stream fish communities', *Transactions of the American Fisheries Society* **113**, 39–55.

Friedland, K. D., Garman, G. C., Bejda, A. J. and Studholme, A. L.: 1988, 'Interannual variation in diet and condition in juvenile bluefish during estuarine residency', *Transactions of the American Fisheries Society* **117**, 474–479.

Fritz, E. S., Meredith, W. H. and Lotrich, V. A.: 1975, 'Fall and winter movements and activity level of the mummichog, *Fundulus heteroclitus*, in a tidal creek', *Chesapeake Science* **16**, 211–215.

Harlin, M. M.: 1995, 'Changes in major plant groups following nutrient enrichment', in: McComb, A. J. (ed.), Eutrophic shallow estuaries and lagoons, CRC Press, Boca Raton, Fl., pp. 173–187

Heck Jr., K. L. and Orth, R. J.: 1980, 'Structural components of eelgrass (*Zostera marina*) meadows in the lower Chesapeake Bay-decapod crustaceans', *Estuaries* **3**, 289–295.

Heck Jr., K. L., Able, K. W., Fahay, M. P. and Roman, C. T.: 1989, 'Fishes and decapod crustaceans of Cape Cod eelgrass meadows: Species composition, seasonal abundance patterns and comparison with unvegetated substrates', *Estuaries* **12**, 59–65.

Heltshe, J. F. and Forrester, N. E.: 1983, 'Estimating species richness using the jackknife Procedure', *Biometrics* **39**, 1–11.

Hoff, J. G. and Ibara, R. M.: 1977, 'Factors affecting the seasonal abundance, composition and diversity of fishes in a southeastern New England estuary', *Estuarine and Coastal Marine Science* **5**, 665–678.

Jordan, F., Coyne, S. and Trexler, J. C.: 1997, 'Sampling fishes in vegetated habitats: Effects of habitat structure on sampling characteristics of the 1-m^2 throw trap', *Transactions of the American Fisheries Society* **126**, 1012–1020.

Karr, J. R.: 1981, 'Assessment of biotic integrity using fish communities', *Fisheries* **6**, 21–27.

Kneib, R. T.: 1997, 'The role of tidal marshes in the ecology of estuarine nekton', *Oceanography and Marine Biology: An Annual Review* **35**, 163–220.

Krebs, C. J.: 1989, *Ecological Methodology*, Harper & Row, New York, NY, 654 pp.

Kushlan, J. A.: 1981, 'Sampling characteristics of enclosure fish traps', *Transactions of the American Fisheries Society* **110**, 557–562.

LaBrecque, E., Fritz, C., Tober, J., Behr, P. J. and Valiela, I.: 1996, 'Abundance and age-specific growth rates in relation to population densities of *Fundulus heteroclitus* in Waquoit Bay estuaries subject to different nitrogen loads', *Biological Bulletin* **191**, 319–320.

Lazzari, M. A., Sherman, S., Brown, C. S., King, J., Joule, B. J., Chenoweth, S. B. and Langton, R. W.: 1999, 'Seasonal and annual variations in abundance and species composition of two nearshore fish communities in Maine', *Estuaries* **22**, 636–647.

Matheson Jr., R. E., Camp, D. K., Sogard, S. M. and Bjorgo, K. A.: 1999, 'Changes in seagrass-associated fish and crustacean communities on Florida Bay mud banks: the effects of recent ecosystem changes?', *Estuaries* **22**, 534–551.

Orth, R. J. and Heck, K. L.: 1980, 'Structural components of eelgrass (*Zostera marina*) meadows in the lower Chesapeake Bay-fishes', *Estuaries* **3**, 278–288.

Pearcy, W. G. and Richards, S. W.: 1962, 'Distribution and ecology of fishes of the Mystic River estuary, Connecticut', *Ecology* **43**, 248–259.

Peterson, J. T. and Rabeni, C. F.: 1995, 'Optimizing sampling effort for sampling warmwater stream fish communities', *North American Journal of Fisheries Management* **15**, 528–541.

Pihl Baden, S. P. and Pihl, L.: 1984, 'Abundance, biomass and production of mobile epibenthic fauna in *Zostera marina* (L.) meadows, western Sweden', *Ophelia* **23**, 65–90.

Pihl, L. and Rosenberg, R.: 1982, 'Production, abundance, and biomass of mobile epibenthic marine fauna in shallow waters, western Sweden', *Journal of Experimental Marine Biology and Ecology* **57**, 273–301.

Raposa, K. B.: 2000, Nekton utilization of tidally restricted, restoring, and reference New England salt marshes, Ph.D. Dissertation, University of Rhode Island, Kingston, RI, 172 pp.

Recksiek, C. W. and McCleave, J. D.: 1973, 'Distribution of pelagic fishes in the Sheepscot River-Back River estuary, Wiscasset, Maine', *Transactions of the American Fisheries Society* **102**, 541–551.

Rey, J. R., Shaffer, J. and Tremain, D.: 1990, 'Effects of re-establishing tidal connections in two impounded subtropical marshes on fishes and physical conditions', *Wetlands* **10**, 27–45.

Rountree, R. A. and Able, K. W.: 1992, 'Fauna of polyhaline subtidal marsh creeks in southern New Jersey: composition, abundance and biomass', *Estuaries* **15**, 171–185.

Rozas, L. P. and Minello, T. J.: 1997, 'Estimating densities of small fishes and decapod crustaceans in shallow estuarine habitats: a review of sampling design with focus on gear selection', *Estuaries* **20**, 199–213.

SAS Institute, Inc.: 1997, SAS user's guide, Version 6, 5th edition, SAS Institute, Inc., Cary, NC.

Sekiguchi, K.: 1995, 'Occurrence, behavior and feeding habits of harbor porpoises (*Phocoena phocoena*) at Pajaro Dunes, Monterey Bay, California', *Aquatic Mammals* **21**, 91–103.

Smith, J. P.: 1997, 'Nesting season food habits of four species of herons and egrets at Lake Okeechobee, Florida', *Colonial Waterbirds* **20**, 198–220.

Smith, K. J. and Able, K. W.: 1994, 'Salt-marsh tide pools as winter refuges for the mummichog, *Fundulus heteroclitus*, in New Jersey', *Estuaries* **17**, 226–234.

Snedecor, W. S. and Cochran, W. G.: 1980, Statistical Methods, Iowa State University Press, Ames, IA, 507 pp.

Sogard, S. M. and Able, K. W.: 1991, 'A comparison of eelgrass, sea lettuce macroalgae, and marsh creeks as habitats for epibenthic fishes and decapods', *Estuarine, Coastal and Shelf Science* **33**, 501–519.

Sokal, R. and Rohlf, F.: 1981, Biometry, Freeman, San Fransisco, CA, 859 pp.

Taylor, D. S., Poulakis, G. R., Kupschus, S. R. and Faunce, C. H.: 1998, 'Estuarine reconnection of an impounded mangrove salt marsh in the Indian River lagoon, Florida: short-term changes in fish fauna', *Mangroves and Salt Marshes* **2**, 29–36.

Tober, J., Fritz, C., LaBrecque, E., Behr, P. J. and Valiela, I.: 1996, 'Abundance, biomass, and species richness of fish communities in relation to nitrogen-loading rates of Waquoit Bay estuaries', *Biological Bulletin* **191**, 321–322.

Valiela, I., Foreman, K. and LaMontagne, M.: 1992, 'Couplings of watersheds and coastal waters: sources and consequences of nutrient enrichment in Waquoit Bay, Massachusetts', *Estuaries* **15**, 443–457.

Varnell, L. M., Havens, K. J. and Hershner, C.: 1995, 'Daily variability in abundance and population characteristics of tidal salt-marsh fauna', *Estuaries* **18**, 326–334.

Vose, F. E. and Bell, S. S.: 1994, 'Resident fishes and macrobenthos in mangrove-rimmed habitats: evaluation of habitat restoration by hydrologic modification', *Estuaries* **17**, 585–596.

Zedler, J.: 1990, A manual for assessing restored and natural coastal wetlands with examples from California, Report #T-CSGCP-021, California Sea Grant, La Jolla CA.

INTERLABORATORY VARIABILITY OF AMPHIPOD SEDIMENT TOXICITY TESTS IN A COOPERATIVE REGIONAL MONITORING PROGRAM

STEVEN M. BAY[1*], ANDREW JIRIK[2], AND STANFORD ASATO[3]

[1]*Southern California Coastal Water Research Project, 7171 Fenwick Lane, Westminster, CA, U.S.A.;*
[2]*Port of Los Angeles, Environmental Management Dept., 425 S. Palos Verdes St., San Pedro, CA, U.S.A;*
[3]*City of Los Angeles, Environmental Monitoring Div., 12000 Vista del Mar, Playa del Rey, CA, U.S.A.;*

[*]*(author for correspondence, e-mail: steveb@sccwrp.org)*

Abstract. Marine sediment toxicity tests are widely applied in monitoring programs, yet relatively little is known about the comparability of data from different laboratories. The need for comparability information is increased in cooperative monitoring programs, where multiple laboratories (often with variable skill levels) perform toxicity tests. An interlaboratory comparison exercise was conducted among seven laboratories in order to document the comparability of sediment toxicity measurements during the Bight'98 regional sediment survey in southern California. Sediments from four stations in Los Angeles and Long Beach Harbors were tested using a 10-day survival test of the amphipod *Eohaustorius estuarius*. All laboratories successfully performed the sediment test and associated reference toxicant test. Statistically significant differences were found in mean amphipod survival rates among some laboratories for the field-collected sediments, but there was little evidence of a consistent bias among laboratories. Although the reference toxicant test indicated a five-fold variation in test sensitivity among laboratories, these results were not accurate predictors of interlaboratory performance for the sediment tests. The laboratories demonstrated excellent concordance (Kendall's W = 0.91) in ranking the field-collected sediments by toxicity. Agreement on classifying the sediments into categories (nontoxic, moderately toxic, and highly toxic) based upon the percent of survival was best for highly toxic sediments. An analysis of test precision based upon the variance among replicates within a test indicated that the measured survival rate for a sample may vary by up to 12 percentage points from the actual response.

Keywords: sediment toxicity, southern California, variability, amphipod, regional monitoring

1. Introduction

Laboratory tests that measure the toxic effects of sediments on benthic organisms, such as amphipod crustaceans, are integral components of national programs to assess sediment quality in coastal areas of the United States (Long, 2000). Amphipod toxicity tests have been used in southern California to examine temporal changes in sediment quality associated with pollution reduction (Swartz *et al.*, 1986; Bay, 1992), identify toxic hot spots (Fairey *et al.*, 1996), and estimate the spatial extent of sediment toxicity (Anderson *et al.*, 1997; Bay *et al.*, 1998). Amphipod toxicity tests were an integral part of the Bight'98 regional survey, a cooperative regional survey that included the testing of 241 sediment samples from the Southern California Bight (Figure 1, Bight '98 Steering Committee, 1998).

Cooperative integrated regional monitoring studies conducted in southern California are distinguished by the extensive participation of local agencies in the collection and analysis of samples. Over 60 stakeholder agencies participated in the

Environmental Monitoring and Assessment **81**: 257–268, 2003.

planning and sampling activities during the Bight'98 regional survey, and seven local laboratories analyzed the sediment samples for toxicity to amphipods. Such a high level of cooperation in these studies provides great advantages in terms of cost sharing, methods standardization, and communication of the results. The active involvement of diverse agencies also presents a concern regarding the integrity of the data collected, however. Stakeholder agencies may have varying degrees of expertise with the range of sample types or analyses utilized in a regional survey, resulting in a greater chance for technical errors or noncomparable data. Consequently, there is an increased need for method standardization and quality assurance programs in cooperative regional surveys. Similar concerns regarding data comparability also arise when the results of environmental studies from diverse programs are combined for higher level analyses to describe spatial patterns or derive sediment quality guidelines (e.g. Long *et al.*, 1995; 1998).

Information regarding the comparability of sediment toxicity data is especially lacking. Although standard amphipod test methods are available (ASTM, 1992; USEPA, 1994a), few interlaboratory comparison studies have been conducted (Mearns *et al.*, 1986; Schlekat *et al.*, 1995; USEPA, 2001). While these studies demonstrate that amphipod toxicity tests are reliable and reproducible, they may not be applicable to data produced by a cooperative regional study for two reasons. First, highly experienced laboratories participated in the previous studies, which may not be the case in a cooperative study. For example, of the seven Bight'98 toxicology laboratories, only one had previously worked with the specific test species used (*Eohaustorius estuarius*),

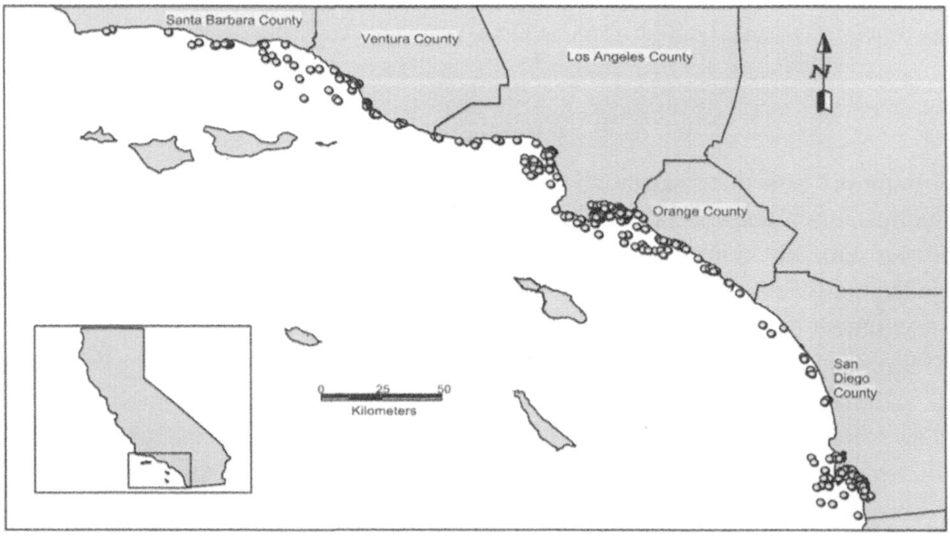

Figure 1. Location of stations analyzed for sediment toxicity during the Bight 98 regional survey of southern California.

and two of the laboratories had not previously conducted sediment toxicity tests with amphipods. Second, most of the existing interlaboratory test data were generated using either spiked or diluted sediment samples. The use of such highly manipulated samples may produce results different from those obtained with field-collected sediments that are handled differently.

This article presents the results of an interlaboratory comparison study that was conducted to accomplish three objectives. The first objective was to determine whether laboratories having limited sediment testing experience were able to perform the 10-day *Eohaustorius* test in accordance with pre-established test acceptability criteria. The second objective was to assess the degree of agreement among laboratories for sediment toxicity samples typical of those encountered in a cooperative regional survey. The third objective was to evaluate whether the participation of multiple laboratories introduced greater variability in test results compared to analyses by a single laboratory.

2. Methods

2.1 SAMPLE COLLECTION AND HANDLING

Sediment samples were collected on May 7, 1998 from four locations in Los Angeles/Long Beach Harbor, California (Figure 2). Station locations were selected based upon prior data to represent a gradient of toxicity ranging from highly toxic (< 50% survival) to moderately toxic (approximately 70% survival). Elevated concentrations of metal and organic contaminants were present at each of the sites,

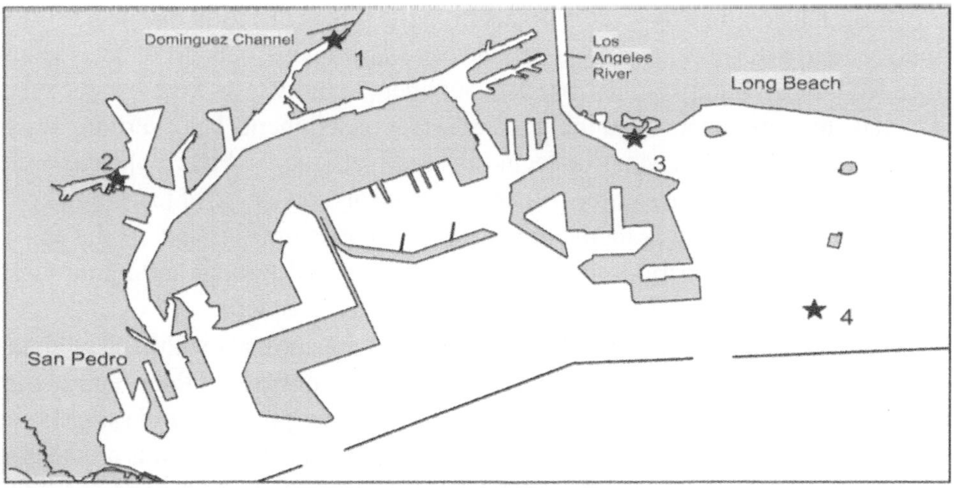

Figure 2. Location of sites sampled in Los Angeles/Long Beach harbor for interlaboratory toxicity comparison.

although the greatest contamination was expected at stations 1 and 3, which receive inputs from large urban stormwater discharges and also from port activities. Station 1 is downstream of a USEPA Superfund site for DDT contamination and also contains elevated concentrations of metals (copper), chlordane, PAHs, and total organic carbon. Station 2 was expected to contain higher concentrations of PCBs (from shipyard operations), relative to the other stations. The principal contaminants at station 3 were metals (lead and zinc) and PAHs. Station 4 was located distant from specific sources of contaminants and was expected to contain the lowest concentrations of contaminants, relative to the other stations. The samples collected for this study were not chemically analyzed.

Sediment samples were collected with a 0.1 m^2 modified Van Veen grab. Multiple grab samples were taken at each station to provide 12 L of surface sediment (top 4 cm of grab). The sediment was placed in polycarbonate containers, chilled on ice, and transported to the laboratory for storage at 4°C in the dark. Each sediment sample was homogenized using an overhead mixer on the following day, split into subsamples, and then distributed to the testing laboratories within 48 hours. The test sediments were kept chilled during transport and storage at the testing laboratories.

2.2 TEST PROCEDURES

Sediment toxicity was determined using a 10-day amphipod survival test (USEPA, 1994a; ASTM, 1992). Test organisms, *Eohaustorius estuarius*, were obtained from Beaver Creek, Oregon. The animals were shipped by overnight courier to each laboratory, where they were acclimated under conditions of 20 g/kg salinity, aeration, constant illumination, and 15°C until the initiation of the test on May 12, 1998. Each laboratory received animals from the same collection batch and sorted the specimens to exclude gravid females and specimens outside of the recommended 3-5 mm length range from use. Sediment toxicity tests were conducted in 1 L glass test containers containing a 2 cm layer of sediment. The overlying water was filtered coastal seawater collected by each laboratory. Five replicates were tested for each sediment sample. A negative control (consisting of test animal collection site sediment) was included in each batch of samples tested. At the end of the 10 day exposure period, the sediment was screened through a 0.5 mm-mesh screen and the number of surviving amphipods was recorded.

A cadmium reference toxicity test was conducted concurrently with each sediment toxicity test. The aqueous phase reference toxicant test consisted of three replicates of five concentrations (0.3, 1.0, 3.2, 5.6, and 10.0 mg Cd/L) plus a control sample. The test concentrations were prepared from a common stock solution that was provided to each laboratory. The number of surviving animals was recorded at the end of 4 d and the LC_{50} (Lethal Concentration, 50% mortality) was calculated. Statistical analysis was conducted using the procedures recommended by USEPA (1994a).

3. Results

All of the participating laboratories met the test acceptability criteria for the sediment and reference toxicant tests, which included (USEPA, 1994a): ≥ 90% survival in controls, water quality parameters within the tolerance range of the test species, and reference toxicant LC_{50} within 2 SD of prior tests. Each laboratory obtained nearly 100% survival in the collection site control sediment and recorded at least 90% survival for the reference toxicant seawater control. The laboratories reported that all of the experiments were conducted within the parameters of the test protocol (including water quality) and that no tests had to be repeated.

3.1 INTERLABORATORY COMPARABILITY

3.1.1 Reference Toxicant

Results of the reference toxicant tests varied approximately five-fold among laboratories. The cadmium LC_{50} for each laboratory (calculated using the Spearman-Karber method) ranged from 1.8 to 9.4 mg/L (Figure 3). The most sensitive test result (lowest LC_{50}) was reported by Laboratory 1; its LC_{50} was significantly less

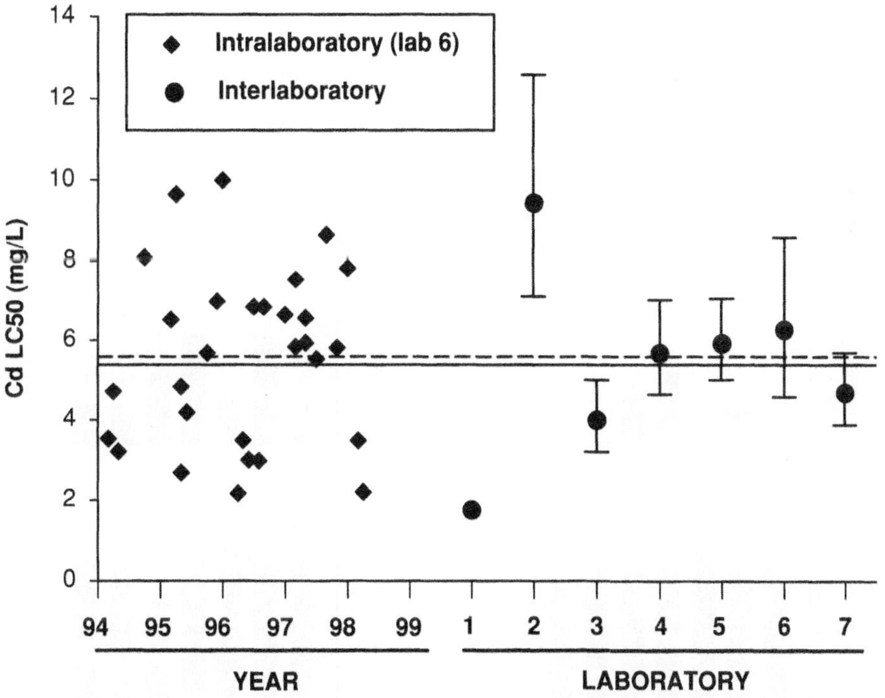

Figure 3. Cadmium reference toxicant test results from the interlaboratory comparison (circles) and from multiple tests conducted by laboratory 6. Error bars indicate the 95% confidence limits. The 95% confidence limits for laboratory 1 are 1.6–1.9 mg/L. Means for the interlaboratory and intralaboratory data are shown by the solid and dashed lines, respectively.

than the values reported by the other laboratories. The LC_{50} for Laboratory 2 was significantly greater than five of the other laboratories. The 95% confidence limits for the results from the other laboratories overlapped one another, indicating that the results were similar. None of the results were classified as an outlier, since all were within two standard deviations of the mean value (USEPA, 1994a). However, inclusion of the data for Laboratory 1 inflated the standard deviation to 2.4. Laboratory 1 would have been classified as an outlier if the standard deviation based on laboratories 2–7 (1.9) was used to establish the control limits.

The reference toxicant test results reported by the interlaboratory study participants were similar to the results of multiple tests conducted by a single laboratory (Laboratory 6) over a period of several years (Figure 3). Both the intralaboratory and interlaboratory LC_{50} data sets had similar means (5.6 mg/L and 5.4 mg/L, respectively) and ranges (1.8–9.4 and 2.2–10, respectively), indicating that testing by multiple laboratories had little effect on the accuracy or precision of the results.

3.1.2 Sediments

Each of the four sediments was consistently identified as toxic by the laboratories. A statistically significant difference in amphipod survival relative to the control sample (t-test, $p < 0.05$) was obtained for all but one of the 28 samples evaluated in the study (4 sediments x 7 laboratories). High mortality (< 35% mean survival among laboratories) was produced by sediments from two stations, 1 and 3 (Figure 4, Table 1). Sediment from station 1 was most toxic; a mean survival of 11%

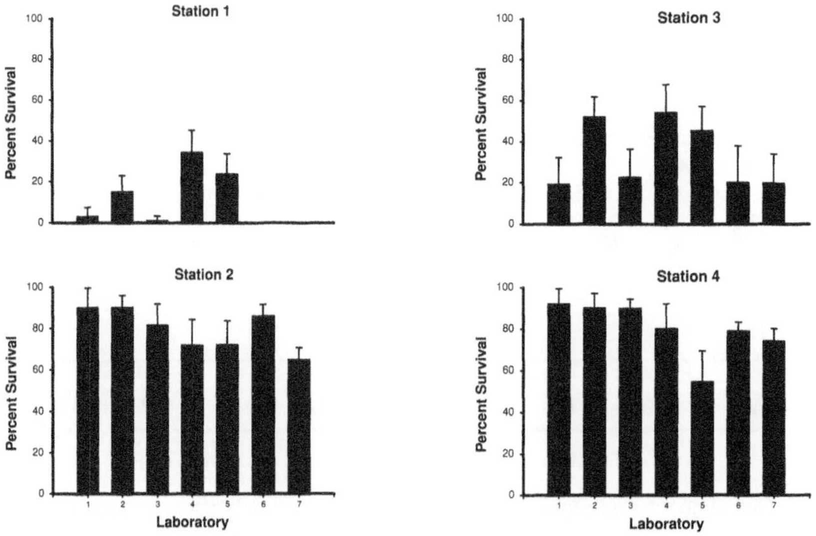

Figure 4. Survival results for *Eohaustorius estuarius* exposed to field sediments. Bars represent the mean (+ 1 standard deviation) of five replicates tested at each laboratory.

Table 1. Summary of toxicity test results for harbor sediment samples.

	Percent Survival			*Variance*	
Station	*Mean*	*Range*	*95% CI*	*Intralab*	*Interlab*
1	11	0–34	±13	41	188
2	79	65–90	±9	80	98
3	33	19–54	±15	180	269
4	80	55–92	±12	74	169

was obtained for this sample and two laboratories reported 0% survival. Low-moderate toxicity was detected in the remaining stations (2 and 4). The laboratories reported 55–92% survival in tests with these samples (Figure 4).

There was little correspondence between the reference toxicant results and the relative response to sediment reported by each laboratory. For example, the reference toxicant results indicated that the amphipod test conducted by Laboratory 1 was significantly more sensitive than the other laboratories' tests, yet the sediment test results for Laboratory 1 were similar to those of most of the other laboratories (Figure 4).

A similar amount of variation among laboratories was present for each sample. Standard deviations for stations 1, 2, 3, and 4 were 14, 10, 16, and 13, respectively. The 95% confidence interval for survival ranged from 9–15% among the stations (Table 1), while an overall confidence interval of ± 12% was calculated based on the grand mean variance of the means for all samples. Thus, the percentage of survival measured by a laboratory is expected to lie within 12 percentage points of the true value 95% of the time.

The variation in the sediment test data attributable solely to interlaboratory vari ability could not be calculated because repeated analyses of the same sample by a single laboratory were not conducted. As an alternative, the pooled variance among replicates within each laboratory was compared to the pooled variance in mean survival among laboratories for each station. The interlaboratory variance was higher than the replicate variance for each of the four sediment samples (Table 1). Station 1 (the most toxic sample) showed the greatest difference, with the interlaboratory variance being 4.5 times greater than the within-replicate variance. The interlaboratory variance was 1.2 to 2.3 times greater than the replicate variance for the other three stations.

The individual sample results among laboratories were compared using three approaches, each reflecting a different method of data interpretation. For the first approach, the percent of survival results among laboratories was compared. Examination of the survival results indicated that some bias may have been present in the data from two laboratories. Laboratory 7 reported the lowest or second lowest survival rate for each of the sediment samples and the results for Laboratory 2 were within the top three for each sample (Table 2, Figure 4). These differences were not statistically

Table 2. Laboratories arranged by order of survival results. Laboratories not significantly different from one another (Tukey pairwise comparison, $p > 0.05$) are underscored.

Station	*Laboratory Number*						
	Highest Survival						*Lowest Survival*
1	4	5	2	1	3	6	7
2	2	1	6	3	5	4	7
3	4	2	5	3	6	7	1
4	1	2	3	4	6	7	5

significant, however. Analysis of variance followed by Tukey pairwise comparisons indicated that the results for Laboratories 2 and 7 were similar to those for most of the other laboratories (Table 2). Significant differences in the percent of survival were detected between laboratories in 25% of the 84 possible pairwise comparisons, but no consistent pattern was observed among laboratories in these differences. Only one case (Laboratory 5 for station 4) was encountered where the results for a laboratory were different from all other laboratories (Table 2).

The second evaluation approach examined the ability of the laboratories to assess the relative toxicity of the four samples. The sediments were assigned ranks based upon the survival results. Each of the seven laboratories ranked stations 1 and 3 as the most toxic and next most toxic samples, respectively (Table 3). Four of the laboratories ranked the sediments in exactly the same order and a fifth only differed in that stations 2 and 4 were tied in the rankings. The Kendall coefficient of concordance based upon these data was 0.91, indicating a high level of agreement ($p < 0.01$) among laboratories.

The final assessment approach examined the ability of the laboratories to classify the degree of sediment toxicity using response thresholds (nontoxic, moderately toxic, and highly toxic) typical of regional monitoring programs. The agreement among laboratories in classifying the samples varied and was dependent upon the magnitude of toxicity. All seven laboratories classified sediment from station 1 (the most toxic sediment) as highly toxic (Table 3). Relatively consistent results were also obtained for station 3; all seven laboratories classified this station as toxic, with five classifying it as highly toxic and two as moderately toxic. The classification results were more variable for stations 2 and 4; approximately half of the laboratories placed these sediments into the nontoxic category (Table 3). The mean survival rate (among laboratories) for these two stations was 80%, the same as the response threshold used to distinguish between nontoxic and moderately toxic samples.

Table 3. Classification of the test sediments based on either relative toxicity (rank) or response-based categories.

	Category by Laboratory[a]							*Category by Laboratory*[b]						
Station	*1*	*2*	*3*	*4*	*5*	*6*	*7*	*1*	*2*	*3*	*4*	*5*	*6*	*7*
1	4	4	4	4	4	4	4	H	H	H	H	H	H	H
2	2	1.5	2	2	1	1	2	N	N	N	M	M	N	M
3	3	3	3	3	3	3	3	H	M	H	M	H	H	H
4	1	1.5	1	1	2	2	1	N	N	N	M	M	M	M

[a] Rank is based on the relative survival percentage for the four stations tested by the laboratory, with a rank of 4 corresponding to the lowest survival (most toxic). In the case of ties (same survival for two stations) the mean of the ranks involved is assigned to each station (e.g., stations 2 and 4 for laboratory 2).

[b] N = nontoxic, M = moderately toxic (survival = 79–50% of control and significantly different from control), H = highly toxic (< 50% of control survival).

4. Discussion

This study has shown that the *E. estuarius* amphipod toxicity test can be conducted with a high degree of success and reproducibility, even among laboratories with varying levels of experience with the method. Each laboratory met the performance criteria specified by the protocol (USEPA, 1994a) and was able to discriminate statistically between the control sediment and moderately toxic field-collected sediments.

Variable results were obtained among some of the laboratories, as shown by statistically significant differences in the results (percent of amphipod survival) for the same sediment sample. However, the variability measured in this study was similar to that found in previous studies (Mearns *et al.*, 1986; Schlekat *et al.*, 1995). This finding is especially noteworthy considering that several of the laboratories participating in this study had limited or no experience conducting the test procedure. Laboratory 5 had no prior sediment testing experience, Laboratory 3 had no amphipod test experience, and Laboratory 1 had less than one year of amphipod test experience. The results for these laboratories were comparable to those from the remaining laboratories, which had at least five years of amphipod test experience. All of the laboratories had multiple years of aquatic toxicity test experience.

No specific cause was identified for the variability in sediment test results observed among laboratories. The lack of a strong bias in the interlaboratory results, indicated by the similarity of station rankings (Table 3), and the similarity of intralaboratory and interlaboratory reference toxicant results, indicates that laboratory-specific differences in organism sensitivity or test methods were not a major factor. The variability may have been due to factors such as changes in contaminant bioavailability due to sediment storage and handling or may reflect the inherent variability of the test organism response.

While these results indicate that precision within one test is greater than the precision within multiple tests conducted by different laboratories, the data cannot be used to determine whether this variability is introduced by multiple tests or by multiple laboratories. Multiple tests within one laboratory may produce a similar level of variability. A measurement of intralaboratory variability, obtained through the repeated measurement of the same sample, is needed to provide a comparison with the interlaboratory variability measured in this study. Such a study is difficult to conduct with sediments because toxicity may be altered by long-term sediment storage. An alternative approach for estimating intralaboratory variability is to simultaneously test splits of the same sample, using different personnel and batches of test organisms (USEPA, 1994b).

A degree of uncertainty must be expected for any study that employs tests that have variability (Mearns *et al.*, 1986). The significance of this variability depends on the way in which the data are used. If the results are used to rank or otherwise describe the relative toxicity of multiple stations, then there is good agreement among laboratories, especially regarding the identification of the most toxic stations ("hot spots"). For regional assessments, thresholds based upon percentage of survival are often used to identify one or more levels of toxic response (Fairey *et al.*, 1996; Long *et al.*, 1998). Our data, representing contaminated field sediments spanning the range of toxicity typically encountered, indicate that survival measurements are likely to vary among laboratories by approximately 12 percentage points from the mean value. This variability may alter the classification of a sediment sample whose true level of toxicity is near a threshold value. This uncertainty is not a unique problem to toxicity tests and can be minimized by stratified sampling designs that utilize the information from multiple sediment samples to characterize the extent and magnitude of toxicity within a region.

Acknowledgments

The authors wish to thank the laboratory personnel of the following organizations who conducted the bioassays for this study: Aquatic Bioassay and Consulting Laboratory (ABCL), University of California Davis Marine Pollution Studies Laboratory (MPSL), City of Los Angeles Environmental Monitoring Division, City of San Diego Ocean Monitoring Program, MEC Analytical Systems, Orange County Sanitation District, and SCCWRP. We thank Darrin Greenstein (SCCWRP) for coordinating the logistics of the exercise and Tim Mikel (ABCL) for assisting with the reference toxicant data analysis. We thank Molly Leecaster (SCCWRP) for her valuable input regarding statistical analysis. Dario Diehl (SCCWRP) as well as the skipper and crew of the *Ocean Sentinel* (County Sanitation Districts of Los Angeles County) ably assisted in the collection of field sediments. We are also grateful to Brian Anderson (MPSL) for sharing his insights into working with *Eohaustorius estuarius* as a test species.

References

ASTM (American Society for Testing and Materials): 1992, Guide for Conducting 10-day Static Sediment Toxicity Tests with Marine and Estuarine Amphipods. E1367-92. ASTM 1992 *Annual Book of Standards*, Vol. 11.04, Philadelphia, PA.

Anderson, B. S., Hunt, J. W., Tudor, S., Newman, J., Tjeerdema, R. S., Fairey, R. S., Oakden, J, Bretz, C., Wilson, C. J., LaCaro, F., Stephenson, M. D., Puckett, H. M., Long, E. R., Fleming, T., Summers, K.: 1997, Chemistry, Toxicity, and Benthic Community Conditions in Selected Sediments of the Southern California Bays and Estuaries. Final Report, State Water Resources Control Board, Sacramento, CA, 157 pp. + app.

Bay, S. M.: 1992, 'Temporal and Spatial Changes in Sediment Toxicity in Santa Monica Bay', in: Cross J. (ed), *Annual Report 1990–91 and 1991–92*, Southern California Coastal Water Research Project, Westminster, CA, pp. 81–87.

Bay, S. M., Greenstein, D. J., Jirik, A. W., Brown, J .S.: 1998, Southern California Bight 1994 Pilot Project: VI. Sediment Toxicity. Southern California Coastal Water Research Project, Westminster, CA, 65 pp.

Bight'98 Steering Committee: 1998, Southern California Bight 1998 regional monitoring survey (Bight'98): Coastal Ecology Workplan. Southern California Coastal Water Research Project, Westminster, CA, 66 pp.

Fairey, R. S., Bretz, C., Lamerdin, S., Hunt, J. W., Anderson, B. S., Tudor, S., Wilson, C. J., LaCaro, F., Stephenson, M .D., Puckett, H. M., Long, E. R.: 1996, Chemistry, Toxicity, and Benthic Community Conditions of Selected Sediments of the San Diego Bay Region. Final Report, State Water Resources Control Board, Sacramento, CA, 179 pp. + app.

Long, E. R.: 2000, 'Degraded Sediment Quality in U.S. Estuaries: A Review of Magnitude and Ecological Implications', *Ecological Applications* **10**, 338–349.

Long, E. R., Field, L. J., MacDonald, D. D.: 1998, 'Predicting Toxicity in Marine Sediments with Numerical Sediment Quality Guidelines', *Environmental Toxicology and Chemistry* **17**, 714–727.

Long, E. R., MacDonald, D., Smith, S., Calder, F.: 1995, 'Incidence of Adverse Biological Effects within Ranges of Chemical Concentrations in Marine and Estuarine Sediments', *Environmental Management* **19**, 81–97.

Mearns, A. J., Swartz, R. C., Cummins, J. M., Dinnel, P. A., Plesha, P., Chapman, P. M.: 1986, 'Interlaboratory Comparison of a Sediment Toxicity Test using the Marine Amphipod *Rhepoxynius abronius*', *Marine Environmental Research* **19**, 13–37.

Schlekat, C. E., Scott, K. J., Swartz, R. C., Albrecht, B., Antrim, L., Doe, K., Douglas, S., Ferretti, J. A., Hansen, D. J., Moore, D. W., Mueller, C., Tang, A.: 1995, 'Interlaboratory Comparison of a 10-day Sediment Toxicity Test Method using *Ampelisca abdita*, *Eohaustorius estuarius*, and *Leptocheirus plumulosus*', *Environmental Toxicology and Chemistry* **14**, 2163–2174.

Swartz, R.C., Cole, F.A., Schultz, D.W., DeBen, W.A.: 1986, 'Ecological Changes in the Southern California Bight near a Large Sewage Outfall: Benthic Conditions in 1980 and 1983', *Marine Ecology Progress Series* **31**, 1–13.

USEPA (U.S. Environmental Protection Agency): 1994a, Methods for Assessing the Toxicity of Sediment-associated Contaminants with Estuarine and Marine Amphipods. EPA/600/R-94/025. U.S. Environmental Protection Agency, Washington, DC.

USEPA (U.S. Environmental Protection Agency): 1994b, Methods for Measuring the Toxicity and Bioaccumulation of Sediment-Associated Contaminants with Freshwater Invertebrates. EPA/600/R-94/024. U.S. Environmental Protection Agency, Duluth, MN.

USEPA (U.S. Environmental Protection Agency): 2001, Methods for Assessing the Chronic Toxicity of Marine and Estuarine Sediment-associated Contaminants with the Amphipod *Leptocheirus plumulosus*. EPA/600/R-01/020. U.S. Environmental Protection Agency, Washington, DC.

MAKING PERFORMANCE-BASED CHEMISTRY WORK: HOW WE CREATED COMPARABLE DATA AMONG LABORATORIES AS PART OF A SOUTHERN CALIFORNIA MARINE REGIONAL ASSESSMENT

RICHARD GOSSETT[1*], RODGER BAIRD[2], KIMBERLY CHRISTENSEN[3], AND STEPHEN B. WEISBERG[4]

[1]*CRG Marine Laboratories, Inc., 2020 Del Amo Blvd., Suite 200, Torrance, CA 90501 USA;*
[2]*Los Angeles County Sanitation Districts, 1965 S. Workman Mill Rd, Whittier, CA 90601 USA;*
[3]*Orange County Sanitation Districts, P.O. Box 8127, Fountain Valley, CA 92728 USA;*
[4]*Southern California Coastal Water Research Project, 7171 Fenwick Lane, Westminster, CA 92683 USA*
*(author for correspondence, e-mail: crglabs@sbcglobal.net)

Abstract. Quality assurance procedures to ensure consistency among chemistry laboratories typically involves the use of standard methods and state certification programs that require laboratories to demonstrate their ability to attain generic performance criteria. To assess whether these procedures are effective for ensuring comparability when processing local samples with potentially complex matrices, seven experienced, state-certified laboratories participated in an intercalibration exercise. Each laboratory was permitted to use their typical methodology for quantifying PAH, PCB and DDT on shared samples collected from Santa Monica Bay and the Palos Verdes Shelf, two sites with a complex mix of constituents. In the initial intercalibration exercise, results from these laboratories differed by as much as an order of magnitude for all three chemical groups. Much, but not all, of the difference was attributable to differences in detection capability. A series of studies was conducted to identify the reasons for the observed differences, which varied among laboratories and included methodological differences, instrument sensitivity differences, and differing interpretations of chromatograms. Following these investigations and resulting modifications to laboratory procedures, the exercise was repeated. The average coefficient of variation among laboratories across all chemical parameters was reduced to less than 30%. Our results suggest that performance-based chemistry can produce comparable results, but the certification processes presently in place that focus on general laboratory procedures and simple matrices are insufficient to achieve comparability.

Keywords: Intercalibration exercise, DDT, PCB, PAH, Performance-based chemistry

1. Introduction

Many environmental assessments require measurement and compilation of sediment chemistry data from multiple laboratories, either to extend temporal records for trends assessment or to extend geographic scale for spatial assessment. Such data compilations assume a degree of comparability among laboratories, even though analytical personnel, methods, and instrumentation may vary. Dissimilarity in sampling and sub-sampling techniques, sample preparation, clean-up procedures, detection capabilities, and instrumental techniques can all lead to differences in analytical results.

Data comparability is enhanced when the chemical analyses are conducted by State-certified laboratories that use standardized methods. The certification process requires laboratories to demonstrate their ability to attain generic performance criteria, but only partially satisfies the presumption of data comparability in real world environmental assessments. Reliance on a priori performance demonstra-

Environmental Monitoring and Assessment **81**: 269–287, 2003.

tion with standardized methods may be unsatisfactory when new compounds are added to the analytical list, when detection goals are set below the demonstrated method capability, when sample matrix components confound the standardized methods, and when methods must be modified or replaced with non-standard procedures in order to meet project-specific goals.

Standard reference materials (SRMs) are often used to help bridge such gaps and demonstrate performance capability among laboratories. However, SRMs fall short of the goal because certified values are typically present only for a subset of the target analytes and may be influenced by the limitations of the methods used to generate the certified results in the first place. Although SRMs are an important component in this process, additional steps may be necessary to insure high quality and comparable data.

To assess whether existing quality assurance criteria and analytical methods are effective for ensuring comparability when processing local samples with complex matrices, seven laboratories participated in an intercalibration exercise conducted as part of the Southern California Bight 1998 Regional Marine Monitoring Survey. All the laboratories were certified by the State of California and had at least ten years of experience. Here we present the results of that intercalibration exercise, identify reasons for observed differences, and discuss procedures important to increasing regional comparability with difficult samples and analytes.

2. Methods

Sediment samples were collected from two locations in Southern California. The first sample was collected from a station on the Palos Verdes Shelf (i.e., LA County Sanitation District Station 7C; Stull *et al.*, 1986). This station was selected because it typically contains concentrations of DDT and PCB at the upper range as compared to other sediments in Southern California. The second sample was collected from Santa Monica Bay (i.e. City of Los Angeles Station E-6; Bascom, 1978). This station is known to have high concentrations of polynuclear aromatic hydrocarbons (PAH). Both stations have complicated organic matrices as a result of decades of ocean disposal of municipal wastewater and various benthic processes. After collection, the sediment samples were thoroughly homogenized, divided, and placed into cleaned glass jars with Teflon lids. They were then stored frozen at -20°C until distributed to the participating laboratories.

After receiving the frozen samples, each participating laboratory used their conventional sample preparation and analytical methods to measure PAH, DDT, and PCB. Extraction methods are presented in Table 1. Each laboratory was also allowed to use its own Gas Chromatographic (GC) Detector providing they could obtain a specified Method Detection Limit (MDL). GC Detectors used for this study included Electron Capture Detectors, Quadrupole Mass Spectrometers, and Ion Trap Mass Spectrometers.

Table 1. Extraction method and instrumentation used by each laboratory.

Laboratory Number	*Extraction Method*	*DDT/PCB Gas Chromatograph Detector*	*PAH Gas Chromatograph Detector*
1	Automated Solvent Extraction	Electron Capture Detector and Ion Trap Mass Spectrometer	Mass Spectrometer
2	Roller Table	Electron Capture Detector	Mass Spectrometer
3	Automated Solvent Extraction	Electron Capture Detector	Mass Spectrometer
4	Solvent Extraction	Electron Capture Detector	Mass Spectrometer
5	Roller Table	Electron Capture Detector	Mass Spectrometer
6	Microwave-Assisted Extraction	Mass Spectrometer (SIM Mode)	Mass Spectrometer
7	Automated Solvent Extraction	Electron Capture Detector	Mass Spectrometer

While laboratories were given considerable methodological freedom, there were several consensus-based restrictions that were placed to ensure consistent MDLs and reporting limits (RLs) among labs. First, a minimum sample size of 10 grams dry weight was specified because some techniques cannot handle larger sample sizes but a minimum of 10 grams was required to reach the MDL required for the survey. Second, the GC column was restricted to the equivalent of a J&W DB-XLB (Frame *et al.*, 1996) for DDT/PCB analysis and to a DB-5 for the analysis of PAH. For electron capture detection, DB-5 was recommended, but not required as the second column for verification purposes. Moreover, each laboratory was required to demonstrate that their chromatographic conditions, e.g. oven temperature ramping rate, separated 40 out of 41 congeners from our custom PCB Calibration Standard.

This study included three phases. The first phase was an initial intercalibration exercise designed to determine how well the laboratories compared using their existing sample preparation and analytical methods. The second phase involved a series of informal studies to identify which laboratory procedures were contributing most to observed differences among laboratories. For example, these laboratory studies included the distribution of one laboratory's extract to other laboratories to determine if the observed differences were due to analysis of the extracts versus the extraction procedure itself. The third and final phase involved redistribution of the original intercalibration samples to assess whether the laboratory improvements identified in the laboratory studies were effective in improving comparability of the chemistry data.

3. Results

For Total Detectable PAH, the differences among laboratories in the first intercalibration exercise were as high as an order of magnitude for both sediment samples (Tables 2 and 3). Most of these differences were attributable to differences in detection limits. For example, Laboratory No. 1 reported detectable values for less than 20% of the target compounds because most were below the detection limit used for the first intercalibration exercise. Whereas Laboratory No. 7 reported measurable concentrations for 100% of these compounds in both sediment samples because their MDL was much lower. However, detection limit differences did not explain all of the discrepancies, as exemplified by the order of magnitude difference for many of the compounds between Laboratories 3 and 4.

After the second intercalibration exercise, the laboratories only differed from the mean PAH value by about 30% in both sediment samples (Tables 4 and 5). The most notable changes were for Laboratories No. 1 and 4 that originally reported significantly lower values, and Laboratory No. 3 that reported higher values than the other laboratories after the first intercalibration exercise. The greater comparability among laboratories after the second intercalibration exercise was also appar-

Table 2. PAH results [ng/dry g or parts per billion (ppb)] for the first intercalibration exercise on sediments collected from station 7C. ND = Not detected, *** = Data not reported, Lab = Laboratory

Compound	*Lab-1*	*Lab-2*	*Lab-3*	*Lab-4*	*Lab-5*	*Lab-6*	*Lab-7*	*Mean*	*SD*	*%CV*
Naphthalene	ND	35	45	ND	31	58	42	42	10	25
2-Methylnaphthalene	ND	57	78	5	54	119	60	62	37	59
1-Methylnaphthalene	ND	23	32	ND	28	66	26	35	18	51
Biphenyl	ND	44	54	17	25	57	21	36	17	48
2,6-Dimethylnaphthalene	28	30	62	ND	39	64	38	44	16	36
Acenaphthylene	25	6	36	11	32	40	28	25	13	50
Acenaphthene	ND	ND	ND	ND	ND	ND	39	ND	***	***
2,3,5-Trimethylnaphthalene	ND	ND	ND	ND	ND	15	18	16	2	13
Fluorene	ND	7	9	ND	ND	20	6	10	6	63
Phenanthrene	ND	36	60	9	64	52	54	46	20	44
Anthracene	ND	ND	48	6	ND	49	34	34	20	58
1-Methylphenanthrene	ND	42	ND	ND	21	ND	15	26	14	55
Fluoranthene	ND	ND	53	12	57	64	63	50	22	43
Pyrene	43	255	374	20	109	108	151	151	124	82
Benz[a]anthracene	ND	ND	79	9	47	49	47	46	25	53
Chrysene	ND	ND	67	9	53	25	63	43	25	58
Benzo[b]fluoranthene	ND	ND	292	14	160	61	168	139	108	77
Benzo[k]fluoranthene	ND	ND	104	10	55	64	83	63	35	56
Benzo[e]pyrene	ND	233	241	19	191	77	198	160	91	57
Benzo[a]pyrene	ND	ND	236	16	186	64	174	135	91	68
Perylene	41	359	312	20	165	138	101	162	129	80
Indeno[1,2,3-c,d]pyrene	ND	ND	26	ND	ND	53	93	57	34	59
Dibenz[a,h]anthracene	ND	ND	ND	ND	ND	ND	40	ND	***	***
Benzo[g,h,i]pyrene	ND	ND	91	ND	112	37	144	96	45	47
Total Detectable PAHs	137	1130	2300	177	1430	1280	1670	1160	781	67

Table 3. PAH results [ng/dry g or parts per billion (ppb)] for the first intercalibration exercise on sediments from station E6. ND = Not detected, *** = Data not reported, LAB = Laboratory

Compound	*LAB-1*	*LAB-2*	*LAB-3*	*LAB-4*	*LAB-5*	*LAB-6*	*LAB-7*	*Mean*	*SD*	*%CV*
Naphthalene	54	171	279	27	139	259	211	163	97	59
2-Methylnaphthalene	129	485	721	59	405	615	653	438	258	59
1-Methylnaphthalene	61	172	272	23	181	222	226	165	91	55
Biphenyl	233	756	1140	97	606	770	650	607	350	58
2,6-Dimethylnaphthalene	131	217	401	37	228	203	356	225	124	55
Acenaphthylene	ND	4	ND	ND	ND	ND	10	7	4	59
Acenaphthene	ND	15	46	ND	ND	ND	7	23	20	89
2,3,5-Trimethylnaphthalene	ND	19	ND	4	15	ND	106	36	47	130
Fluorene	ND	38	75	2	24	69	26	39	28	72
Phenanthrene	ND	137	469	9	109	112	95	155	160	103
Anthracene	ND	ND	111	13	19	18	39	40	41	102
1-Methylphenanthrene	ND	154	ND	ND	51	ND	31	79	66	84
Fluoranthene	76	ND	495	26	87	108	173	161	171	106
Pyrene	91	ND	1120	28	79	111	165	266	421	158
Benz[a]anthracene	ND	ND	284	30	65	38	100	103	105	101
Chrysene	60	ND	320	31	83	46	136	113	108	96
Benzo[b]fluoranthene	ND	ND	672	19	205	38	178	222	264	119
Benzo[k]fluoranthene	ND	ND	205	18	77	41	68	82	73	89
Benzo[e]pyrene	ND	ND	367	11	171	63	195	161	138	85
Benzo[a]pyrene	ND	ND	409	13	162	ND	189	193	163	85
Perylene	ND	249	183	5	72	32	59	100	95	95
Indeno[1,2,3-c,d]pyrene	ND	ND	ND	ND	69	23	155	82	67	81
Dibenz[a,h]anthracene	ND	ND	ND	ND	ND	38	42	40	3	8
Benzo[g,h,i]pyrene	ND	ND	60	ND	109	30	163	90	58	64
Total Detectable PAHs	835	2420	7630	453	2960	2840	4030	3020	2380	79

Table 4. PAH results [ng/dry g or parts per billion (ppb)] for the second intercalibration exercise on sediments collected from station 7C. ND = Not detected, *** = Data not reported, LAB = Laboratory

Compound	*LAB-1*	*LAB-2*	*LAB-3*	*LAB-4*	*LAB-5*	*LAB-6*	*LAB-7*	*Mean*	*SD*	*%CV*
Naphthalene	***	32	28	34	27	11	***	27	9	35
2-Methylnaphthalene	45	54	63	56	51	54	***	54	6	11
1-Methylnaphthalene	***	19	29	20	29	ND	***	24	5	22
Biphenyl	49	26	39	47	27	33	***	37	10	26
2,6-Dimethylnaphthalene	82	31	75	79	27	28	***	53	27	54
Acenaphthylene	***	15	42	87	43	12	***	40	30	76
Acenaphthene	***	ND	ND	7	ND	ND	***	ND	***	***
2,3,5-Trimethylnaphthalene	***	ND	ND	19	ND	14	***	17	4	22
Fluorene	***	3	21	13	ND	ND	***	13	9	72
Phenanthrene	ND	71	53	58	69	66	***	63	8	12
Anthracene	***	20	39	44	24	21	***	29	11	38
1-Methylphenanthrene	***	25	ND	ND	ND	23	***	24	1	6
Fluoranthene	51	68	55	39	72	75	***	60	14	23
Pyrene	144	215	137	138	172	168	***	162	30	18
Benz[a]anthracene	48	46	61	56	30	60	***	50	12	24
Chrysene	36	58	63	78	74	63	***	62	15	24
Benzo[b]fluoranthene	***	163	267	103	53	95	***	136	83	61
Benzo[k]fluoranthene	***	49	80	63	50	95	***	67	20	29
Benzo[e]pyrene	97	124	193	131	62	113	***	120	43	36
Benzo[a]pyrene	76	141	203	109	67	52	***	108	57	52
Perylene	127	259	227	237	140	142	***	189	58	31
Indeno[1,2,3-c,d]pyrene	***	35	ND	ND	88	79	***	67	28	42
Dibenz[a,h]anthracene	***	18	ND	ND	ND	ND	***	ND	***	***
Benzo[g,h,i]pyrene	28	99	75	ND	75	91	***	74	28	38
Total Detectable PAHs	***	1570	1750	1420	1180	1300	***	1440	224	15

Note: Laboratory No. 7 did not report any results for the second intercalibration exercise.

Table 5. PAH results [ng/dry g or parts per billion (ppb)] for the second intercalibration exercise on sediments from station E6. ND = Not detected, *** = Data not reported, LAB = Laboratory

Compound	*LAB-1*	*LAB-2*	*LAB-3*	*LAB-4*	*LAB-5*	*LAB-6*	*LAB-7*	*Mean*	*SD*	*%CV*
Naphthalene	173	162	170	191	139	193	***	171	20	12
2-Methylnaphthalene	388	435	480	532	336	525	***	449	78	17
1-Methylnaphthalene	***	145	185	166	153	144	***	159	17	11
Biphenyl	650	644	850	800	535	796	***	712	121	17
2,6-Dimethylnaphthalene	365	212	255	343	214	269	***	276	65	23
Acenaphthylene	***	8	ND	ND	ND	ND	***	ND	***	***
Acenaphthene	***	ND	25	15	ND	ND	***	20	7	35
2,3,5-Trimethylnaphthalene	***	22	ND	119	47	ND	***	62	50	81
Fluorene	ND	25	49	40	39	52	***	41	11	26
Phenanthrene	114	131	145	130	142	141	***	134	11	7
Anthracene	***	33	34	58	41	29	***	39	11	29
1-Methylphenanthrene	ND	62	27	68	73	128	***	71	36	51
Fluoranthene	183	280	150	135	146	183	***	179	53	30
Pyrene	211	196	155	230	125	185	***	184	38	21
Benz[a]anthracene	93	126	145	118	37	114	***	105	38	36
Chrysene	115	88	120	152	127	145	***	124	23	18
Benzo[b]fluoranthene	***	164	330	179	60	92	***	165	105	63
Benzo[k]fluoranthene	***	63	103	167	60	90	***	97	43	45
Benzo[e]pyrene	117	115	155	183	51	115	***	123	44	36
Benzo[a]pyrene	94	109	195	191	52	65	***	118	62	52
Perylene	ND	91	78	110	70	26	***	75	31	42
Indeno[1,2,3-c,d]pyrene	***	44	ND	ND	88	66	***	66	22	33
Dibenz[a,h]anthracene	***	26	ND	ND	ND	ND	***	ND	***	***
Benzo[g,h,i]pyrene	34	100	ND	ND	80	97	***	78	30	39
Total Detectable PAHs	***	3280	3650	3930	2610	3450	***	3390	494	15

Note: Laboratory 7 No. did not report any results for the second the intercalibration exercise.

ent for individual compounds as indicated by the median Coefficient of Variation (CV) that was greater than 60% after the first exercise (Tables 2 and 3), but was less than 30% after the second exercise (Tables 4 and 5).

Similar patterns in the chemistry results for Total Detectable PCB and Total Detectable DDT (Tables 6–9) were found when comparing data from the first intercalibration exercise with those from the second intercalibration exercise. After the first exercise, there was about a factor of 3 difference among laboratories for both sediment samples. After the second intercalibration exercise, this difference was generally less than a factor of 2 between the highest and lowest values reported. The largest change was in the 4,4'-DDE values because each laboratory experienced different problems with measuring these compounds at such high concentrations found in these sediments. For example, one laboratory found that they were not diluting the sample extract sufficiently to bring the peak size within the linear range of the detector. Another laboratory had to modify its extraction and clean-up procedures to adjust to the higher concentrations.

4. Discussion

This study was a precursor to a regional survey of sediment chemical concentrations in southern California, similar to that of Schiff (2000). It was conducted to determine whether multiple laboratories using different methods could produce data of sufficient similarity, or if it was necessary instead to have a single laboratory conduct all analyses. While there were considerable differences prior to the intercalibration exercise, it was possible to at least partially resolve these differences and achieve more comparable results using a performance-based approach.

As part of the National Status and Trends Program, the National Institute of Standards and Technology (NIST) was funded by the National Oceanic and Atmospheric Adminstration (NOAA), and later by the U.S. EPA, to organize trace organic intercalibration exercises using marine matrix materials. To determine whether all our laboratories had achieved comparability, we established performance criteria based on the NIST intercalibration exercises. To indicate acceptable performance according to the NIST criteria, results from each laboratory needed to be within 40% of the mean value for 80% of the sentinel compounds. We were able to achieve this goal, even with samples that were selected to be more challenging than average. Still it is important to recognize that we did observe larger differences for some compounds. We concluded that it is unrealistic to expect that that all compounds will be measured within 40% of the mean even within the most rigorous performance-based exercise.

Resolving pre-existing differences among laboratories resulted from many factors, including adjustments to sample extraction methods such as adding additional drying agents, improvement of clean-up procedures such as changing or increasing adsorbents, changing sample sizes, adjusting GC oven parameters for im-

Table 6. DDT and PCB results [ng/dry g or parts per billion (ppb)] for the first intercalibration exercise on sediments from station 7C. ND = Not detected, *** = Data not reported, LAB = Laboratory

Compound	*LAB-1*	*LAB-2*	*LAB-3*	*LAB-4*	*LAB-5*	*LAB-6*	*LAB-7*	*Mean*	*SD*	*%CV*
4,4'-DDE	6570	12900	***	17100	16300	5930	4830	10600	5500	52
4,4'-DDD	410	886	***	1090	8	384	285	510	401	79
4,4'-DDT	547	683	***	1490	480	185	483	645	447	69
2,4'-DDE	820	1870	***	2060	769	769	801	1180	610	52
2,4'-DDD	113	413	***	363	328	73	143	239	146	51
2,4'-DDT	7	26	***	ND	352	12	23	84	150	179
Total Detectable DDTs	8460	16800	***	21900	19500	7350	6560	13400	6760	50
PCB 18	ND	18	***	23	12	10	13	15	5	34
PCB 28	25	49	***	45	33	12	42	35	14	41
PCB 52	46	91	***	98	61	55	88	73	22	30
PCB 49	30	65	***	64	39	17	60	46	20	44
PCB 44	56	80	***	70	46	31	60	57	17	30
PCB 37	ND	230	***	ND	163	10	3	102	113	111
PCB 74	29	57	***	54	34	35	58	44	13	30
PCB 70	78	150	***	104	71	52	137	99	39	39
PCB 66	61	129	***	73	79	25	90	76	34	45
PCB 101	12	95	***	73	302	50	73	101	102	102
PCB 99	24	58	***	49	22	31	52	39	15	39
PCB 119	ND	ND	***	ND	ND	ND	2	ND	***	***
PCB 87	18	ND	***	28	440	24	44	111	184	166
PCB 110	24	335	***	ND	90	46	85	116	125	108
PCB 81	ND	ND	***	ND	ND	10	38	24	20	83
PCB 151	ND	ND	***	30	20	12	14	19	8	43
PCB 77	ND	ND	***	8	24	8	2	10	10	91

continued on the following page

Table 6—*Continued from previous page.* DDT and PCB results [ng/dry g or parts per billion (ppb)] for the first intercalibration exercise on sediments from station 7C.
ND = Not detected, *** = Data not reported, LAB = Laboratory

Compound	*LAB-1*	*LAB-2*	*LAB-3*	*LAB-4*	*LAB-5*	*LAB-6*	*LAB-7*	*Mean*	*SD*	*%CV*
PCB 149	24	54	***	50	12	28	40	35	16	47
PCB 123	ND	ND	***	9	3	11	ND	8	4	58
PCB 118	60	102	***	94	10	48	87	67	35	52
PCB 114	ND	ND	***	ND	ND	9	***	ND	***	***
PCB 168/153	23	97	***	83	37	34	99	62	35	55
PCB 105	ND	85	***	62	21	32	49	50	25	50
PCB 138	ND	117	***	ND	314	26	975	358	429	120
PCB 158	ND	ND	***	ND	7	21	6	11	8	74
PCB 187	9	22	***	17	9	14	22	15	6	38
PCB 183	ND	10	***	8	6	8	11	8	2	21
PCB 126	ND	ND	***	ND	ND	8	ND	ND	***	138
PCB 128	ND	20	***	14	7	12	16	14	5	34
PCB 167	ND	ND	***	5	ND	6	3	5	1	29
PCB 177	ND	12	***	ND	4	9	10	9	3	37
PCB 200	ND	3	***	18	ND	5	***	8	8	94
PCB 156	ND	ND	***	7	ND	13	15	12	4	36
PCB 157	ND	ND	***	ND	ND	7	1	4	4	97
PCB 180	11	40	***	1	18	19	48	23	18	78
PCB 170	ND	19	***	ND	13	14	21	17	4	24
PCB 169	ND	ND	***	ND	13	9	ND	7	6	88
PCB 189	ND	5	***	3	ND	6	2	4	2	50
PCB 194	ND	9	***	4	6	12	2	7	4	57
PCB 206	ND	ND	***	6	4	6	21	9	8	81
Total Detectable PCBs	529	1950	***	1100	1920	785	1390	1280	586	46

Note: Laboratory No. 3 did not report any results for PCB or DDT in the first intercalibration exercise.

Table 7. DDT and PCB results [ng/dry g or parts per billion (ppb)] for the first intercalibration exercise on sediments from station E6. ND = Not detected, *** = Data not reported, LAB = Laboratory

Compound	*LAB-1*	*LAB-2*	*LAB-3*	*LAB-4*	*LAB-5*	*LAB-6*	*LAB-7*	*Mean*	*SD*	*%CV*
4,4'-DDE	207	316	***	474	123	188	151	254	123	48
4,4'-DDD	17	26	***	35	21	10	26	24	8	36
4,4'-DDT	2	67	***	79	42	17	10	34	30	89
2,4'-DDE	25	54	***	30	23	13	21	27	14	51
2,4'-DDD	26	56	***	107	67	5	14	46	38	84
2,4'-DDT	1	ND	***	ND	6	16	4	7	6	97
Total Detectable DDTs	277	519	***	725	283	249	226	378	182	48
PCB 18	ND	7	***	12	6	11	8	9	2	26
PCB 28	18	17	***	22	20	ND	27	21	4	18
PCB 52	26	24	***	39	29	31	40	30	7	22
PCB 49	15	18	***	25	17	3	24	17	7	42
PCB 44	45	26	***	27	19	14	34	26	11	42
PCB 37	ND	29	***	21	15	22	ND	22	5	25
PCB 74	ND	16	***	17	12	13	25	17	5	27
PCB 70	24	29	***	37	30	15	52	31	12	37
PCB 66	35	42	***	40	32	9	39	33	12	37
PCB 101	50	52	***	44	84	29	64	53	17	33
PCB 99	19	16	***	15	14	13	35	20	8	41
PCB 119	ND	ND	***	ND	ND	ND	2	ND	***	***
PCB 87	16	26	***	26	22	23	33	24	5	22
PCB 110	18	63	***	7	35	33	66	38	22	58
PCB 81	ND	ND	***	ND	ND	14	23	19	7	35
PCB 151	ND	ND	***	17	15	9	13	13	3	25
PCB 77	ND	ND	***	8	ND	7	2	6	4	64

continued on the following page

Table 7—*Continued from previous page.* DDT and PCB results [ng/dry g or parts per billion (ppb)] for the first intercalibration exercise on sediments from station E6. ND = Not detected, *** = Data not reported, LAB = Laboratory

Compound	*LAB-1*	*LAB-2*	*LAB-3*	*LAB-4*	*LAB-5*	*LAB-6*	*LAB-7*	*Mean*	*SD*	*%CV*
PCB 149	25	31	***	40	14	25	37	29	9	30
PCB 123	ND	ND	***	6	4	12	ND	7	4	61
PCB 118	48	50	***	56	8	32	74	45	21	46
PCB 114	ND	ND	***	ND	ND	11	***	ND	***	***
PCB 168/153	27	62	***	61	21	32	100	50	27	55
PCB 105	ND	36	***	35	15	21	29	27	8	30
PCB 138	36	50	***	49	8	33	80	45	23	51
PCB 158	ND	6	***	ND	5	11	67	22	30	135
PCB 187	ND	16	***	15	5	15	22	15	6	37
PCB 183	ND	9	***	7	4	9	11	8	2	29
PCB 126	ND	ND	***	ND	ND	19	1	10	13	134
PCB 128	ND	17	***	15	7	12	16	13	4	29
PCB 167	ND	ND	***	3	ND	8	3	4	3	61
PCB 177	ND	9	***	8	5	8	12	8	2	25
PCB 200	ND	4	***	21	ND	6	***	10	9	90
PCB 156	ND	ND	***	5	ND	14	12	9	4	44
PCB 157	ND	ND	***	ND	ND	10	1	4	5	121
PCB 180	10	32	***	1	43	18	43	25	16	65
PCB 170	ND	22	***	24	5	17	22	17	7	41
PCB 169	ND	ND	***	ND	ND	11	ND	ND	***	***
PCB 189	ND	4	***	1	ND	8	2	3	3	86
PCB 194	ND	13	***	7	7	11	3	8	3	42
PCB 206	ND	14	***	4	ND	6	9	8	4	53
Total Detectable PCBs	411	739	***	713	502	595	1030	657	200	30

Note: Laboratory No. 3 did not report any results for PCB or DDT in the first round of the intercalibration exercise.

Table 8. DDT and PCB results [ng/dry g or parts per billion (ppb)] for the second intercalibration exercise on sediments from station 7C. ND = Not detected, *** = Data not reported, LAB = Laboratory

Compound	*LAB-1*	*LAB-2*	*LAB-3*	*LAB-4*	*LAB-5*	*LAB-6*	*LAB-7*	*Mean*	*SD*	*%CV*
4,4'-DDE	***	11800	14800	11800	6680	8610	***	10700	3150	29
4,4'-DDD	***	791	994	693	382	398	***	652	262	40
4,4'-DDT	***	879	562	334	326	434	***	507	229	45
2,4'-DDE	***	1720	2780	1210	908	1230	***	1570	736	47
2,4'-DDD	***	300	383	317	102	264	***	273	105	38
2,4'-DDT	***	13	ND	ND	6.0	11	***	10	4	35
Total Detectable DDTs	***	15500	19500	14400	8410	10900	***	13800	4280	31
PCB 18	***	13	18	17	10	12	***	14	3	24
PCB 28	***	35	54	34	24	29	***	35	11	32
PCB 52	***	72	91	73	47	71	***	71	16	22
PCB 49	***	49	53	46	33	43	***	45	8	17
PCB 44	***	54	66	53	37	47	***	51	11	21
PCB 37	***	156	36	9	11	14	***	45	63	139
PCB 74	***	47	73	38	40	47	***	49	14	28
PCB 70	***	93	83	74	65	82	***	79	11	13
PCB 66	***	95	98	72	57	60	***	76	19	25
PCB 101	***	101	77	63	51	78	***	74	19	25
PCB 99	***	54	68	49	33	34	***	48	15	31
PCB 119	***	ND	ND	ND	ND	ND	***	ND	***	***
PCB 87	***	29	63	35	32	34	***	39	14	36
PCB 110	***	40	269	220	61	64	***	131	106	81
PCB 81	***	ND	ND	ND	ND	ND	***	ND	***	***
PCB 151	***	ND	22	23	8	16	***	17	7	41
PCB 77	***	ND	ND	ND	5	6	***	6	1	24

continued on the following page

Table 8—*Continued from previous page.* DDT and PCB results [ng/dry g or parts per billion (ppb)] for the second intercalibration exercise on sediments from station 7C. ND = Not detected, *** = Data not reported, LAB = Laboratory

Compound	*LAB-1*	*LAB-2*	*LAB-3*	*LAB-4*	*LAB-5*	*LAB-6*	*LAB-7*	*Mean*	*SD*	*%CV*
PCB 149	***	45	49	34	31	39	***	40	8	19
PCB 123	***	ND	9	6	8	7	***	7	1	19
PCB 118	***	94	85	57	60	77	***	77	14	18
PCB 114	***	ND	ND	ND	19	ND	***	19	***	***
PCB 168/153	***	71	71	59	56	54	***	62	8	13
PCB 105	***	53	44	46	51	50	***	49	4	8
PCB 138	***	126	22	ND	62	66	***	69	43	62
PCB 158	***	10	ND	ND	6	4	***	6	3	49
PCB 187	***	15	21	12	15	18	***	16	3	21
PCB 183	***	8	10	5	7	9	***	8	2	23
PCB 126	***	ND	ND	ND	ND	ND	***	ND	***	***
PCB 128	***	18	19	12	9	20	***	15	5	31
PCB 167	***	ND	ND	2	8	8	***	6	3	58
PCB 177	***	ND	11	9	8	8	***	9	1	16
PCB 200	***	ND	ND	4	ND	4	***	4	0.4	10
PCB 156	***	ND	11	10	ND	21	***	14	6	44
PCB 157	***	ND	ND	1	ND	ND	***	ND	***	***
PCB 180	***	28	35	16	31	33	***	28	8	27
PCB 170	***	15	21	10	11	15	***	14	4	29
PCB 169	***	ND	ND	ND	ND	ND	***	ND	***	***
PCB 189	***	2	ND	ND	ND	ND	***	ND	***	***
PCB 194	***	4	18	ND	7	10	***	10	6	62
PCB 206	***	ND	ND	ND	4	5	***	5	0.4	9
Total Detectable PCBs	***	1330	1500	1100	901	1080	***	1180	232	20

Note: Laboratory No. 1 and 7 did not report any results for PCB or DDT in the second round of the intercalibration exercise.

Table 9. DDT and PCB results [ng/dry g or parts per billion (ppb)] for the second intercalibration exercise on sediments from station E6. ND = Not detected, *** = Data not reported, LAB = Laboratory

Compound	*LAB-1*	*LAB-2*	*LAB-3*	*LAB-4*	*LAB-5*	*LAB-6*	*LAB-7*	*Mean*	*SD*	*%CV*
4,4'-DDE	***	390	332	345	164	219	***	290	94	33
4,4'-DDD	***	24	24	28	3	18	***	19	10	51
4,4'-DDT	***	49	69	70	45	25	***	52	19	36
2,4'-DDE	***	57	32	20	17	17	***	29	17	60
2,4'-DDD	***	55	64	78	49	14	***	52	24	46
2,4'-DDT	***	3	ND	ND	19	6	***	9	9	92
Total Detectable DDTs	***	579	521	541	297	299	***	447	138	31
PCB 18	***	12	11	8	9	9	***	10	2	15
PCB 28	***	17	24	19	19	18	***	19	3	13
PCB 52	***	29	46	33	28	31	***	34	7	21
PCB 49	***	18	20	20	15	16	***	18	2	13
PCB 44	***	23	26	23	18	18	***	22	4	17
PCB 37	***	35	16	6	20	20	***	19	11	55
PCB 74	***	15	15	18	12	16	***	15	2	14
PCB 70	***	32	30	34	27	28	***	30	3	10
PCB 66	***	33	33	35	28	23	***	30	5	16
PCB 101	***	59	88	40	87	46	***	64	22	35
PCB 99	***	31	69	23	15	15	***	31	22	73
PCB 119	***	ND	ND	ND	ND	ND	***	ND	***	***
PCB 87	***	19	26	24	24	19	***	22	3	14
PCB 110	***	15	56	64	23	38	***	39	21	53
PCB 81	***	ND	ND	ND	ND	ND	***	ND	***	***
PCB 151	***	16	ND	16	4	10	***	11	5	48
PCB 77	***	ND	ND	ND	ND	ND	***	ND	***	***

continued on the following page

Table 9—*Continued from previous page.* DDT and PCB results [ng/dry g or parts per billion (ppb)] for the second intercalibration exercise on sediments from station E6.
ND = Not detected, *** = Data not reported, LAB = Laboratory

Compound	*LAB-1*	*LAB-2*	*LAB-3*	*LAB-4*	*LAB-5*	*LAB-6*	*LAB-7*	*Mean*	*SD*	*%CV*
PCB 149	***	29	34	32	7	23	***	25	11	44
PCB 123	***	ND	5	5	4	8	***	5	2	28
PCB 118	***	46	46	51	10	40	***	39	17	43
PCB 114	***	ND	ND	ND	ND	ND	***	ND	***	***
PCB 168/153	***	47	50	53	39	29	***	44	10	22
PCB 105	***	27	24	33	28	22	***	27	4	16
PCB 138	***	41	14	47	33	40	***	35	13	36
PCB 158	***	6	ND	ND	8	8	***	8	1	16
PCB 187	***	14	16	14	11	15	***	14	2	12
PCB 183	***	11	7	7	7	12	***	9	3	30
PCB 126	***	ND	ND	ND	ND	ND	***	ND	***	***
PCB 128	***	23	17	14	6	9	***	14	7	50
PCB 167	***	ND	ND	3	ND	4	***	4	1	38
PCB 177	***	7	7	7	4	8	***	7	2	23
PCB 200	***	ND	ND	6	ND	ND	***	6	***	***
PCB 156	***	ND	6	7	ND	3	***	5	2	39
PCB 157	***	ND	ND	1	ND	ND	***	ND	***	***
PCB 180	***	24	22	ND	16	17	***	20	4	20
PCB 170	***	15	14	15	6	18	***	14	4	32
PCB 169	***	ND	ND	ND	***	ND	***	ND	***	***
PCB 189	***	ND	ND	ND	ND	ND	***	ND	***	***
PCB 194	***	10	9	8	4	7	***	8	2	29
PCB 206	***	7	ND	11	ND	5	***	8	3	33
Total Detectable PCBs	***	678	732	688	513	562	***	635	93	15

Note: Laboratory No. 1 and 7 did not report any results for PCB or DDT in the second round of the intercalibration exercise.

proved separation, and switching from older instrumentation to newer instrumentation with greater sensitivity. Another major factor affecting the results was interpretation of the chromatograms themselves. For example, in some cases a laboratory ignored a peak that other laboratories were identifying as a target analyte. In other cases, peaks were switched, or the procedure for drawing baselines used for integrating the peaks needed to be refined to provide consistency between laboratories.

A detailed description of the specific analytical factors identified that contributed to the variability has not been provided here since they are numerous and varied for each laboratory. Some of the larger factors controlling the variation were described earlier. The specific changes each laboratory made are not critical because another set of laboratories would probably encounter different issues and the critical factor for improving comparability among laboratories is the process itself. Equally important was providing the laboratories with experience handling extremely difficult matrices and a communication process that started with consensus on sample size, maximum MDL, and chromatographic separation, but extended to nearly every decision about sample processing alternatives. A good performance-based approach fosters communication among laboratories that would not normally occur in its absence. Another important factor in reducing inter-laboratory variability was the recognition among participants that a performance-based approach conducted on samples collected locally was worth the commitment. For many of the laboratories, the studies associated with achieving comparability consumed more effort than processing the actual samples from the regional survey that this exercise was intended to support. We believe that the increased knowledge and staff education gained through participation offset the extra cost of the time invested.

Our results suggest that performance-based chemistry can produce comparable results, even given a wide range of approaches and instrumentation. However, we also found that the certification process presently in place, which focuses only on general laboratory procedures and simple matrices, is insufficient to achieve comparability for all samples encountered in the field. While project-specific intercalibration exercises will probably always be necessary to achieve the level of comparability that we achieved here, there are several changes to the certification process that would improve performance-based chemistry on an ongoing basis. First, additional methods flexibility is needed since certified methods presently lag current technology. For instance, California's certified methods for PCB are still based on Aroclors, which presents a challenge when projects such as this one are based on congeners. Additionally, there needs to be a wider array of reference materials. All of the laboratories that participated in this study had previously passed NIST's SRM exercise and were State-certified in California. However, the available SRM material does not contain certified values for DDT and the matrix was limited to dried samples that did not test the laboratories ability to handle wet samples.

Acknowledgments

We would like to thank Dr. Eddy Zeng for his direction and outstanding efforts towards the success of this project and to the chemists at the seven participating laboratories for their hard work and dedication toward improving the quality of chemistry data produced in Southern California.

References

Bascom, W.: 1978, 'Life in the bottom'. SCCWRP Annual Report 1978. p. 57–80.

Frame, G. Cochran, J.W., and Bowadt, S.: 1996, 'Complete PCB Congener distribution for 17 Aroclor mixtures determined by three HRGC systems optimized for comprehensive, quantitative, congener analysis', *Journal of High Resolution Chromatography*, **19**, 657–688.

Schiff, K.C.: 2000, 'Sediment chemistry on the mainland shelf of the Southern California Bight', *Marine Pollution Bulletin* **40**, 268–276.

Stull, J.K., Baird, R.B., and Heesen, T.C.: 1986, 'Marine sediment core profiles of trace constituents offshore of a deep wastewater outfall', *Journal of the Water Pollution Control Federation* **58**, 985–991.

CHARACTERIZATION AND STATISTICAL MODELING OF BACTERIAL (*ESCHERICHIA COLI*) OUTFLOWS FROM WATERSHEDS THAT DISCHARGE INTO SOUTHERN LAKE MICHIGAN

GREG A. OLYPHANT[1*], JUDITH THOMAS[1], RICHARD L. WHITMAN[2], AND DENVER HARPER[3]

[1]*Department of Geological Sciences, Indiana University, Bloomington, IN, U.S.A.;*
[2]*United States Geological Survey, Lake Michigan Ecological Research Station, Porter, IN, U.S.A.;*
[3]*Indiana Geological Survey, Bloomington, IN, U.S.A.*
[*]*(author for correspondence, e-mail: olyphant@indiana.edu)*

Abstract. Two watersheds in northwestern Indiana were selected for detailed monitoring of bacterially contaminated discharges (*Escherichia coli*) into Lake Michigan. A large watershed that drains an urbanized area with treatment plants that release raw sewage during storms discharges into Lake Michigan at the outlet of Burns Ditch. A small watershed drains part of the Great Marsh, a wetland complex that has been disrupted by ditching and limited residential development, at the outlet of Derby Ditch. Monitoring at the outlet of Burns Ditch in 1999 and 2000 indicated that *E. coli* concentrations vary over two orders of magnitude during storms. During one storm, sewage overflows caused concentrations to increase to more than 10,000 cfu/100 mL for several hours. Monitoring at Derby Ditch from 1997 to 2000 also indicated that *E. coli* concentrations increase during storms with the highest concentrations generally occurring during rising streamflow. Multiple regression analysis indicated that 60% of the variability in measured outflows of *E. coli* from Derby Ditch (n = 88) could be accounted for by a model that utilizes continuously measured rainfall, stream discharge, soil temperature and depth to water table in the Great Marsh. A similar analysis indicated that 90% of the variability in measured *E. coli* concentrations at the outlet of Burns Ditch (n = 43) during storms could be accounted for by a combination of continuously measured water-quality variables including nitrate and ammonium. These models, which utilize data that can be collected on a real-time basis, could form part of an *Early Warning System* for predicting beach closures.

Keywords: *E. coli* bacteria, storm-period monitoring, statistical forecasting equations

1. Introduction

Public beaches in many coastal regions that have experienced urban and suburban development are threatened by high levels of fecal bacteria and associated contaminants. Analyses of water samples collected from the surf zone of public beaches in northwestern Indiana have revealed concentrations of *Escherichia coli* (*E. coli*) bacteria that exceed the federal limit for safe swimming (235 colony forming units (cfu) per 100 mL; U.S. EPA, 1986). In order to protect the health of beach visitors, health officials and beach managers are obligated to close affected beaches to swimming and bathing when the waters are deemed unsafe. Unfortunately, the standard methods for determining *E. coli* concentrations in water samples require at least 24 hours for culturing and counting of colony forming units. As a consequence, one can only know if the water was safe for swimming yesterday—not today.

An obvious source of bacterially contaminated lake water is discharges from streams that receive point and non-point pollution from contributing areas within their watersheds. In northern Indiana, there are three large watersheds and three

Environmental Monitoring and Assessment **81**: 289–300, 2003.

small watersheds that drain directly into Lake Michigan. The large watersheds contain urban, suburban and rural areas that contribute bacterial contamination to the stream network through a variety of point and non-point sources. Of particular concern in the large watersheds are periodic combined sewage overflows (CSO) that discharge large volumes of raw sewage into the drainage system. These overflows move down the drainage system as contaminant plumes and eventually discharge into the lake where they are dispersed by currents and waves. Under certain conditions, the contaminated water can reach public beaches along the shoreline and cause a serious threat to human health. The smaller watersheds drain portions of the once extensive Great Marsh. The Great Marsh is a large wetland complex situated between past and present shorelines of Lake Michigan. Over the past century much of the Great Marsh was drained by ditching, but a program of restoration in now in effect. Possible sources of bacterial contamination in the watersheds of the Great Marsh include residential developments that utilize onsite sewage treatment systems (septic fields and dry wells) and a population of mammals (mostly deer and raccoon) that has increased in the last few decades.

In 1995, an Interagency Technical Task Force on *E. coli* (referred to here as the "Task Force") was formed to begin addressing the problem of bacterial contamination that has led to numerous closings of public beaches along the Lake Michigan shoreline of Indiana. As a first step in developing an Early Warning System for unsafe bathing conditions, members of the Task Force have conducted studies of *E. coli* contamination in two streams that drain into Lake Michigan from Indiana (Figure 1). One of the streams (Burns Ditch) drains the largest watershed in the area and is subject to numerous CSOs. The other stream (Derby Ditch) drains part of the Great Marsh. There are no obvious point sources or septic systems that enter into the stream but *E. coli* concentrations are consistently high and the primary sources are assumed to be non-point sources such as riparian wildlife (Whitman *et al.*, 1991). The primary goal of this study was to characterize the variability of *E. coli* concentrations in streamflow during storm runoff events and to explore the possibility of developing empirical equations that can successfully predict outflows and concentrations of *E. coli* as a function of continuously measured hydrologic and water-quality variables. If such equations can be successfully developed, then outflows of bacterially contaminated streams could be predicted on a real-time basis. Such predictions could lead to improved forecasts of the threat to human health at nearby bathing beaches.

2. Monitoring Sites, Instrumentation, and Methods

Derby Ditch was studied first. This site was selected because the watershed was scheduled for restoration efforts (mainly ditch plugging with some internal drainage diversions), and data on *E. coli* outflows were acquired before and while the restoration took place. The drainage area of Derby Ditch is only about 25 km^2, and all of the drainage shown in Figure 1 is artificial. Although there are a few pockets of residential development within the watershed (Town of Beverly Shores), most of the watershed is within the Indiana Dunes National Lakeshore. When the study commenced, the parklands consisted mostly of ditched and degraded wetlands, and a very dense population of animals (especially deer) was found in the undeveloped areas. The outlet of Derby Ditch and its adjacent beaches were being monitored by the Task Force. Water samples from the outlet indicated that *E. coli* concentrations in the streamflow were commonly above the 235 cfu/100 mL threshold. In May 1997, a monitoring station was established at the upstream end of a culvert that discharges streamflow from the Derby Ditch watershed into Lake Michigan. The culvert is 90 cm in diameter. A Doppler water current meter and pressure transducer were placed inside the culvert to measure water current depth and velocity. From these measurements and the geometry of the culvert, volumetric discharge was computed. Electronic sensors for measuring specific conductance, water temperature, and turbidity were also placed within the water current by attaching them to the base of the corrugated pipe within the culvert. Water samples were collected once per week during the spring, summer, and autumn months of 1997 through 2000. These grab samples were collected by members of the Task Force and analyzed for their *E. coli* concentrations. Additional samples were collected using an auto sampler during ten storms that occurred in the spring and summer months of 1997 through 2000. Samples were collected in sterile containers. The samples were stored on ice until delivery to a laboratory within 6 hours of collection. The methods for measuring *E. coli* followed the membrane filtration (MF) techniques outlined in the American Public Health Associations volume of Standard Methods (APHA, 1998).

An additional monitoring station was established in the Great Marsh approximately 800 m upstream from the outlet. The monitoring station contained a monitoring well fitted with an electronic pressure transducer for measuring water levels, a tipping-bucket rain gauge, and an electronic sensor for measuring soil temperature.

As shown in Figure 1, Burns Ditch at its outlet discharges drainage from a 1,000 km^2 area that is mostly urbanized in its western portion but still mostly agricultural and undeveloped in its eastern portion. Also, as shown on the map,there are several locations within the watershed where Task Force members collected water samples weekly between April and October 1997–2000, for determination of *E. coli* concentrations. Water quality is generally poor within the stream system and CSOs occur on a regular basis when storm runoff from source areas within the watershed temporarily overwhelms the sewage system.

In April 1999, a monitoring station was established on the end of a dock at a private marina (Lefty's Landing) near the outlet of Burns Ditch. The station was designed to facilitate continuous monitoring of streamflow characteristics (river stage and velocity) and water quality parameters and to allow samples of streamflow to be collected using a hand-held depth sampling device and an auto-sampler. The

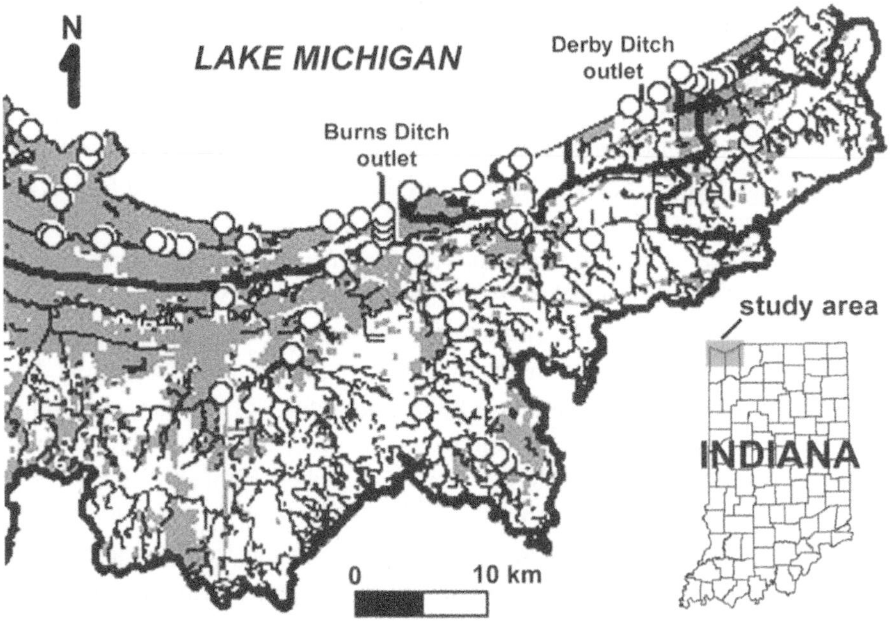

Figure 1. Map showing watersheds in northwestern Indiana that contribute streamflow to southern Lake Michigan. Watershed boundaries are indicated by thick black lines. White circles indicate locations where Task Force personnel collected water samples on a weekly basis for bacterial analysis, and grey shading represents areas of urban development. Burns Ditch outlet receives drainage from the large and partly urbanized watershed encompassing most of the mapped area. Derby Ditch outlet receives drainage from a much smaller and only moderately developed watershed that is part of the previously more extensive Great Marsh.

hand-held sampler could be adjusted to collect 100 mL samples simultaneously at 1/4, 1/2, and 3/4 of the total flow depth. Prior to each sampling, the total flow depth was determined using a scaled metal probe. The position of each sample intake was adjusted and the apparatus was placed into the stream for sample collection. The samples were then combined into a single composite sample prior to analysis for bacterial concentrations. The auto-sampler collected samples from a fixed point approximately 2 meters below the mean water surface of the stream. During the period of monitoring at Lefty's Landing, samples were collected for bacterial analysis during three separate storm periods. A major CSO occurred during one of those storm periods. During that period samples collected by the hand-held sampler and by the auto-sampler were analyzed for *E. coli* concentrations. This allowed a comparison to be made between the two sampling methods.

Stream velocity was measured using a Doppler current meter, and river stage was measured using an electronic pressure transducer that was suspended in the stream at a depth of approximately 2 meters below the mean water surface level. A detailed survey of the river cross-section at the point of measurement allowed development of a rating curve for the cross-sectional area of flow, and the area determinations were combined with the velocity measurements to provide a continuous record of volumetric discharge. A water-quality multiprobe was also suspended in the stream at a depth of 2 m. The multiprobe was fitted with sensors for measuring water temperature (Tw, °C), specific conductance (Spc, mS/cm), pH, oxidation/reduction potential (ORP, mV), dissolved oxygen (DO, mg/L), ammonium as nitrogen ($NH_4 - N$, mg/L), nitrate as nitrogen ($NO_3 - N$, mg/L), and turbidity (τ, NTU).

All electronic data presented in this paper were originally collected as hourly averages or totals. The hourly values are based on raw measurements taken at one-minute intervals for initial processing by programmable digital data loggers. In some cases the hourly data were further processed (averaged or totalized) to obtain daily values.

3. Results

3.1 STORM-PERIOD VARIABILITY OF *E. COLI* CONCENTRATIONS

The data plotted in Figure 2 exhibit the typical response of a small wetland watershed to a pronounced and short-term pulse of rainfall. The data were collected at the outlet of Derby Ditch when an isolated thunderstorm produced 1.5 cm of rainfall in one hour. Stream discharge increased sharply, then began a long slow decline. Samples of streamflow indicated that bacterial concentrations increased sharply during the rising phase of the discharge, but declined and exhibited only small variations once the discharge had peaked (Figure 2a). As shown in Figure 2b, the rise and fall of bacterial concentrations was accompanied by a similar fluctuation of turbidity, but specific conductance actually declined somewhat as the *E. coli* concentration peaked. These trends indicate that storm-period outflows are associated with increases of suspended solids (turbidity being a surrogate for total suspended solids) and decreases of dissolved solids (specific conductance being a surrogate for total dissolved solids). During this storm, the concentrations of *E. coli* in the streamflow dropped off quickly. Possible causes of such a response include die-off, and settling into bottom sediments. Since bacteria in streamflow tend to be associated with sediment, the reduction of sediment load (as inferred by the trend of turbidity) may have been responsible for the decline of bacteria in the outflow.

The other storms that we monitored at Derby Ditch showed similar relationships between discharge, *E. coli*, turbidity, and specific conductance. However, many of the other storms were less pronounced, with the rainfall occurring sporadically over a longer duration so hydrographs showing the relationships are less convincing.

The data presented in Figure 3 were collected at Lefty's Landing when a major storm, producing more than 4 cm of rainfall, caused a CSO in the interior of the watershed. *E. coli* concentrations were already above 2,000 cfu/100 mL when the sampling was initiated. By the end of the 24-hour period of sampling (which roughly corresponded to the period of peak discharge), the concentration of *E. coli* had increased to more than 25,000 cfu/100 mL. Turbidity increased sharply along with *E. coli* concentrations, peaked with the stream discharge, and then began to slowly decline (Figure 3b). Water temperature decreased during the storm and increased as the discharge receded. As shown in Figure 3c, nitrate and ammonium exhibited an inverse response, with nitrate increasing and ammonium decreasing during the main part of the storm. The decline of ammonium may have been a result of oxidation as the contaminant plume was transported down stream in the turbulent flow.

The extended periods of high turbidity and nitrate, both of which show a clear relationship to *E. coli* concentration during the rising phase of the storm discharge, indicated that the contaminant plume continued to travel through the study reach (albeit with lesser concentration) for a long time after sampling for bacterial concentrations had ceased. The other two storms that we monitored were much smaller in intensity and had lower *E. coli* concentrations, though during the latter storms bacterial concentrations were still well above the 235 cfu/100mL threshold. During these storms, ammonium concentrations remained below 0.5 mg/L, nitrate concentrations fluctuated between 2 and 6 mg/L, and turbidity never exceeded 40 NTU. Taken as a whole, these findings indicate that the electronically monitored water-quality variables are effectively tracking the variability of *E. coli* concentrations in the streamflow of Burns Ditch.

3.2 PREDICTIVE MODEL FOR *E. COLI* OUTFLOWS FROM DERBY DITCH

During the period of monitoring in the Derby Ditch watershed, the Task Force collected water samples for bacterial analysis every Thursday morning between the months of April and October. Simultaneous sampling by the Task Force and electronic data logging occurred on 88 occasions. Data shown in Figure 4 show how *E. coli* concentrations varied in relation to stream discharge, soil temperature

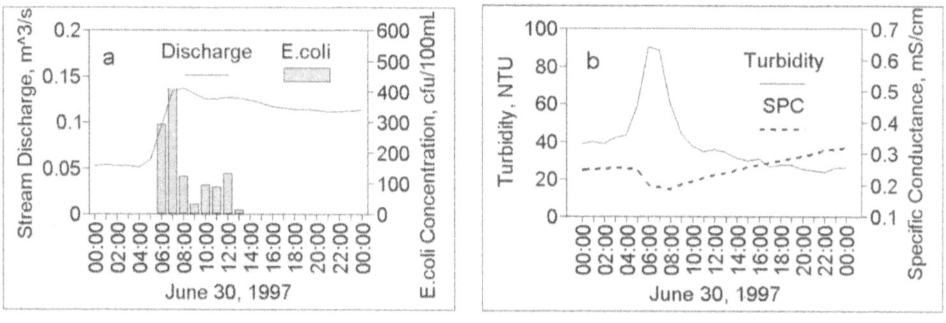

Figure 2. Graphs showing (a) stream discharge and *E. coli* concentrations in Derby Ditch streamflow, and (b) turbidity and specific conductance of streamflow during a brief but intense rainstorm in spring, 1997.

and water levels in the Great Marsh during the 1998 field season. During the period of study, there was a consistent tendency for bacterial concentrations to be relatively low in the spring, then increase in summer and remain high until autumn. This trend was similar to that of soil temperature as measured in the Great Marsh. However, the concentrations could spike at any time. Some of the spikes in Figure 4 are associated with storm discharges (flushing events), but others seem to be more closely related to periods when the water table in the Great Marsh dropped and the soil temperature increased (population explosions).

Because the variations in *E. coli* concentrations seemed to associate with a number of storm- and non-storm-related factors, a general model that included both kinds of variables was developed for the total bacterial outflows (product of water discharge and bacterial concentration) from Derby Ditch by Olyphant.

The hypothesized statistical regression model is of the following form:

$$lnE_t = b_0 + b_1 Tg_t + b_2 lnQ_t + b_3 Dtw_t + b_4 P_t + \rho e_{t-1} + v_t \qquad (1)$$

where E_t is the average daily outflow of E.coli in cfu/s, Tg_t is the average daily ground temperature (°C) as measured at the station in the Great Marsh, Q_t is the average daily discharge at the outlet (L/s), Dtw_t is average daily depth to water measured at the

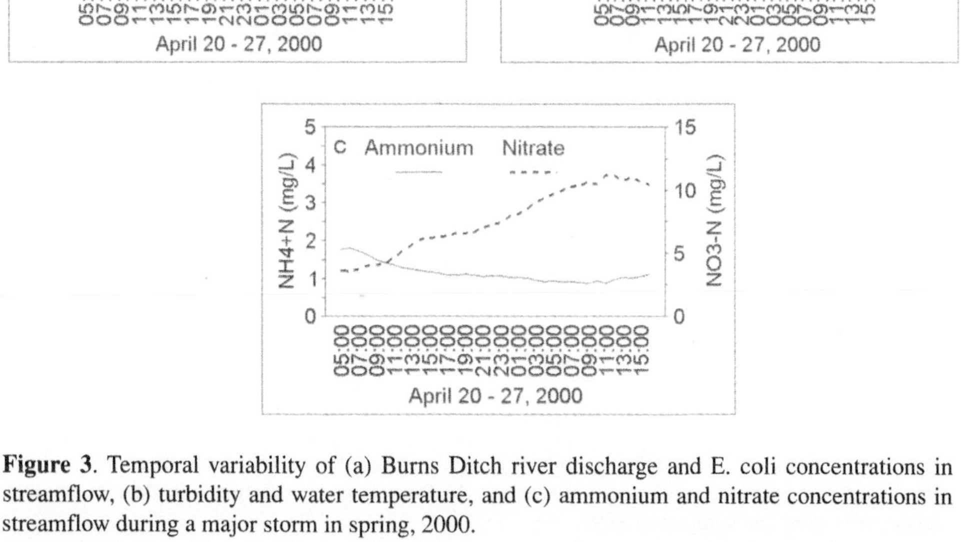

Figure 3. Temporal variability of (a) Burns Ditch river discharge and E. coli concentrations in streamflow, (b) turbidity and water temperature, and (c) ammonium and nitrate concentrations in streamflow during a major storm in spring, 2000.

station in the Great Marsh (m), and P_t is the total daily precipitation (cm) measured at the station in the Great Marsh. All of the independent variables are averages or totals for the 24-hour period preceding the time that the water samples were collected. The regression parameters $b_0 - b_4$ were estimated using generalized least-squares, as was the temporal autocorrelation coefficient ρ, because the data were collected as a (weekly) time series and are subject to serial autocorrelation. The last two terms in Equation 1 represent temporally correlated and random errors, respectively. An iterative procedure developed by Cochrane and Orcutt (1949) was utilized to converge on the best-fit parameters satisfying Equation 1.

The results of the regression analysis are summarized in Table 1. All of the regression parameters are statistically significant (95% confidence level) and the model accounts for 69% of the variability in the calculated mean daily outflows of *E. coli* determined from measurements taken over the four-year period of monitoring.

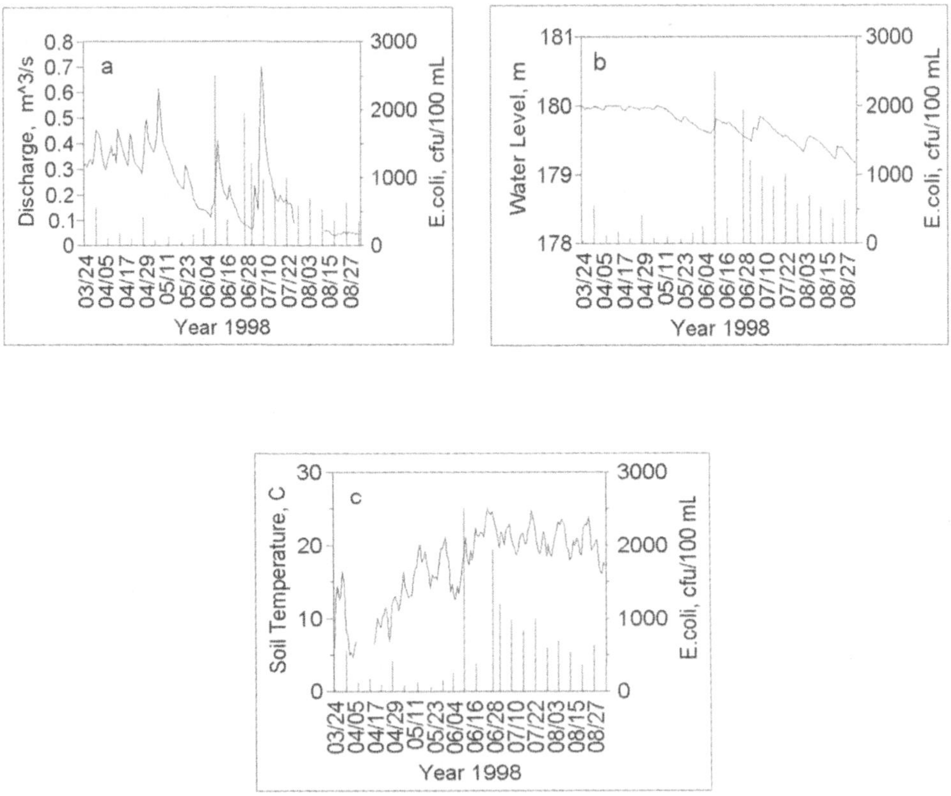

Figure 4. Graphs showing concentrations of E. coli in streamflow (bars) determined from weekly grab samples in relation to: (a) stream discharge, (b) water table elevation, and (c) soil temperature during the spring and summer of 1998.

Table 1. Parameter Estimates Associated with Equation 1

Parameter	b_0	b_1	b_2	b_3	b_4
Estimate	5.444	0.116	1.114	1.117	0.857
Standard Error	1.025	0.023	0.177	0.391	0.165
t-ratio	5.131**	5.051**	6.285**	2.855**	5.204**

Estimated autocorrelation coefficient, $\rho = 0.40$. Multiple correlation coefficient, $R = 0.83$. Total sample size, $N = 88$.

*******Estimated regression parameter is statistically different from zero at the 95 percent confidence level.*

3.3 PREDICTIVE MODEL FOR *E. COLI* CONCENTRATIONS AT THE OUTFALL OF THE LITTLE CALUMET RIVER AT BURNS DITCH

Since Burns Ditch at its outlet is the largest of all the streams that drain into Lake Michigan from Indiana, and because the water quality at the outlet is affected by CSOs, members of the Task Force wanted to explore the possibility of tracking contaminant outfalls in real time using surrogate variables. To this end, a multiple regression equation was developed that predicts *E. coli* concentrations in streamflow as a linear combination of the water-quality variables that are measured electronically with the multiprobe. The hypothesized model is of the following form:

$$lnEc_t = b_0 + b_1Tw_t + b_2pH_t + b_3ORP_t + b_4[NH_4^+N]_t + b_5[NO_3{-}N]_t + b_6\tau_t + \rho e_{t-1} + v_t \qquad (2)$$

where Ec_t is the *E. coli* concentration in streamflow at time t and the independent variables are hourly average values of the electronically measured water-quality variables. Dissolved oxygen (DO) was not included in the regression as it basically mimics the behavior of oxidation/reduction potential (ORP) in streamflow. The regression parameters $b_0 - b_6$ were estimated using generalized least-squares, as was the temporal autocorrelation coefficient ρ, because the data used in the regression analysis were collected during the three storm periods as hourly time-series. The last two terms in Equation 2 represent temporally correlated and random errors respectively. An iterative procedure developed by Cochrane and Orcutt (1949) was utilized to converge on the best-fit parameters satisfying Equation 2. The results of the regression analysis are presented in Table 2.

As indicated by the calculated t-ratios (Table 2), four of the six regression coefficients ($b_1 - b_6$) are statistically different from zero despite the relatively small degrees of freedom resulting from the limited sample size (N = 43). The multiple correlation coefficient (R = 0.95) is large indicating that the combination of electronically measured water-quality variables account for 90 percent of the observed variability in the measured *E. coli* concentrations. Water temperature and pH were not found to contribute significantly to the regression at the 90 percent confidence level (t-ratios < 1.69; Table 2). The coefficient relating ORP to *E. coli* concentra-

Table 2. Parameter Estimates Associated with Equation 2a

Parameter	b_0	b_1	b_2	b_3	b_4	b_5	b_6
Estimate	−81.221	0.311	7.165	0.055	1.475	1.003	0.021
Standard Error	46.406	0.189	4.267	0.028	0.587	0.238	0.004
t-ratio	−1.750*	1.644	1.679	1.986*	2.513**	4.208**	5.314**

Estimated autocorrelation coefficient, $\rho = -0.15$. Multiple correlation coefficient, $R = 0.95$. Total sample size, $n = 43$.

* *Estimated regression parameter is statistically different from zero at the 90 percent confidence level (two-tailed test).*
** *Estimated regression parameter is statistically different from zero at the 95 percent confidence level (two-tailed test).*

tions is significantly different from zero at the 90 percent confidence level, but not at the 95 percent confidence level. On the other hand, regression coefficients relating nitrate, ammonium, and turbidity to *E. coli* concentrations are all significantly different from zero at the 95 percent confidence level. Nitrate and ammonium are commonly associated with sewage, and turbidity (which is associated with suspended solids) has been shown to have a good correlation with bacterial concentrations in numerous other studies of contaminated streamflow.

4. Discussion

The results of this study provide some support to the frequently reported conclusion that bacterial concentrations in streamflow increase during storms (e.g. Kunkle and Meiman, 1967; Matson *et al.*, 1978; McDonald and Kay, 1981). At the outlet of Burns Ditch, *E. coli* concentrations rose exponentially during the rising phase of the water discharge when a major CSO occurred up-drainage in April, 2000. Based on the continuously measured water-quality variables, the bacterial concentrations in the streamflow may have remained high for a very long time after the discharge peaked. In contrast, the storm monitored at the outlet of Derby Ditch in June, 1997 exhibited a much shorter peak of bacterial concentration with a sharp decline that preceded the recession of the water discharge. This behavior is similar to that of streams in which bacterial loads are derived from bed scour and inflows from adjacent streambanks (cf. Kunkle, 1970; McDonald *et al.*, 1982).

Long-term monitoring at the outlet of Derby Ditch indicated that storms were not the only cause of high *E. coli* concentrations in streamflow. Indeed some of the highest concentrations of bacteria at the outlet of this relatively undeveloped watershed occurred near the end of protracted dry periods when the water table in the Great Marsh dropped and the soil temperature increased. This indicates that bacterial populations in source areas of the watershed may experience significant growth at such times, and also supports the contention that bacteria can survive (and even multiply) in the seemingly unfavorable stream environment (Kittrell and Furfari, 1963; Toranzos, 1991; Solo-Gabriele *et al.*, 2000).

A statistical model that predicts bacterial outflows as a function of rainfall, stream discharge, soil temperature and depth to water in the Great Marsh accounts for almost 70 percent of the variability of measured bacterial outflows over a 4-year-long study at the outlet of the Derby Ditch watershed. The four variables, which were all statistically significant predictors, are surrogates for more difficult to measure controls such as bed scour and lateral inflows of bacterial laden water, as well as bacterial population growth within watershed source areas.

A time series regression model aimed at predicting *E. coli* concentrations in stormflows as a function of continuously measured water-quality variables accounted for more than 90 percent of the observed variability in hourly bacterial concentrations during three storm periods of intensive sampling at Burns Ditch. The most important independent variables were turbidity, nitrate and ammonium. Turbidity is highly correlated with suspended sediment load in Burns Ditch (Thomas, 2001), and elevated levels of nitrate and ammonium are clearly associated with human waste (Robertson, *et al.*, 1991). Although two of the potentially useful predictors of *E. coli* concentration (Tw and pH) did not yield statistically significant t-ratios in the present study, significance may be achieved with larger sample sizes.

Despite the relatively small samples used to generate the statistical models discussed in this paper, the use of such models in the future may represent the best available method for real-time forecasting of bacterial concentrations in streams and their outflows to receiving water bodies. More studies in a wider range of settings and conditions will be needed to fully realize the utility of statistically based predictive models for bacterial concentrations and outflows, and to establish their validity. This should be a worthwhile endeavor, because predictive modeling represents a better approach to understanding the source and nature of *E. coli* contamination of beaches and it may be a superior public health warning system, when compared to the presently practiced microbiological monitoring and reporting systems.

Acknowledgments

Funding to support this study was provided through Grants from the U.S. Environmental Protection Agency, The Indiana Department of Environmental Management, and the Great Lakes Commission. Support in the form of a fellowship to Judith Thomas was provided by the Illinois-Indiana Sea Grant Program. Bacterial data used in the Derby Ditch regression analysis were provided by Task Force personnel. Analyses of samples collected during storms were supervised by Dr. G. T. Asteriadis (Purdue University North Central), and Dr. A. V. Gochee (Indiana University, Northwest).

References

American Public Health Association: 1998, 'Standand Methods for the Examination of Water and Waste Water.' 20th Ed. APHA, Washington, D.C.

Cochrane, D. and Orcutt, D. H.: 1949, 'Application of Least Squares Regression to Relationships Containing Autocorrelated Error Terms', *Journal of the American Statistical Association* **44**, 23–61.

Kittrell, F. W. and Furfari, S. A.: 1963, 'Observations of Coliform Bacteria in Streams', *Journal of the Water Pollution Control Federation* **35**, 1361–1384.

Kunkle, S. H.: 1970, 'Sources and Transport of Bacterial Indicators in Rural Streams', Proceedings of the Symposium on Interdisciplinary Aspects of Watershed Management, August 3–7, 1970, Montana State University, Bozeman, MT, pp. 106-132.

Kunkle, S. H. and Meiman, J. R.: 1967, 'Water Quality of Mountain Watersheds', Hydrology Paper No. 21, Colorado State University, Ft. Collins, CO.

Matson, E. A., Horner, S. G. and Buck, J. D.: 1978, 'Pollution Indicators and other Micro-organisms in River Sediment', *Journal of the Water Pollution Control Federation* **50**, 13-19.

McDonald, A. and Kay, D.: 1981, ' Enteric Bacterial Concentrations in Reservoir Feeder Streams: Baseflow Characteristics and Response to Hydrograph Events', *Water Research* **15**, 961–968.

McDonald, A., Kay, D. and Jenkins, A.: 1982, 'Generation of Fecal and Total Coliform Surges by Stream Flow Manipulation in the Absence of Normal Hydrometeorological Stimuli', *Applied and Environmental Microbiology* **44**(2), 292–300.

Robertson, W. D., Cherry, J. A. and Sudicky, E. A.: 1991, 'Ground-Water Contamination from Two Small Septic Systems on Sand Aquifers, *Ground Water* **29**(1), 82–92.

Solo-Gabriele, H. M., Wolfert, M. A., Desmarias, T. R. and Palmer, C. J.: 2000, 'Sources of Escherichia coli in a Coastal Subtropical Environment', *Applied Environmental Microbiology* **66**, 230–237.

Thomas, J, C.: 2001, 'Relationships Between Suspended Solids, Water Quality Parameters, and E.coli Concentrations in Streamflow Discharging into Southern Lake Michigan, at the Outlet of Burns Ditch, Portage, Indiana, Unpublished Master's Thesis, Indiana University, Bloomington, IN, 45 pp.

Toranzos, G. A.: 1991, 'Current and Possible Alternate Indicators of Fecal Contamination in Tropical Waters: A Short Review', *Environmental Toxicology. Water Qual.* **6**, 121–130.

U.S. EPA.: 1986, Ambient Water Quality Criteria for Bacteria—1986. Janurary 1986. EPA440/5-84-002. 17 pp.

Whitman, R. L., Sobat, T. A., Dustman, W. and May, S. W.: 1991, 'Fecal Contamination Source Determination of Derby Ditch', Unpublished Report to the Indiana Dunes National Lakeshore, 31 pp.

COMPARISON OF BEACH BACTERIAL WATER QUALITY INDICATOR MEASUREMENT METHODS

RACHEL T. NOBLE[1*], STEPHEN B. WEISBERG[2], MOLLY K. LEECASTER[3]
CHARLES D. MCGEE[4], KERRY RITTER[2], KATHY O. WALKER[5], AND PATRICIA M. VAINIK[6]

[1]*University of North Carolina at Chapel Hill, Institute for Marine Sciences, Morehead City, NC, USA;*
[2]*Southern California Coastal Water Research Project, Westminster, CA, USA;*
[3]*Idaho National Engineering and Environmental Laboratories, Idaho Falls, ID, USA;*
[4]*Orange County Sanitation District, Fountain Valley, CA, USA;* [5]*Sanitation Districts of Los Angeles County, Carson, CA, USA;* [6]*City of San Diego Metropolitan Wastewater Department, San Diego, CA, USA;*
[*]*(author for correspondence, e-mail: rtnoble@email.unc.edu)*

Abstract. Three methods (membrane filtration, multiple tube fermentation, and chromogenic substrate technology kits manufactured by IDEXX Laboratories, Inc.) are routinely used to measure indicator bacteria for beach water quality. To assess comparability of these methods, quantify within-laboratory variability for each method, and place that variability into context of variability among laboratories using the same method, 22 southern California laboratories participated in a series of intercalibration exercises. Each laboratory processed three to five replicates from thirteen samples, with total coliforms, fecal coliforms or enterococci measured depending on the sample. Results were generally comparable among methods, though membrane filtration appeared to underestimate the other two methods for fecal coliforms, possibly due to clumping. Variability was greatest for the multiple tube fermentation method. For all three methods, within laboratory variability was greater than among laboratories variability.

Keywords: microbiology, intercalibration, variability, bacterial indicators, beach water quality

1. Introduction

Coastal beaches are the subject of extensive water quality monitoring to detect fecal contamination from human activities, such as wastewater discharge, industrial input, and surface runoff. Included in many of these monitoring programs is measurement of indicator bacteria, such as total coliforms, fecal coliforms, and enterococci. While indicator bacteria are not necessarily pathogenic, they are found abundantly in human wastes, where pathogenic organisms may exist. Bacterial indicators are used in preference to direct tests for pathogens because bacteria indicator measurements are less expensive and correlate with the incidence of illness in swimmers (Cabelli, 1983; Haile *et al.*, 1999).

Indicator bacteria have historically been measured using either membrane filtration (MF) or multiple tube fermentation (MTF), though chromogenic substrate methods, such as those manufactured by IDEXX Laboratories, Inc. (IDEXX), have recently been gaining popularity. The three methods each are based upon measuring different products of bacterial growth. MF enumerates bacterial colonies on a specific growth substrate. MTF measures metabolic activity as determined by fermentation and the production of gas. Chromogenic methods measure the ability of organisms to metabolize a specific labeled substrate, thereby releasing a chromogen. These differences in analytical endpoint provide the potential for differing results among methods.

Environmental Monitoring and Assessment **81**: 301–312, 2003.

There have been a number of studies comparing the response of MF and MTF, but only a few studies have compared these methods to IDEXX (Edberg *et al.*, 1990; Stasiak and Cheng, 1991; Green *et al.*, 1997; Eckner, 1998; Francy and Darner 2000). No study has quantified response among all three methods, nor placed differences among methods into context of variability among laboratories that use the same methods. Furthermore, no study has compared measurement precision among the three methods. IDEXX kits have the advantages of being less expensive and faster than the historically used methods, but these advantages are of little value if the results produced by IDEXX are not comparable to that from the historic methods.

Here we present a series of intercalibration studies that were conducted among 22 southern California laboratories. We use these studies to assess comparability of results among the three bacterial indicator measurement methods, quantify within laboratory variability for each method and place that variability into context of variability among laboratories using the same method.

2. Methods

Thirteen experiments were conducted on five separate occasions (Table 1). The first set of experiments involved quantification of total coliforms in transport media at three bacterial indicator concentrations. The second set of experiments involved quantification of fecal coliforms [or *Escherichia coli (E. coli)*, to which the IDEXX method, Colilert®, is specific] in transport media at three bacterial indicator concentrations. The third set of experiments involved quantification of enterococci in transport media at three bacterial indicator concentrations. The fourth set of experiments involved quantification of total coliforms and fecal coliforms (or *E. coli*) at a single concentration in seawater and fecal coliforms (or *E. coli*) in transport medium. The fifth set of experiments involved quantification of fecal coliforms (or *E. coli*) from natural seawater spiked with treated wastewater effluent (Table 1).

Ten of the thirteen experiments were performed by seeding 24 hour-old stock cultures of *E. coli* (ATCC 75922) or *Enterococcus faecalis* (ATCC 29212) into 10-liter carboys of NYSDH-1 transport medium (Toombs and Conner 1980). Transport media was prepared prior to the day of the experiment in two-liter volumes and sterilized. Carboys were sterilized separately. Bacteria were added to the transport media and mixed for twenty minutes on a magnetic mixer prior to dispensing the first sample. Targeted concentrations were 100–1,000 (low), 1,000–10,000 (medium) and 10,000–100,000 bacteria/100 mL (high, Table 1). The amount of stock culture necessary to achieve the target concentrations was based on MF analyses from the preceding day.

In experiments 5 and 10, stock cultures of *E. coli* (ATCC 75922) were used to inoculate natural seawater collected from Palos Verdes, California. In experiment

Table 1. Median bacterial indicator count and standard deviation in each experiment. Numbers with the same letter code are not significantly different. MF = Membrane Filtration; and MTF = Multiple Tube Fermentation.

Experiment Number	*Date*	*Indicator*		*Matrix*	*Method*	*Median Count*	*Standard Deviation*
1	04/21/98	Fecal Coliforms	Bacterial Culture	Transport Media	MF	306[b]	55[a]
					MTF	967[a]	431[a]
					IDEXX	996[ab]	1,131[a]
2	04/21/98	Fecal Coliforms	Bacterial Culture	Transport Media	MF	340[b]	115[b]
					MTF	2,610[a]	1,652[a]
					IDEXX	3,964[a]	1,626[ab]
3	04/21/98	Fecal Coliforms	Bacterial Culture	Transport Media	MF	65,400[b]	9,422[a]
					MTF	173,750[a]	45,066[a]
					IDEXX	160,000[ab]	35,777[a]
4	06/09/98	Fecal Coliforms	Bacterial Culture	Seawater	MF	696[b]	125[a]
					MTF	1,945[a]	1,306[b]
					IDEXX	1,601[ab]	407[a]
5	06/09/98	Fecal Coliforms	Bacterial Culture	Transport Media	MF	848[b]	124[b]
					MTF	1,928[a]	2,454[a]
					IDEXX	1,528[ab]	379[ab]
6	06/23/98	Fecal Coliforms	Effluent Wastewater	Seawater	MF	764[a]	195[a]
					MTF	1,093[a]	1,389[a]
7	03/31/98	Total Coliforms	Bacterial Culture	Transport Media	MF	179[a]	74[a]
					MTF	265[a]	231[a]
					IDEXX	172[a]	51[a]
8	03/31/98	Total Coliforms	Bacterial Culture	Transport Media	MF	1,797[a]	280[a]
					MTF	2,088[a]	1,638[a]
					IDEXX	1,613[a]	318[a]
9	03/31/98	Total Coliforms	Bacterial Culture	Transport Media	MF	21,920[a]	3,098[a]
					MTF	19,564[a]	13,061[a]
					IDEXX	16,680[a]	6,202[a]
10	06/09/98	Total Coliforms	Bacterial Culture	Seawater	MF	1,676[a]	230[a]
					MTF	2,617[a]	2,325[b]
					IDEXX	1,601[a]	47[a]
11	04/28/98	Enterococci	Bacterial Culture	Transport Media	MF	177[a]	29[a]
					MTF	246[a]	164[a]
					IDEXX	164[a]	39[a]
12	04/28/98	Enterococci	Bacterial Culture	Transport Media	MF	2,600[a]	773[a]
					MTF	2,540[a]	1,792[a]
					IDEXX	1,967[a]	506[a]
13	04/28/98	Enterococci	Bacterial Culture	Transport Media	MF	28,571[a]	10,260[a]
					MTF	41,600[a]	23,249[a]
					IDEXX	18,429[a]	3,742[a]

6, filtered primary-treated wastewater from the Orange County Sanitation District Plant #1 was added to natural seawater (collected from Newport Beach, California, Table 1). Before inoculating the seawater sample, the primary treated wastewater was filtered through Whatman Grade 415 filter paper to remove fine particulates.

On the morning of each experiment, the contents of the carboys were aliquotted into subsamples, which were packed in ice and distributed to the participating laboratories. All laboratories began their analyses at the same time, approximately five hours after the stock sample was prepared. The originating laboratory analyzed the first and last sample dispensed from each carboy by MF and MTF to validate homogeneity of samples aliquotted from the carboy.

Each laboratory used its own standard operating procedures. Methods used by participants included 9221B, C and E, 9222B and D, 9230B and C in Standard Methods for the Examination of Water and Wastewater, APHA, AWWA, WEF, 19th edition, 1995 and EPA method 1600 (APHA 1995). Colilert® and Enterolert™ (IDEXX Laboratories, Inc, Westbrook, ME) were used in both 15-tube MTF format (only one labortory) and 51 well Quanti-Tray® format. Three to five replicates were performed for each indicator at each concentration. Several laboratories used more than one analytical method, which resulted in more than 22 sets of results for some experiments.

To test the hypothesis that the median within-laboratory values were the same among methods, we performed an ANOVA on ranks, separately for each experiment (Conover and Iman, 1981). A Bonferoni adjustment to significance level was used to account for pair-wise testing of methods. Similar analyses were performed to test for difference in within-laboratory variance among methods.

To assess uncertainty for individual laboratories reporting counts based on a single sample, 60% and 95% confidence limits were estimated using data from all laboratories, pooled across experiments. After log transforming the data, least squares regression was used to model the within laboratory variances as a function of within laboratory means. Confidence intervals were then back-transformed to original scale. Separate regressions were performed for each indicator.

3. Results

Comparability of bacterial densities measured among methods was indicator specific (Figure 1; Table 1). For enterococci and total coliforms, we saw no significant differences among methods in any of the individual experiments, though we did observe that MTF results were higher in five of the seven experiments. We also observed that IDEXX yielded the lowest median count in all three of the experiments conducted with enterococci. For fecal coliforms, MF produced the lowest counts in all six tests, and in several experiments the median count was less than 50% of the other methods. MF results were significantly lower than MTF results in many of the individual tests, but differed significantly from results from IDEXX only in experiment 2.

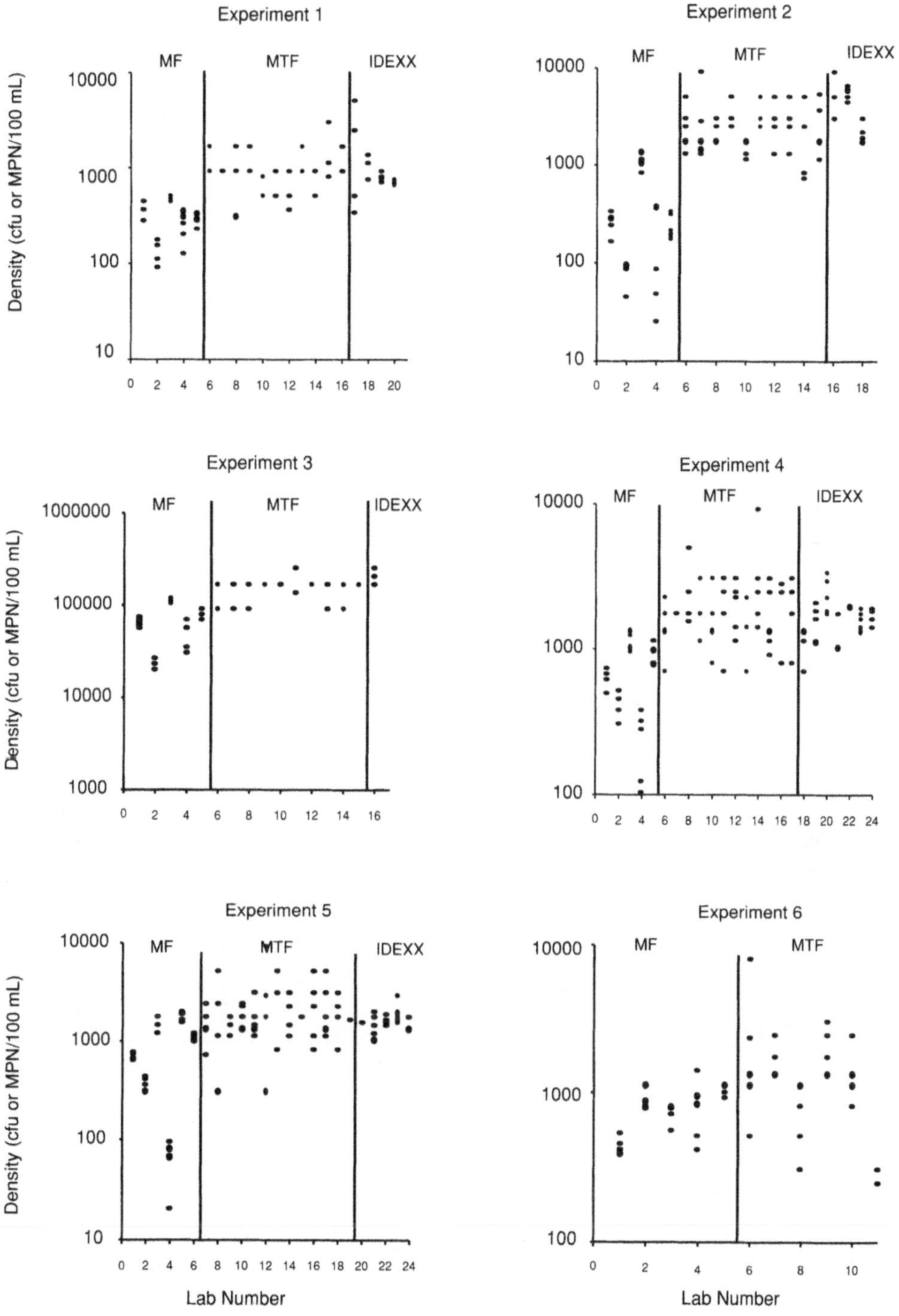

Figure 1. Log-transformed bacterial density vs. Laboratory (Lab) number. Laboratory numbering was random and arranged by method. Experiments 1–6 are for Fecal Coliforms. MF = Membrane Filtration Method, MTF = Multiple Tube Fermentation Method, IDEXX = Colilert® or Enterolert™.

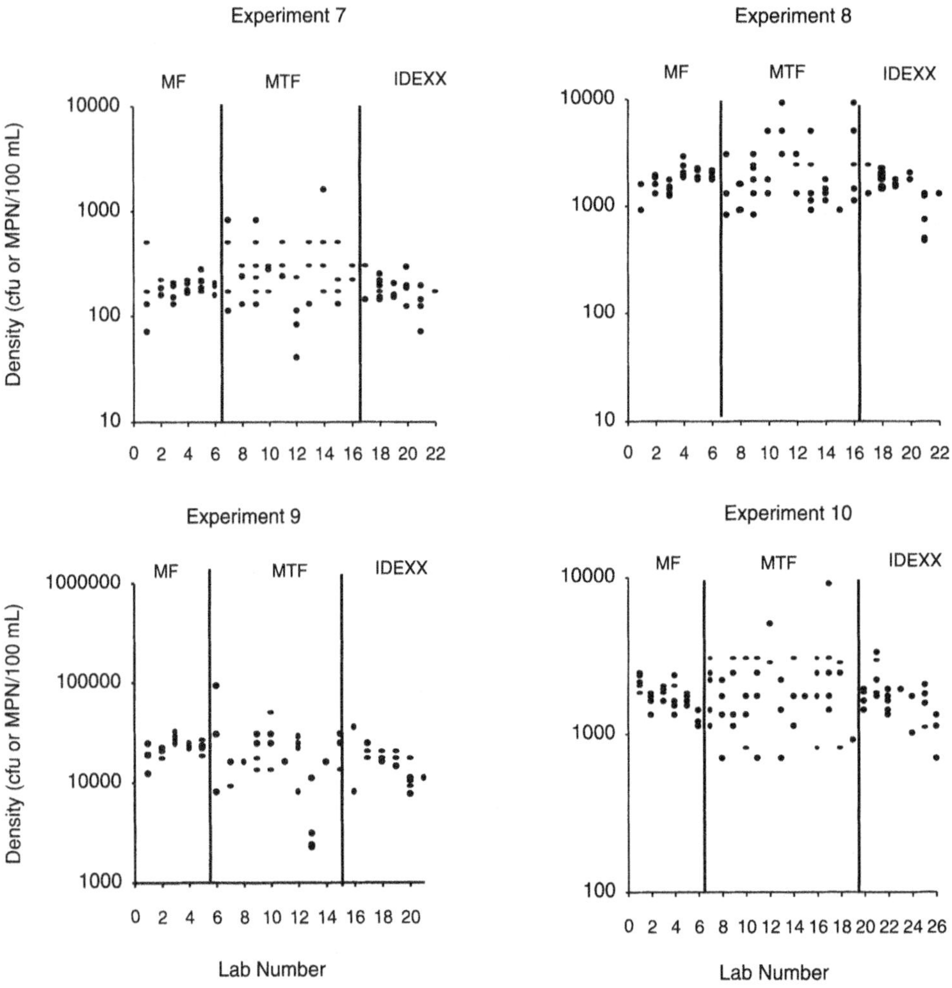

Figure 1—*Continued.* Log-transformed bacterial density vs. Laboratory (Lab) number. Laboratory numbering was random and arranged by method. Experiments 7–10 are for Total Coliforms.

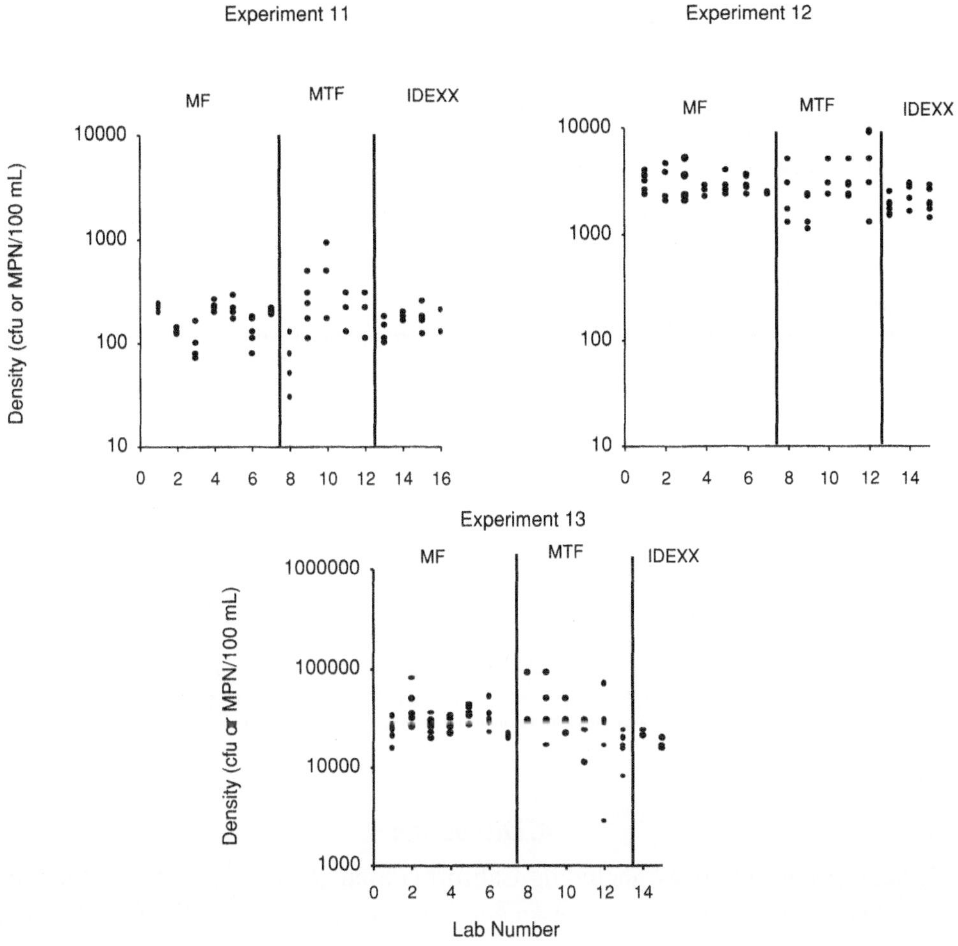

Figure 1—*Continued*. Log-transformed bacterial density vs. Laboratory (Lab) number. Laboratory numbering was random and arranged by method. Experiments 11–13 are for Enterococci.

Variability was consistently highest for MTF, and in several cases was almost an order of magnitude higher (Table 1). The variability associated with MF was generally a little lower than that with IDEXX, but not significantly different in any of the individual experiments. The coefficient of variation appeared to be independent of which bacterial indicator was measured.

The confidence interval for MTF was about twice as large as for the other two methods (Table 2). In most cases, the upper 95% confidence interval was more than twice the threshold value. The confidence intervals are all asymmetric around the threshold, reflecting the lognormal distribution of the data. Also, the confidence intervals presented in Table 2 are all specific to the threshold concentration, as the data displayed a significant variance: mean ratio.

Table 2. Upper and lower confidence limits around California's AB411 single-sample thresholds. AB411 thresholds: total coliforms 10,000 cfu or MPN/100 mL, fecal coliforms 400 cfu or MPN/100 ml, enterococci 104 cfu or MPN/100 mL.

		Lower 95% CI	Lower 60% CI	Upper 60% CI	Upper 95% CI
Total coliforms	MTF	3,482	5,901	16,946	28,717
	MF	7,947	8,915	11,218	12,583
	IDEXX	4,339	6,587	15,181	23,048
Fecal coliforms	MTF	122	221	724	1310
	MF	188	274	584	851
	IDEXX	164	256	625	976
Enterococci	MTF	36	61	176	299
	MF	70	86	126	153
	IDEXX	63	81	134	173

4. Discussion

Several recent initiatives, including California State Assembly Bill 411, the federal Beaches Environmental Assessment, Closure, and Health (BEACH) Act and the World Health/USEPA Expert Consultation of Safety of Recreational Waters, have been catalysts for increased beach monitoring and greater consistency in standards on which to base public health warning decisions. One price for increased consistency, though, can be loss in flexibility. For instance, California, which has the most beach monitoring in the United States (Schiff *et al.*, 2001), recently mandated that warnings be based on any sample that exceeds a single sample standard, whereas agencies previously had flexibility to collect confirmation samples. Examination of data from Los Angeles County reveals that a majority of the single sample standard exceedences for the five-year period between 1995 and 2000 (Schiff *et al.*, In Press) were within measurement error for MTF. Conversely, there were similar numbers of measurements that were less than the standard, but within the lower confidence bounds. These findings suggest there is a great deal of uncertainty associated with warnings based on a single sample standard.

The magnitude of our variability estimates, as well as our finding of higher variability for MTF, is consistent with that of previous studies (Fleisher, 1990). MTF is a most probable number technique, which is a statistical estimate based on the percentage of test tubes eliciting a positive response to the presence of bacteria. As the estimate is based on a binomial distribution, its variance is primarily a function of the number of tubes used. IDEXX methods are also based on a most probable number technique, but they employ a prepackaged 51-well tray that is logistically easier to use than test tubes. As a result of the greater number of wells, this method yielded a variance similar to that of MF.

Although our variability estimates were similar to that of previous studies, our finding of lower fecal coliform values using MF differed from previous studies which generally have found comparable results for MF and MTF. It also differs from our finding of similar results for total coliforms and enterococci using the two methods. Several laboratories reported filtration difficulty and colony growth that was extremely patchy, suggesting clumping of the bacteria. This could be an artifact of our using a laboratory strain inoculated into transport media. To investigate this possibility, we used a seawater matrix in experiment four and effluent as an inoculant in experiment six. We saw the same pattern of lower MF values in experiment four, but method differences were not apparent when using effluent as the inoculant. While use of a laboratory strain might increase the likelihood of clumping, discussions with our participating laboratories revealed that they loosely follow the standard method protocol which calls for shaking the sample at least 25 times to enhance homogeneity, instead typically shaking the sample only about five times. As a plating method, MF is more susceptible to clumping and our results emphasize the need for strict adherence to quality control guidelines regarding shaking and sample dilution.

Many environmental assessments require compilation of data from multiple laboratories, either to extend temporal records for trends assessment or to enhance geographic scale to accomplish regional/national assessments. Such data compilations assume a degree of comparability among laboratories, even though analytical personnel, methods, and instrumentation may vary. Our study suggests that differences among methods are small relative to inherent measurement error, though our results with fecal coliform suggest that this could be of concern if laboratory procedures are not followed precisely. We also found that differences among laboratories using the same methods were generally small and less than the normal variability encountered using a single method in a single laboratory (Figure 1). Overall, the increase in variability among measurements from pooled data was only about 30% higher than that obtained using a single analytical method performed at a single laboratory.

Our study also included participation by volunteer monitoring organizations which are becoming an increasingly large component of beach monitoring in some areas. The volunteer organizations involved in this study produced data comparable to that of the certified professional laboratories. While the volunteers involved in our study were more experienced than most, having conducted their own monitoring activities for many years and having benefited from EPA-sponsored training, our study demonstrates that data from properly trained volunteer organizations can be an equal addition to beach quality assessments.

The use of IDEXX kits has become increasingly popular in the past few years because they are less expensive and require less formal training than historically used methods. The use of IDEXX kits requires minimal space and equipment, making them ideal for volunteer organizations or for agencies without microbiology laboratories. Our study is among the first to compare this method to others and is the first test to quantify its variability. We found it to be an acceptable, and perhaps preferable, method. The median values were similar to that produced by MTF, without evidence of the clumping that seemed to compromise the MF measurements. Moreover, measurement variability was considerably less than MTF, which can be partially attributed to the use of the Quanti-Tray® 51-well format compared to the MTF procedure which typically involves a 15-tube format. Measurement variability for IDEXX was comparable to that of MF. While these findings support the use of IDEXX for marine water quality testing, they are not comprehensive. The bacteria measured in all but one of our experiments were laboratory strains, with no background bacteria to compete or interfere with the analyses. While we saw little difference in results between our experiments using laboratory cultures and wastewater effluent as inoculant, our experiments were also conducted on samples that contained low levels of suspended solids, which can potentially interfere with colony growth. Side-by-side testing of samples from the natural environment, particularly during high turbidity conditions, is a logical next step in evaluating these methods.

Acknowledgements

The authors thank all of the participants in the study who spent laboratory time and also attended the many meetings in preparation for the exercises; Algalita Marine Research Foundation, Aliso Water Management Authority and Southeast Regional Reclamation Authority, Aquatic Bioassay and Consulting Laboratories, City of Long Beach Department of Health & Human Services, City of Oceanside, City of Oxnard, City of Los Angeles Environmental Monitoring Division, City of San Diego, City of Santa Barbara, City of Ventura, Encina Wastewater Authority, Goleta Sanitation District, Instituto de Investigaciones Oceanalogicas (UABC), Los Angeles County Department of Health Services, County Sanitation Districts of Los Angeles County, Orange County Public Health Laboratory, Orange County Sanitation District, San Diego County Department of Environmental Health, San

Elijo Joint Powers Authority, Santa Barbara Health Care Services, Southern California Marine Institute, and Surfrider Foundation. We especially thank the County Sanitation Districts of Los Angeles County and the Orange County Sanitation District for preparing the samples.

References

American Public Health Association, *Standard Methods for the Examination of Water and Wastewater*: 1995, 18th ed, Eaton, A.D., Clesceri, L.S., and Greenberg, A.E. (eds.), Washington, D.C.

APHA. See American Public Health Association.

Conover, W.J., and Iman, R.L.: 1981, 'Rank Transform as a Bridge Between Parametric and Nonparametric Statistics,' *The American Statistician* **35**, 125.

Cheung, W.H.S., Chang, K.C.K., and Hung, R.P.S.:1991, "Variations in microbial indicator densities in beach waters and health-related assessment of bathing water quality,' *Epidemiol. Infect.* **106**, 329–344.

Eckner, K.F.: 1998, 'Comparison of membrane filtration and multiple-tube fermentation by the Colilert and Enterolert methods for detection of waterborne coliform bacteria, Escherichia coli, and enterococci used in drinking and bathing water quality monitoring in southern Sweden,' *Appl. Environ. Microbiol.* **64**, 3079–3083.

Edberg, S.C., Allen, M.J. Smith, D.B. and Kriz, N.J.: 1990, 'Enumeration of total coliforms and Escherichia coli from source water by the defined substrate technology,' *Appl. Environ. Microbiol.* B, 366–369.

Fleisher, J.M.: 1990, 'The effects of measurement error on previously reported mathematical relationships between indicator organism density and swimming-associated illness: a quantitative estimate of the resulting bias,' *Internat. J. of Epidemiol.* **19**, 1100–1106.

Francy, D.S. and Darner, R.A.: 2000, 'Comparison of methods for determining Escherichia coli concentrations in recreational waters' *Wat. Res.* **24**, 2770–2778.

Green, B.L., Clausen, E.M. and Litsky, W.: 1977, 'Two-temperature membrane filter method for enumerating fecal coliform bacteria from chlorinated effluents', *Appl. Environ. Microbiol.* **33**, 1259–1264.

Haile, R.W., J.S. Witte, M. Gold, R. Cressey, C. McGee, R.C. Millikan, A. Glasser, N. Harawa, C. Ervin, P. Harmon, J. Harper, J. Dermand, J. Alamillo, K. Barrett, M. Nides and Wang, G.: 1999, 'The health effects of swimming in ocean water contaminated by storm drain runoff,' *Epidemiol.* **10**, 355–363.

Messer, J.W., and Dufour, A.P.: 1998, 'A rapid, specific membrane filtration procedure for enumeration of enterococci in recreational water,' *Appl. Environ. Microbiol.* **64**, 678–680.

Natural Resources Defense Council.: 1998, 'Testing the waters, Has your vacation beach cleaned up its act?' Vol. 8. p. 145. New York, NY.

Schiff, K.C., Weisberg, S.B. and Dorsey, J.H.: 2001, 'Microbiological monitoring of marine recreational waters in southern California,' *Environ. Management.* **27**:149–157.

Schiff, K.C., Morton, J. and Weisberg, S.B.: In press, 'Retrospective evaluation of shoreline water quality along Santa Monica Bay beaches,' *Mar. Env. Res.*

Stasiak, M.C., and Cheng, S.H.: 1991, 'Coliform detection using membrane filtration and Colilert, and *E. coli* detection using nutrient AGAR Plus MUG and EC Plus MUG,' pp. 741–748 in: Proceedings of the Water Quality Technology Conference, Orlando. American Water Works Association, Denver, CO.

Toombs, R.W., and Conner, D.A.: 1980, 'Proficiency test sample media for single and mixed pure cultures of water pollution indicator bacteria,' *Appl. Environ. Microbiol.* **40**, 883–887.

MOLECULAR APPROACHES TO MICROBIOLOGICAL MONITORING: FECAL SOURCE DETECTION

KATHARINE G. FIELD[1*], ANNE E. BERNHARD[2], AND TIMOTHY J. BRODEUR[1]

[1]*Department of Microbiology, Oregon State University, Corvallis, OR 97331;*
[2]*Department of Civil and Environmental Engineering, University of Washington, Seattle, WA 98195-2700*
*(author for correspondence, e-mail: kate.field@orst.edu)

Abstract. Molecular methods are useful both to monitor natural communities of bacteria, and to track specific bacterial markers in complex environments. Length-heterogeneity polymerase chain reaction (LH-PCR) and terminal restriction fragment length polymorphism (T-RFLP) of 16S rDNAs discriminate among 16S rRNA genes based on length polymorphisms of their PCR products. With these methods, we developed an alternative indicator that distinguishes the source of fecal pollution in water. We amplify 16S rRNA gene fragments from the fecal anaerobic genus *Bacteroides* with specific primers. Because *Bacteroides* normally resides in gut habitats, its presence in water indicates fecal pollution. Molecular detection circumvents the complexities of growing anaerobic bacteria. We identified *Bacteroides* LH-PCR and T-RFLP ribosomal DNA markers unique to either ruminant or human feces. The same unique fecal markers were recovered from polluted natural waters. We cloned and sequenced the unique markers; marker sequences were used to design specific PCR primers that reliably distinguish human from ruminant sources of fecal contamination. Primers for more species are under development. This approach is more sensitive than fecal coliform assays, is comparable in complexity to standard food safety and public health diagnostic tests, and lends itself to automation and high-throughput. Thus molecular genetic markers for fecal anaerobic bacteria hold promise for monitoring bacterial pollution and water quality.

Keywords: Microbiological monitoring, fecal contamination, fecal source discrimination, *Bacteroides*, anaerobic bacteria, LH-PCR, T-RFLP, bacterial communities, molecular markers

1. Introduction

Fecal contamination of aquatic environments afflicts many regions of the U.S., with associated human health risks and environmental damage. Human fecal pollution spreads many dangerous bacterial pathogens and viruses such as hepatitis, while other pathogens such as *Cryptosporidium parvum*, *Giardia lamblia*, *Salmonella* spp., and *Escherichia coli (E. coli)* O157:H7 are associated with animal fecal pollution. Fecal bacteria, pathogenic protista and viruses may be highly concentrated in sediments (Pommepuy *et al.*, 1992) and in shellfish (DePaola *et al.*, 1990). In 1995, 4.9 million acres of shellfish-growing waters were restricted nationally due to water pollution. Additionally, in 1996, over 2,500 beaches in the U.S. were posted or closed due to microbial contamination.

According to the Clean Water Action Plan of the U.S. EPA, "after 25 years, the clean water program has addressed many of the major pipe discharges of sewage and industrial waste. By far the predominant source of remaining water polluting problems is from urban and agricultural lands and facilities such as animal feeding operations..."

Environmental Monitoring and Assessment **81**: 313–326, 2003.

Often the source of fecal contamination in water cannot be determined. For example, failing septic systems, overloads at sewage treatment facilities, and run-off from non-point sources such as farm manure may all be candidates. In addition, although sediments are known to be a source of human pathogens, the contribution of bacterial pollution "stored" in sediments and re-suspended during storm events is unknown relative to other sources. In order to adequately assess human health risks and develop management plans for watersheds providing drinking water and supporting recreation and fisheries, it is necessary to know the sources of fecal contamination.

The standard method of measuring fecal pollution does not distinguish between human and animal sources. The method requires growing fecal coliforms or *Escherichia coli* from water samples to estimate their concentrations. Because these bacteria are found in a variety of warm-blooded animals and even reptiles (Harwood *et al.*, 1999), this method does not distinguish between human and animal sources. In addition, coliform-based methods can be confounded by growth of coliforms in receiving water and sediments (Gerba and McLeod, 1976; Hood and Ness, 1982; Howell *et al.*, 1996; LaLiberte and Grimes, 1982). Coliforms adhere to sediments and are transported in and out of the water column, affecting the measure of fecal pollution in unknown ways (Weiskel *et al.*, 1996; Gary and Adams, 1985; Gerba and McLeod, 1976; Hood and Ness, 1982; Howell *et al.*, 1996; LaLiberte and Grimes, 1982; Sherer *et al.*, 1988). It is remarkable that although coliforms are the gold standard for measuring fecal pollution, the influences of these important factors are still so poorly understood.

2. Methods of detecting the source of fecal pollution

Several methods of discriminating the source of fecal pollution have been proposed, but each has drawbacks. The ratio of fecal streptococci to fecal coliforms has been used to differentiate human fecal contamination from that of other animals (Sinton *et al.*, 1993). But because strains of coliforms and streptococci have different survival rates, the ratio changes in complex ways over time, making it unreliable.

Antibiotic resistance patterns of fecal streptococci or coliforms can distinguish among sources of fecal pollution (Andrews *et al.*, 1997; Wiggins, 1996; Wiggins *et al.*, 1999). Because this method requires culturing a large number of fecal streptococcal isolates, screening them for resistance against a panel of antibiotics, and analyzing the results by discriminant analysis, it is labor-intensive and time-consuming. When sets of isolates from a single type of feces have been analyzed, rates of correct classification have varied from about 64 to 87%. However, when individual isolates from mixed fecal sources were analyzed, rates of correct classification were lower. Thus the accuracy of this method may be unacceptably low.

E. coli ribotyping (Carson *et al.*, 2001; Parveen *et al.*, 1999) and rep-PCR DNA fingerprinting (Dombek *et al.*, 2000) are genotyping methods that depend on growing a large number of isolates. Ribotyping correctly discriminates between human and non-human fecal isolates with a high rate of success, but is less successful at identifying animal sources (Carson *et al.*, 2001; Parveen *et al.*, 1999). A ribotyping analysis is estimated to take two weeks to complete (M. Samadpour, pers. comm.). DNA fingerprinting utilizes complex, tedious-to-interpret patterns that don't always correctly group isolates, even after statistical analysis (Dombek *et al.*, 2000). Both methods require massive local collections of strains for comparison. Although such collections are underway in several locales, the expense of maintaining them often puts them out of reach of anyone except private consulting companies.

Phage can be used as indicators of fecal pollution (Goyal *et al.*, 1980; Havelaar *et al.*, 1993; Osawa *et al.*, 1981; Palmateer *et al.*, 1991; Paul *et al.*, 1997; Singh and Gerba, 1983; Tartera and Jofre, 1987; Tartera *et al.*, 1989). Coliphage abundance is well correlated with sewage effluent (Paul *et al.*, 1997); coliphage are also associated with non-point discharge (Paul *et al.*, 1997). RNA phages, which constitute only a minor fraction of human coliphages, show specificity in host distribution (Osawa *et al.*, 1981). Although coliphage may indicate the presence of human enteric viruses in polluted water (Havelaar *et al.*, 1993; Palmateer *et al.*, 1991; Sobsey, 1989), their use may be limited since only a small portion of fecal bacteria may contain phages (Osawa *et al.*, 1981), and virus concentrations in sewage samples are highly variable (Havelaar *et al.*, 1993).

All of these methods of fecal source discrimination are time consuming, labor intensive, require specialized equipment or extensive strain collections, are subject to culture bias, and may have unacceptably low rates of accuracy. A method is needed that:

- Is quick and reliable;
- Does not require culturing isolates;
- Does not require a large library of reference strains;
- Is flexible for easy field handling and storage of samples; and
- Requires a minimum of specialized equipment.

Modern methods of identifying microbes in the environment, foods, and clinical diagnosis rely on genetic markers, not growing the microbes. We adapted these methods to source discrimination of fecal pollution.

3. Markers from Fecal *Bacteroides* to Distinguish the Source of Fecal Pollution

To distinguish the source of fecal pollution in fresh and marine waters, we have developed a new Polymerase Chain Reaction (PCR)-based indicator system, utilizing molecular markers from the *Bacteroides* group of fecal anaerobic bacteria (Bernhard and Field, 2000a; Bernhard and Field, 2000b). The method differs from existing methods of detecting fecal pollution in that it detects genetic marker sequences that are not only specific to fecal bacteria, but are also specific to the host species that produced the feces, allowing discrimination among different potential sources. Furthermore, the method does not require isolating and growing the indicator bacteria, and it is rapid and accurate.

Bacteroides is a genus of non-spore-forming, obligate anaerobes that make up one-third of the human fecal flora, far outnumbering coliforms (Holdeman *et al.*, 1976; Moore and Holdeman, 1974; Salyers, 1984). The most abundant *Bacteroides* species in human feces are either human-specific, or present only at very low levels in other species (Allsop and Stickler, 1985; Straub and Dixon, 1997). In addition, *Bacteroides* and its close relatives in the genus *Prevotella* are found exclusively in feces, animal rumens, the mouth and teeth, and other cavities within animals and humans (Paster *et al.*, 1994). When found in water, they are invariably diagnostic of pollution. Because they are strict anaerobes, they are thought not to survive very long once released into receiving waters (Avelar *et al.*, 1998; Kreader, 1998). *Bacteroides* has been identified from environmental water samples for several days after dispersal in water, (Kreader, 1995; Straub and Dixon, 1997). It is detectable by PCR for up to 14 days at 4°C, even in the presence of predators, while at higher temperatures more typical of natural waters (14°C), *Bacteroides* is detectable for 4–5 days (Kreader, 1998).

In the past, *Bacteroides* has had limited use as an indicator because of the relative complexity of growing strict anaerobes. We circumvent this by using molecular, rather than culture-based, methods to detect the bacteria. Molecular approaches make it possible to use difficult-to-grow, but diagnostic, microbes in fecal pollution assays. They also avoid cultivation bias, which is the tendency to preferentially grow weedy, easy-to-grow species and strains, thus not getting an accurate representation of the actual population.

To establish diagnostic markers, we used recently developed technologies that discriminate among mixtures of bacterial gene sequences by detecting sequence length differences in a particular gene fragment (Avaniss-Aghajani *et al.*, 1994; Brunk *et al.*, 1996). Length Heterogeneity Polymerase Chain Reaction (LH-PCR; Rappé *et al.*, 1997; Suzuki *et al.*, 1997) and Terminal Restriction Fragment Length Polymorphism (T-RFLP; Bruce, 1997; Clement *et al.*, 1998; Liu *et al.*, 1997) analyze differences in lengths of gene fragments due to insertions and deletions and point mutations affecting restriction recognition sites, and estimate the relative abundance of each fragment.

To find the *Bacteroides* marker sequences that discriminate fecal sources, we use the Polymerase Chain Reaction (PCR, a method of making millions of copies of a specified gene in a test tube) to amplify specific *Bacteroides* 16S rDNA markers from DNAs extracted from feces. Amplified fragments were cut with restriction enzymes and screened by LH-PCR and T-RFLP using GeneScan on an automated DNA sequencer (Bernhard and Field, 2000a). These methods allowed us to identify specific markers that distinguish cow from human feces (for examples, see Figure 1). Once reliable markers were found, their signals could be followed easily and quickly from many samples.

We also detected these host-specific *Bacteroides* markers in polluted water samples from rivers and Tillamook Bay, Oregon (Figure 2), by filtering water samples and extracting DNA from the filter. The *Bacteroides* signal was easily detectable in all fecal samples and river and bay water samples.

T-RFLP and LH-PCR are expensive methods that require an automated DNA sequencer, and are thus not accessible to many laboratories. To solve this problem,

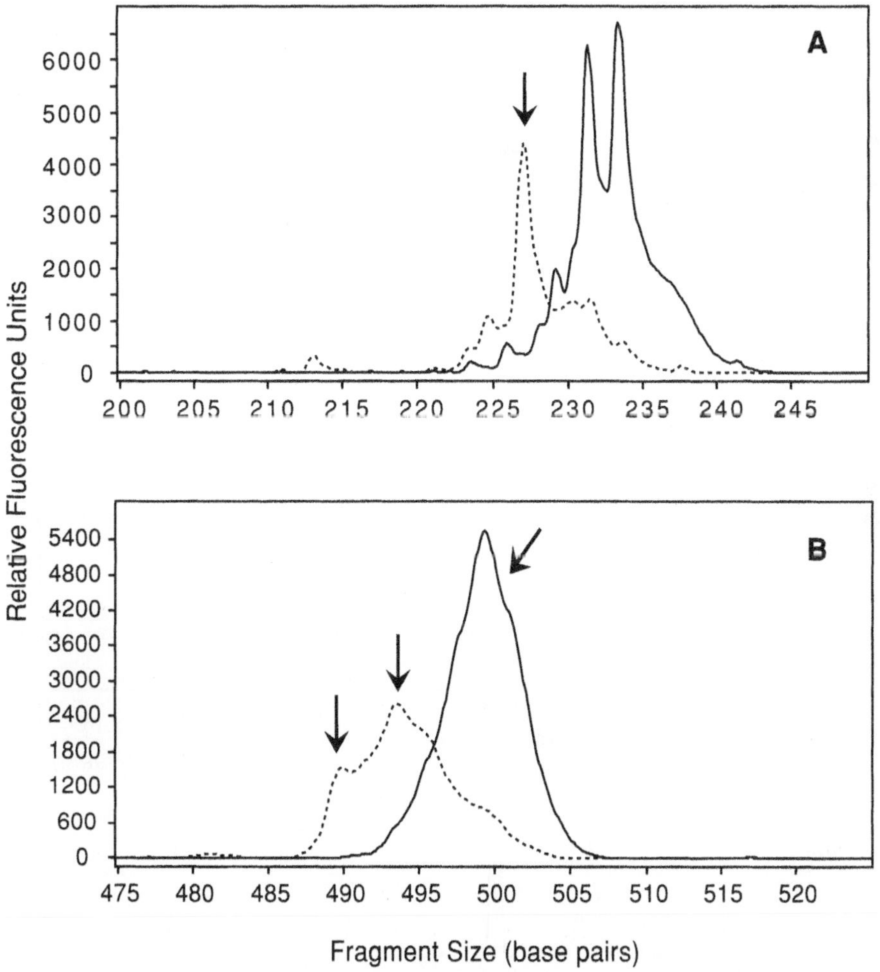

Figure 1. Example of T-RFLP analysis of *Bacteroides* 16S rDNA fragments. Arrows indicate ruminant- and human-specific markers. Fecal DNAs were amplified with PCR primers Bac32F-FAM and Bac708R and cut with restriction enzyme *Aci*I. Solid lines are from human fecal DNA; dotted lines are from cow fecal DNA. Panel A shows all fragments from 200 to 250 base pairs; panel B shows fragments from 475 to 525 base pairs.

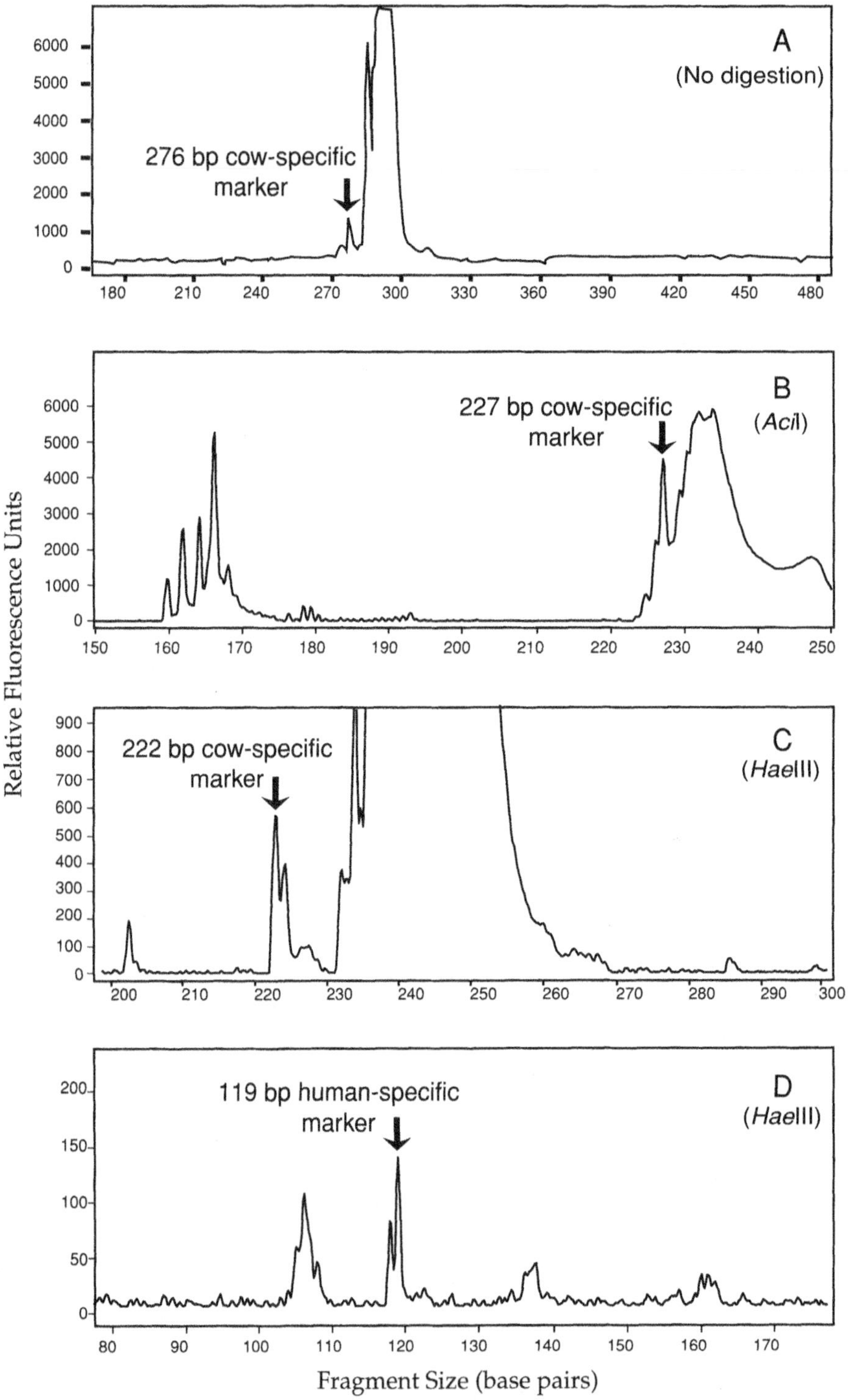

Figure 2. Arrows indicate examples of ruminant specific (A, B, C) or human-specific (D) *Bacteroides* 16S rDNA markers from water samples collected from Tillamook Bay, amplified with primers Bac32F and Bac303R (A) or Bac708R (B, C, D) and digested with *Aci*I (B) or *Hae*III (C and D).

we prepared *Bacteroides* 16S rDNA clones libraries from DNAs extracted from human and cow fecal samples, identified the clones which comprised the unique T-RFLP and LH-PCR peaks, and designed PCR primers specific to these sequences (Bernhard and Field, 2000b). Primers HF8 and HF10, when paired with a general *Bacteroides* primer (Bac32F), specifically amplify signals from human fecal DNA (Figure 3). Primers CF123 and CF151, when paired with a general *Bacteroides* primer (Bac32F), specifically amplify signals from cow fecal DNA (Figure 4). With these primers, we found that we could reliably and specifically amplify host-specific markers from both feces and polluted water samples.

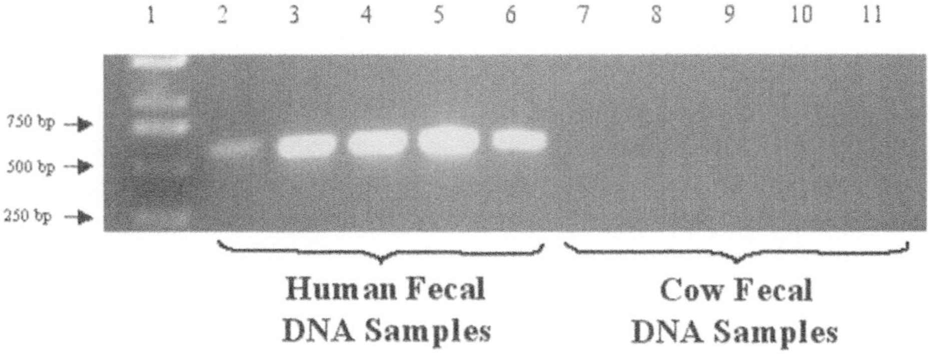

Figure 3. Specific amplification of a human fecal DNA marker with *Bacteroides* 16S rDNA PCR primers HF10 and Bac32F.

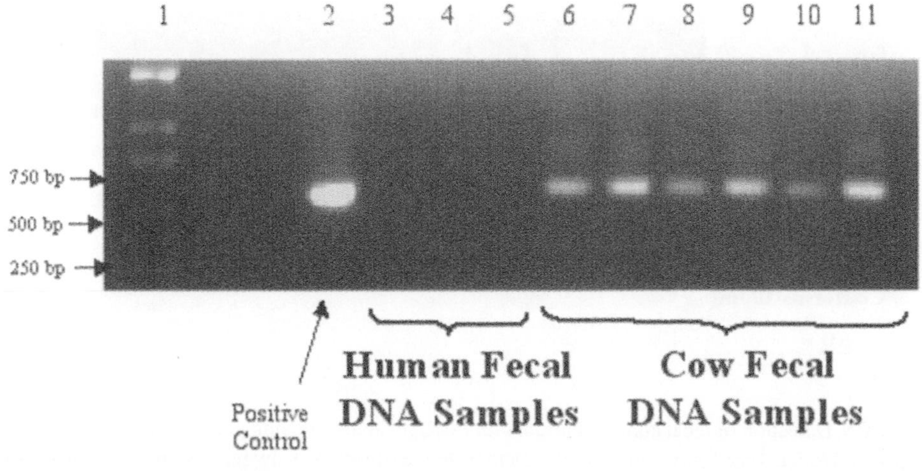

Figure 4. Specific amplification of a cow fecal DNA marker with *Bacteroides* 16S rDNA PCR primers CF151 and Bac32F.

Using general *Bacteroides* PCR primers, fecal pollution can be detected with greater sensitivity than by using membrane filtration of fecal coliforms. Figure 5 shows cow fecal dilutions analyzed by PCR of *Bacteroides* markers and membrane filtration. *Bacteroides* markers were detected to well below the 14 CFU/100ml cutoff, above which shellfish cannot be harvested. However, the irregularity of the PCR bands at the lowest dilutions emphasizes that fecal material in the dilutions was somewhat unevenly dispersed.

The source-specific *Bacteroides* primers are somewhat less sensitive than the general *Bacteroides* primers, because the bacterial groups they sample make up only part of the total *Bacteroides* present. Nevertheless, they are of comparable sensitivity to standard coliform tests (Bernhard and Field, 2000b).

4. Are Genetic Markers from Fecal *Bacteroides* Universal?

The utility of this method of fecal source discrimination will depend on the range of applicability of the markers. Markers represent particular related groups of fecal bacteria, found in one host organism but not another. How widespread will these groups be? Three lines of evidence suggest that the markers will be universally distributed among members of each host group.

First, there is no overlap between cow and human *Bacteroides* sequences. Although LH-PCR and T-RFLP *Bacteroides* peaks from human and cow fecal DNAs largely overlapped, implying that the two hosts share many species, analysis of a large number of *Bacteroides* gene sequences from cow and human feces found no sequence overlap between the two host species (Figure 6) (Mintie *et al.*, 2000 (abstract). Sequences from the human-specific markers were closely related to

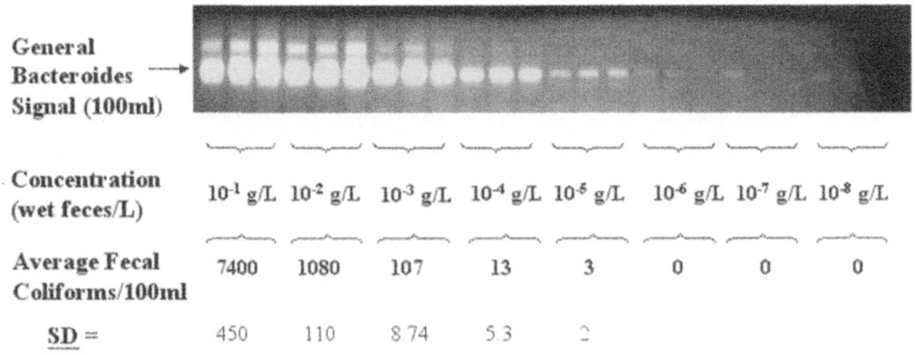

Figure 5. Detection of fecal contamination with PCR/gel analysis using general *Bacteroides* primers compared to fecal coliform counts (CFU) by membrane filtration. Cow feces were mixed with water and diluted. Triplicate subsamples of each dilution were used for membrane filtration of fecal coliforms and DNA extraction followed by PCR detection. 1g wet fecal mass = 1.4 x 10-1 g dry fecal mass.

known human *Bacteroides* isolates such as *B. uniformis*. Sequences from the cow-specific markers formed a previously unknown lineage within the *Cytophaga/Flexibacter/Bacteroides* (CFB) division. Marker clones recovered from water samples were members of the same lineages as the human and cow fecal markers.

Second, these markers occur throughout a broad phylogenetic group (ruminants), suggesting that they co-evolved with host groups. To determine the host specificity of the primers, we tested them with fecal DNAs collected from a broad range of animals (Bernhard and Field, 2000b). Although the human primers only amplified

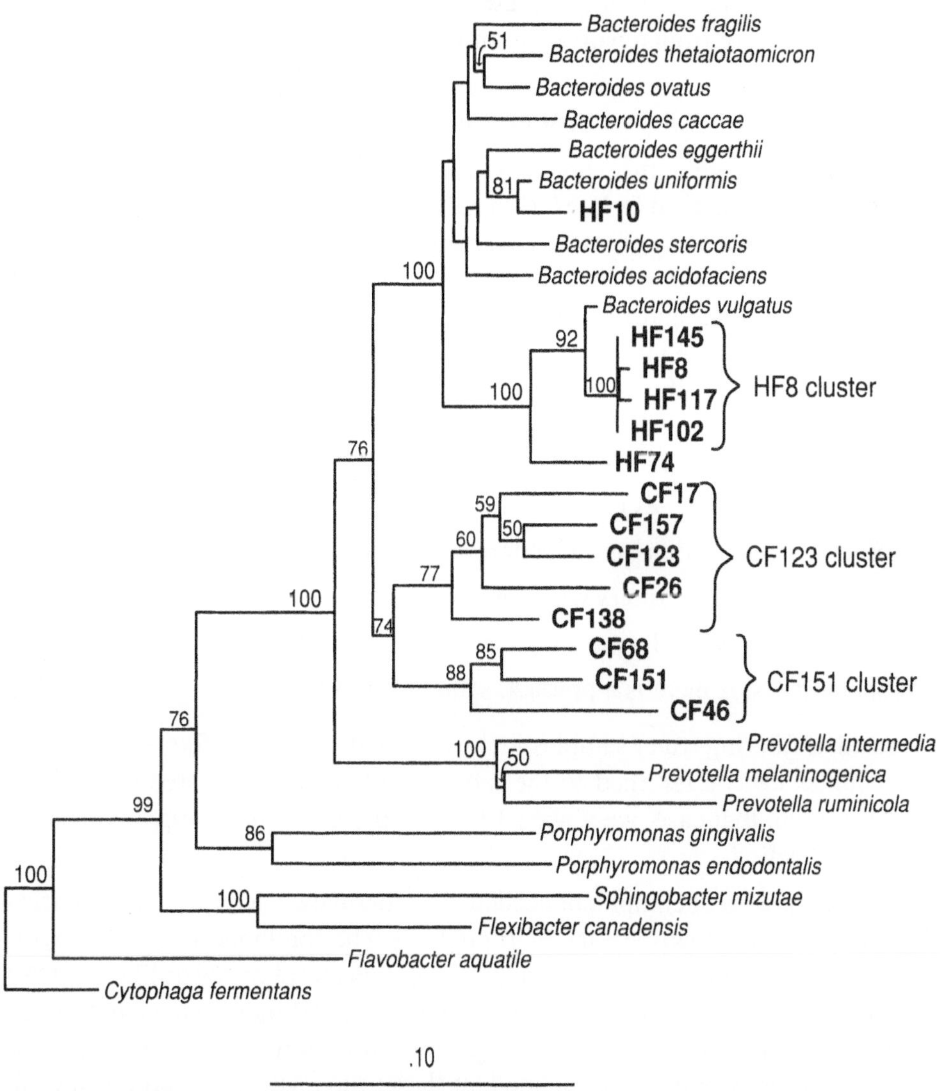

Figure 6. Phylogenetic relationships among partial 16S rDNA sequences of clones from human (HF) and cow (CF) feces. The tree was inferred by neighbor joining. Numbers above internal branches are bootstrap values; values below 50% are not shown.

markers from human fecal DNA, the cow primers amplified markers from all ruminant animals tested, included deer, elk, goat, and sheep, and even the pseudoruminant, llama. This suggests that the unique group of *Bacteroides* (Figure 6) found in cows evolved in ruminants in general. If all ruminants have this unique group, then it is virtually certain that cows in different geographic areas will have it. Depending on choice of primer, either very general or very specific *Bacteroides* groups can be differentiated.

Third, fecal samples from widespread areas of the United States have proven to have the marker bacterial sequences. For example, primers developed for Oregon cow and human feces also amplify DNA markers from the Northeastern U. S. (Figures 3 and 4 are from a Northeastern U.S. study).

5. Protocol for Source Detection

A principal advantage of this approach to source discrimination is that it does not require culturing bacterial isolates. As a result, the method can be very rapid, with a turnaround time of a day or less. Conversely, samples can be stored before analysis if necessary or convenient, since the analysis does not require living cells. This allows for convenient and flexible field handling. In addition, scoring does not require complex pattern interpretation. Either the specific primers amplify a band, in which case that specific type of fecal pollution is present in the sample, or there is no band, in which case the specific source of fecal pollution is absent or present at a level below detection. A summary of the protocol is as follows:

1. Collect water samples (50 mL);
2. Filter samples at time of collection, store filters in lysis buffer;
3. Extract DNA from filters with commercial kit;
4. Perform PCR with diagnostic primers; and
5. Score PCRs: if there is a product, sample is positive.

Instead of filtering water samples in the field and storing the filters in preservative, samples can be returned to the lab, filtered and either stored or analyzed immediately. In that case, we always filter and process water samples within six hours of collection.

The major pieces of equipment needed are a thermal cycler, for PCR amplification, and equipment to inspect the PCR products. The commonest and least expensive method of visualizing PCR products is gel electrophoresis utilizing a DNA stain. This requires a gel box, power source, light or UV light source, and camera or gel documentation system. Also required is filtration equipment (disposable syringes and filters in the field, or vacuum flasks and a vacuum source in the lab) and a microfuge and lab ware for DNA extraction.

This method of source discrimination can be conveniently adapted to high-throughput analysis. For example, DNA extraction, PCR, and scoring (by fluorescence) can all be done by robot.

6. Future Experimental Work

We are currently developing primers for other animal species, and for distinguishing wild from domestic ruminants. In addition, in both lab and field trials, we are establishing the comparative sensitivity of each primer, relative to general *Bacteroides* primers, each other, and coliforms and *E. coli*-based methods. Finally, we are developing a quantitative assay, using real-time quantitative PCR.

Acknowledgements

We are grateful for assistance from Caragwen Bracken. This work was partially supported by Grant NA76RG0476 (project no. R/ECO-04) from the National Oceanic and Atmospheric Administration to the Oregon Sea Grant College Program, by appropriations made by the Oregon State Legislature, by Grant R827639-01-0 from the U. S. Environmental Protection Agency, and by Grant 00-51130-9818 from the U. S. Department of Agriculture.

References

Allsop, K., and Stickler, J.D.: 1985, 'An assessment of *Bacteroides fragilis* group organisms as indicators of human faecal pollution', *J. Appl. Bacteriol.* **58**, 95–99.

Andrews, R. W., Conway, R. A., Corr, C. L., Dobratz, E. J., Dougherty, D. P., Eppard, J. R., Knupp, S. R., Limjoco, M. C., Mettenburg, J. M., Rinehardt, J. M., Sonsino, J., Torrijos, R. L., Zimmerman, M. E., and Wiggins, B. A.: 1997, 'Classification of fecal streptococci isolated from potential sources of fecal pollution using discriminant analysis: supporting evidence from large datasets', Abstracts, American Society of Microbiology General Meeting, Miami Beach, Florida.

Avaniss-Aghajani, E. A., Jones, K., Chapman, D., and Brunk, C.: 1994, 'A molecular technique for identification of bacteria using small subunit ribosomal RNA sequences', *BioTechniques* **17**, 144–149.

Avelar, K. E. S., Morales, S. R., Pinto, L. J. F., Silva e Sousa, R. M. C. P. D., and Ferreira, M. C. d. S.: 1998, 'Influence of stress conditions of *Bacteroides fragilis* survival and protein profiles', *Zent. bl. Bakteriol.* **287**, 399–409.

Bernhard, A. E., and Field, K.G.: 2000a, 'Identification of nonpoint sources of fecal pollution in coastal waters by using host-specific 16S ribosomal DNA genetic markers from fecal anaerobes', *Appl. Environ. Microbiol.* **66**, 1587–1594.

Bernhard, A. E., and Field, K. G.: 2000b, 'A PCR assay to discriminate human and ruminant feces on the basis of host differences in *Bacteroides-Prevotella* genes encoding 16S rRNA', *Appl. Environ. Microbiol.* **66**, 4571–4574.

Bruce, K. D.: 1997, 'Analysis of *mer* gene subclasses within bacterial communities in soils and sediments resolved by fluorescent-PCR-restriction fragment length polymorphism profiling', *Appl. Environ. Microbiol.* **63**, 4914–4919.

Brunk, C. F., Avaniss-Aghajani, E., and Brunk, C. A.: 1996, 'A computer analysis of primer and probe hybridization potential with bacterial small-subunit rRNA sequences', *Appl. Environ. Microbiol.* **62**, 872–879.

Carson, C. A., Shear, B. L., Ellershiek, M. R., and Asfaw, A.: 2001, 'Identification of fecal *Escherichia coli* from humans and animals by ribotyping', *Appl. Environ. Microbiol.* **67**, 1503–1507.

Clement, B. G., Kehl, L. E., DeBord, K. L., and Kitts, C. L.: 1998, 'Terminal restriction fragment patterns (TRFPs), a rapid, PCR-based method for the comparison of complex bacterial communities', *J. Microbiol. Methods* **31**, 135–142.

DePaola, A., Hophins, L. H., Peeler, J. T., Wentz, B., and McPhearson, R. M.: 1990, 'Incidence of *Vibrio parahaemolyticus* in U. S. coastal waters and oysters', *Appl. Environ. Microbiol.* **56**, 2299–2302.

Dombek, P. E., Johnson, L. K., Zimmerley, S. T., and Sadowsky, M. J.: 2000, 'Use of repetitive DNA sequences and the PCR to differentiate *Escherichia coli* isolates from human and animal sources', *Appl. Environ. Microbiol.* **66**, 2572–2577.

Gary, H. L., and Adams, J. C.: 1985, 'Indicator bacteria in water and stream sediments near the snowy range in southern Wyoming', *Water, Air, Soil Pollut.* **25**, 133–144.

Gerba, C. P., and McLeod, J. S.: 1976, 'Effect of sediments on the survival of *Escherichia coli* in marine waters', *Appl. Environ. Microbiol.* **32**, 114–120.

Goyal, S. M., Zerda, K. S., and Gerba, C. P.: 1980, 'Concentration of coliphage from large volumes of water and wastewater', *Appl. Environ. Microbiol.* **59**, 2956–2962.

Harwood, V. J., Butler, J., Parrish, D., and Wagner, V.: 1999, 'Isolation of fecal coliform bacteria from the diamondback terrapin (*Malaclemys terrapin centrata*)', *Appl. Environ. Microbiol.* **65**, 865–867.

Havelaar, A. H., van Olphen, M., and Drost, Y. C.: 1993, 'F-specific RNA bacteriophages are adequate model organisms for enteric viruses in fresh water', *Appl. Environ. Microbiol.* **59**, 2956–2962.

Holdeman, V., Cato, E. P., and Moore, W. E. C.: 1976, 'Human fecal flora: variation in bacterial composition within individuals and a possible effect of emotional stress', *Appl. Environ. Microbiol.* **31**, 359–375.

Hood, M. A., and Ness, G. E.: 1982, 'Survival of *Vibrio cholerae* and *Escherichia coli* in estuarine waters and sediments', *Appl. Environ. Microbiol.* **43**, 578–584.

Howell, J. M., Coyne, M. S., and Cornelius, P. L.: 1996, 'Effect of sediment particle size and temperature on fecal bacteria mortality rates and the fecal coliform/fecal streptococci ratio', *J. Environ. Qual.* **25**, 1216–1220.

Kreader, C. A.: 1995, 'Design and evaluation of *Bacteroides* DNA probes for the specific detection of human fecal pollution', *Appl. Environ. Microbiol.* **61**, 1171–1179.

Kreader, C. A.: 1998, 'Persistence of PCR-detectable *Bacteroides* distasonis from human feces in river water', *Appl. Environ. Microbiol.* **64**, 4103–4105.

LaLiberte, P., and Grimes, D. J.: 1982, 'Survival of *Escherichia coli* in lake bottom sediment', *Appl. Environ. Microbiol.* **43**, 623–628.

Liu, W.-T., Marsh, T. L., Cheng, H., and Forney, L. J.: 1997, 'Characterization of microbial diversity by determining terminal restriction fragment length polymorphisms of genes encoding 16S rRNA', *Appl. Environ. Microbiol.* **63**, 4516–4522.

Mintie, A., Bernhard, A. E., Trippet, D., Cooper, B., Buccafurni, D., and Field, K. G.: 2000 (abstract), 'Phylogenetic diversity of *Bacteroides-Prevotella* 16S rRNA genes from cow and human feces', American Society for Microbiology, Los Angeles, CA.

Moore, W. E. C., and Holdeman, L. V.: 1974, 'Human fecal flora: the normal flora of 20 Japanese-Hawaiians', *Appl. Microbiol.* **27**, 552–555.

Osawa, S., Furuse, K., and Watanabe, I.: 1981, 'Distribution of ribonucleic coliphages in animals', *Appl. Environ. Microbiol.* **41**, 164–168.

Palmateer, G. A., Dutka, E. M., Jantzen, E. M., Meissner, S. M., and Sakeellaries, M. G.: 1991, 'Coliphage and bacteriophage as indicators of recreational water quality', *Water Research* **25**, 355–357.

Parveen, S., Portier, K. M., Robinson, K., Edmiston, L., and Tamplin, M.: 1999, 'Discriminate analysis of ribotype profiles of *Escherichia coli* for differentiating human and nonhuman sources of fecal pollution', *Appl. Environ. Microbiol.* **65**, 3142–3147.

Paster, B. J., Dewhirst, F. E., Olsen, I., and Fraser, G. J.: 1994, 'Phylogeny of *Bacteroides*, *Prevotella*, and *Porphyromonas* spp. and related bacteria', *J. Bacteriol.* **176**, 725–732.

Paul, J. H., Rose, J. B., Jiang, S. C., London, P., Xhou, X., and Kellogg, C.: 1997, 'Coliphage and indigenous phage in Mamala Bay, Oahu, Hawaii', *Appl. Environ. Microbiol.* **63**, 133–138.

Pommepuy, M., Guillaud, J. F., Dupray, E., Derien, A., Le Guyader, F., and Cormier, M.: 1992, 'Enteric bacteria survival factors', *Water Sci. Technol.* **12**, 93–103.

Rappé, M. S., Kemp, P. F., and Giovannoni, S. J.: 1997, 'Phylogenetic diversity of marine coastal picoplankton 16S rRNA genes cloned from the continental shelf off Cape Hatteras, North Carolina', *Limnol. Ocean.* **42**, 811–826.

Salyers, A. A.: 1984, '*Bacteroides* of the human lower intestinal tract', Ann. Rev. *Microbiol.* **38**, 293–313.

Sherer, B. M., Miner, J. R., Moore, J. A., and Buckhouse, J. C.: 1988, 'Resuspending organisms from a rangeland stream bottom', *Trans. Amer. Soc. Ag. Eng.* **31**, 1217–1222.

Singh, S. N., and Gerba, C. P.: 1983, 'Concentration of coliphages from water and sewage with charge-modified filter aid.', *Appl. Environ. Microbiol.* **45**, 232–237.

Sinton, L. W., Donnison, A. M., and Hastie, C. M.: 1993, 'Faecal streptococci as faecal pollution indicators: a review. Part II: Sanitary significance, survival, and use', *New Zealand J. Marine Freshwater Res.* **27**, 117–137.

Sobsey, M. D.: 1989, 'Inactivation of health-related microorganisms in water by disinfection processes', *Water Sci. Technol.* **21**, 179–195.

Straub, D. V., and Dixon, B.: 1997, '*Bacteroides vulgatus*, an alternative indicator for the assessment of fecal contamination of shellfish and estuarine waters', 38th Annual Western Fish Disease Workshop, Bodega Marine Laboratory, Bodega Bay, CA.

Suzuki, M. T., Rappé, M. S., Haimberger, Z. W., Winfield, H., Adair, N., Ströbel, J., and Giovannoni, S. J.: 1997, 'Bacterial diversity among small-subunit rRNA gene clones and cellular isolates from the same seawater sample', *Appl. Envir. Microbiol.* **63**, 983–989.

Tartera, C., and Jofre, J.: 1987, 'Bacteriophages active against *Bacteroides fragilis* in sewage polluted waters', *Appl. Environ. Microbiol.* **53**, 1632–1637.

Tartera, C., Lucena, F., and Jofre, J.: 1989, 'Human origin of *Bacteroides fragilis* bacteriophages present in the environment', *Appl. Environ. Microbiol.* **55**, 2696–2701.

Weiskel, P. K., Howes, B. L., and Heufelder, G. R.: 1996, 'Coliform contamination of a coastal embayment: sources and transport pathways', *Environ. Sci. Technol.* **30**, 1872–1881.

Wiggins, B. A.: 1996, 'Discriminant analysis of antibiotic resistance patterns in fecal streptococci, a method to differentiate human and animal sources of fecal pollution in natural waters', *Appl. Environ. Microbiol.* **62**, 3997–4002.

Wiggins, B. A., Andrews, R. W., Conway, R. A., Corr, C. L., Dobratz, E. J., Dougherty, D. P., Eppard, J. R., Knupp, S. R., Limjoco, M. C., Mettenburg, J. M., Rinehardt, J. M., Sonsino, J., Torrijos, R. L., and Zimmerman, M. E.: 1999, 'Use of antibiotic resistance analysis to identify nonpoint sources of fecal pollution', *Appl. Environ. Microbiol.* **65**, 3483–3486.

CHARACTERIZATION OF MICROBIAL COMMUNITIES FROM COASTAL WATERS USING MICROARRAYS

O. COLIN STINE[1*], AMY CARNAHAN[1], RUBY SINGH[2], JAN POWELL[1], JON P. FURUNO[1], ALICIA DORSEY[1], ELLEN SILBERGELD[1], HENRY N. WILLIAMS[2], AND J. GLENN MORRIS[1]

[1]*Department of Epidemiology and Preventive Medicine, School of Medicine;*
[2]*School of Dentistry, University of Maryland Baltimore, Baltimore, Maryland, U.S.A.*
**(author for correspondence, email: ostin001@umaryland.edu)*

Abstract. Molecular methods, including DNA probes, were used to identify and enumerate pathogenic *Vibrio* species in the Chesapeake Bay; our data indicated that *Vibrio vulnificus* exhibits seasonal fluctuations in number. Our work included a characterization of total microbial communities from the Bay; development of microarrays that identify and quantify the diversity of those communities; and observation of temporal changes in those communities. To identify members of the microbial community, we amplified the 16S rDNA gene from community DNA isolated from a biofilm sample collected from the Chesapeake Bay in February, 2000. The resultant 75 sequences were 95% or more similar to 7 species including two recently described *Shewanella* species, *baltica* and *frigidimarina*, that have not been previously isolated from the Chesapeake. When the genera of bacteria from biofilm after culturing are compared to those detected by subcloning amplified 16S fragments from community DNA, the cultured sample exhibited a strong bias. In oysters collected in February, the most common bacteria were previously unknown. Based on our 16S findings, we are developing microarrays to detect these and other microbial species in these estuarine communities. The microarrays will detect each species using four distinct loci, with the multiple loci serving as an internal control. The accuracy of the microarray will be measured using sentinel species such as *Aeromonas* species, *Escherichia coli*, and *Vibrio vulnificus*. Using microarrays, it should be possible to determine the annual fluctuations of bacterial species (culturable and non-culturable, pathogenic and non-pathogenic). The data may be applied to understanding patterns of environmental change; assessing the "health" of the Bay; and evaluating the risk of human illness associated with exposure to and ingestion of water and shellfish.

Keywords: Microarrays, Chesapeake Bay, non-culturable bacteria, DNA sequencing, microbial ecology

1. Introduction

The complex microbial flora of the Chesapeake Bay is poorly studied due to the technical difficulties involved in isolation and identification of the diverse culturable and non-culturable populations present. Even among culturable bacteria, characterization of this diverse community requires an enormous amount of labor to identify and enumerate the bacteria at a single time point (Mallory *et al.*, 1977). Studies of single species have revealed substantial seasonal variation. However, advances in molecular genetics, particularly, the availability of high-throughput sequencing and microarrays, provide, for the first time, the basis for a comprehensive characterization of the ecology of the Bay's bacterial flora. To do so requires a DNA sequence database of the bacteria from the Bay and the development of microarrays based on those sequences (Stine, 2001).

Previous characterizations of the seasonal variation of microbial flora have focused on single species of bacteria. Identification of a selected species often in-

Environmental Monitoring and Assessment **81**: 327–336, 2003.

volves growth on a semi-selective medium followed by hybridization utilizing a specific DNA probe. In the Chesapeake region this methodology has been applied to *Aeromonas hydrophila* (Kaper *et al.*, 1981), *Bdellovibrio* (Williams *et al.*, 1980),*Vibrio parahaemolyticus* (Colwell *et al.*, 1977) and *V. vulnificus* (Wright *et al.*, 1996). These studies have shown marked seasonal variation in the studied species, with all being more common in the summer (and generally non-detectable during the winter months). *Bdellovibrio*-like organisms (BLO) are observed in the Chesapeake Bay most commonly in the spring and, geographically, in the middle (from N to S) regions (Williams *et al.*, 1982). These studies were dependent on the ability to culture BLO on *V. parahaemolyticus*. Similarly, the distribution of *V. vulnificus* in the Chesapeake Bay has been assessed using cultures and a species-specific DNA oligonucleotide probe (VVAP) derived from the hemolysin gene (Wright *et al.*, 1993). *V. vulnificus* was found to constitute about 10% of the culturable bacteria between May and December. In contrast, it was not detected in February and March even though the total bacterial count was similar all year long. The results were similar in samples taken from oysters except the total bacterial count went down in the February and March. *V. vulnificus* was most commonly isolated from the middle regions of the Bay (Wright *et al.*, 1996).

The identities of the bacteria in the Chesapeake Bay have not been completely ascertained. Of the single species studies, *V. vulnificus* (one of the most common species in the Bay) at its maximal abundance only accounts for 10% of the culturable bacteria. As their name implies, the unculturables are difficult to study by conventional methods. Their presence has been demonstrated by isolating "community" DNA from environmental samples, and then amplifying 16S ribosomal RNA genes (rDNA) using "universal primers," followed by subcloning and sequencing of the subclones. These sequences are then compared to those in Genbank as a means of identifying whether they are known or unknown. Looking in extreme environments, this method has revealed novel divisions of bacteria (Barns *et al.*, 1994; Hugenholtz *et al.*, 1996). A comparison of cultured and uncultured bacteria from the North Sea demonstrated that the uncultured bacteria constituted the vast majority (99%) of the total bacteria (Eilers *et al.*, 2000). The previously unknown and uncultured bacteria may be known to us only by the sequences fragment of the 16S rDNA gene and the origin of the sample in which they are found (Field *et al.*, 1997). The methods that currently permit the monitoring of unculturable species, FISH and chromosome painting, are dependent upon the availability of probes specific for each species and are labor intensive, particularly if more than one species is to be monitored. Microarrays have the potential to monitor multiple species simultaneously with less labor.

16S rDNA is the paradigm for studying evolutionary change in bacteria. The phylogenetic relatedness of even unculturable bacteria can be determined from 16S sequences (Maidek *et al.*, 1999). Ribosomal RNA is found in every known bacterium because it is necessary for the translation of messenger RNA into protein. Although some small (20–30 base pairs) regions of the 16S sequence do not vary among bacteria, the majority of the molecule (c. 1400 base pairs) has a different sequence from one bacterial species to the next. The more distantly related the species are, the larger the number of changes between their respective sequences of the 16S gene. However, the strong conservation of the 16S genes means that in some genera, species level identification cannot be made using 16S.

The evolutionary paradigm extends to sequence differences among all homologous genes. Homology refers to genes that are genetically related and are found in different organisms. The major difference is that other genes tend to vary more rapidly over evolutionary time than the 16S gene. Thus, other loci can be used to distinguish between closely related species and strains within species. An example is the *rpoB* (RNA polymerase beta subunit) gene (Mollet *et al.*, 1997; Dahloff *et al.*, 2000). There are no known "universal" primers for other genes. We have designed primers that will amplify genes across the boundaries of genus and species, but to cover the entire bacterial kingdom for another gene, even a gene that is highly conserved and found universally, multiple primer sets will be needed.

In microarrays, a capture DNA sequence is fixed to the substrate surface and the sample DNA is labeled and placed in solution. The advantage provided by the microarray methodology is that many genes may be probed simultaneously. Commercial microarrays are treated glass slides with up to 8,000 genes or probes. Microarrays are just beginning to be used for bacterial identification. Simple microarrays based on 16S rDNA sequences have already been demonstrated (Bavykin *et al.*, 2001; Edman *et al.*, 2000). However, 16S genes have only limited ability to discriminate within and between species. In some cases, the differences are only a single nucleotide, a level of distinction that is difficult to resolve using microarrays. Sequence information from other genes has the potential to discriminate within and between species and can provide internal controls against potential cross-hybridization. We are developing the databases necessary to distinguish which species are present in the Chesapeake Bay.

2. Microbial Communities in the Chesapeake Bay

To identify the bacterial species inhabiting the Chesapeake Bay, we collected community DNA samples from biofilms in 2000; and from oysters in 2001. These samples were either cultured or subjected to direct isolation of "community" DNA using a modification of the bead beating technique (Hugenholtz *et al.*, 1998). The modifications involve eliminating the organic extractions in favor of salt precipitations to remove proteins and recover nucleic acids, primarily DNA. This "community" DNA was amplied by PCR initiated using "universal" primers for 16S rDNA. The amplified product was subcloned using TA cloning kit (Invitrogen) that takes advantage of the propensity of *Taq* polymerase to add an overhanging A, when completing the synthesis of a new fragment. The subclones were cycle sequenced using a Big Dye Terminator Kit (Applied Biosystems, Inc.). The resulting fragments were separated using an ABI 3700 Automatic Sequencer. As an internal check and to maximize the quality of the data, each subclone was sequenced in both directions. The raw data in the form of trace files from both forward and reverse sequencing reactions were analyzed by the programs PHRED and PHRAP (Ewing and Green, 1998; Ewing *et al.*, 1998; Green, 1998) and viewed using CONSED (Gordon *et al.*, 1998). Low quality sequences from the ends of the sequence were removed. The sequences were aligned using CLUSTALX (Jeanmougin *et al.*, 1998). The genetic relatedness between strains was compared using the Neighbor Joining and Bootstrapping algorithms of PAUP (Swofford, 2000). The sequences were compared to known sequences using BLASTN against Genbank (Altschul *et al.*, 1997).

Community DNA was isolated from a cultured sample of Chesapeake Bay biofilm collected in February, 2000. The 75 subcloned rDNA sequences were compared to known sequences in Genbank and were 95% or more similar to 7 species, all in the gamma-Proteobacteria. The species were: *Aeromonas jandaei*, *Shewanella baltica*, *Aeromonas salmonicida*, *Shewanella frigidmarina*, *Escherichia coli*, *Aeromonas trota* and *Vibrio vulnificus*. The most commonly detected species was *A. jandaei* followed by *S. baltica*. However, the identities of the *Aeromonas* species should be marked with an asterisk, since *Aeromonas* is the paradigmatic example of a genus in which the 16S sequence is indicative of the genus, but does not identify the species within the genus. Two biochemically distinct species often have strains that share an identical sequence and a single biochemically distinct species may have strains with slightly different sequences. Thus, species identification based on DNA sequences must depend upon sequence from other loci (Carnahan *et al.*, 2001; Carnahan, 2001).

The two *Shewanella* sequences detected in the cultured sample of Chesapeake Bay biofilm collected in February correspond to species which were only recently proposed based on specimens from the Baltic Sea and Antarctica (Ziemke *et al.*, 1998, Reid *et al.*, 1999). Although not previous known in the Chesapeake Bay, their presence in a winter sample may be significant.

Cultivation of bacteria from biofilms is likely to produce a biased sample of the bacteria present. For example, the genera of bacteria growing in a biofilm were identified with and without culturing. Biofilm was removed from standard polyvinyl tubing in plumbing fed by a municipal water supply. An aliquot was cultured on the recommended 'nonselective' media (R2A medium). All 55 colonies were picked and identified using 16S rDNA sequence, 15 genera were identified. The two most common, in order, were *Afipia* and *Sphingomonas*. Another aliquot was subjected to direct isolation of community DNA. After PCR amplification and subcloning, 168 sequences revealed the presence of 50 genera. The six most common, in order, were *Leptospira*, *Sphingomonas*, *Escherichia*, *Bacillus*, *Pseudomo-*

nas and *Afipia*. Thus, the cultured and the community DNA samples estimate different relative frequencies for commonly found genera, perhaps indicative of a bias induced by the culture conditions. In addition, several genera were detected that require special conditions for growth. These included *Bdellovibrio*, a predator, the bacterial endosymbiant of *Acanthameoba* (a eukaryotic predator), *Geobacter*, an anaerobe, *Ketogulonigenium*, another anaerobe, and *Legionella*, a potential pathogen. Thus, the community sample detects bacteria that cannot be grown under a single set of culture conditions that bias the sample.

Analysis of community DNA from oysters collected from the Chesapeake Bay in February revealed that the most common bacteria are unknown. All 57 subcloned sequences were very similar, differing by less than 3% of the nucleotides. Whether these sequences represent a single bacterial species or a single genus, as in the case of *Aeromonas*, requires analyses of other loci. The closest 16S rDNA sequence in Genbank differs at 18 to 20% of the nucleotides. The closest 10 sequences are all from uncultured bacteria, i.e. bacteria that are known only by their 16S sequence and their sample origin. Twenty percent differences between 16S sequences can be seen between different orders and different divisions of bacteria. Thus, the taxonomy of the bacteria in oysters is unclear.

Additional genes can be analyzed directly from community DNA. For example, bacterial tetracycline resistance genes can be identified using specific primers and PCR amplification. When community DNA from the oyster sample is amplified using *tetL* specific primers, an agarose gel reveals a clear band at the same position as the amplified product from cultured strains known to contain the *tetL* genes. Thus, this population harbors the *tetL* gene and some of the members are presumably tetracycline resistant. These data are not totally unexpected, as tetracycline has been detected in Bay sediments (Meyer, personal communication).

3. Genomic Changes in Bacterial Species

Identifying strains of a single bacterial species or differentiating among closely related species can be done using multilocus sequencing typing (MLST) (Maiden *et al.*, 1998). We have developed primers for *recA* for *Vibrios*. The primers amplify the base sequence from 813-1598 (numbering based on the *V. cholerae* sequence (accession # X71969)) and were selected from two regions conserved between *V. cholerae* and *V. anguillarum* (accession # M80525). Unpublished data also demonstrate that these primers successfully amplify the *recA* locus from *E. coli* and *Salmonella* spp. The genetic relatedness based on recA sequence from strains representing 5 species of the genus *Vibrio* is shown in Figure 1. The neighbor joining tree of selected strains from *V. cholerae*, *V. parahaemolyticus*, *V. mimicus*, *V. fluvialis* and *V. vulnificus*, all of which are found in the Chesapeake Bay, is consistent with a relatedness estimated using biochemical methods. In particular, *V. cholerae*, and *V. mimicus* are sibling species as expected. Bootstrapping reveals that these sequences are distinct 100% of the time in 1000 replicates. Biochemically, the distinct (by bootstrap analysis) clusters labeled *V. mimicus*, *V. fluvialis*, "5", and the one containing A-5 and NRT-36S all have members that are biochemically identified as *V. cholerae*. Thus, either there is a substantial amount of genetic divergence within *V. cholerae* or the biochemical tests may be more variable within the divergent species than previously thought (Stine *et al.*, 2000).

The biochemical assays and the DNA sequence based systems of taxonomy for bacteria are both based on how bacterial genomes are evolving. Bacteria have very small genomes and live in highly diverse environments. The diversity of environments requires a selected set of enzymes, not all of which can be found in any single, small bacterial genome. Thus, bacteria genomes have evolved through a process of gaining and losing genes. The most remarkable example of this process is the potentially deadly bloody diarrhea causing *E. coli* O157:H7. It is thought that O157 evolved from harmless *E. coli* strains like K12. The complete sequence of each of these 2 strains reveals they differ by the presence of over 1500 out of 5000 genes that are in one, but not both genomes (Perna *et al.*, 2001). Any one of these genes may be used to differentiate between the strains. Although investigators were surprised by the magnitude of the number of genes that distinguish the two *E. coli* strains, the principle that distinct strains will have a number of genetic differences was expected.

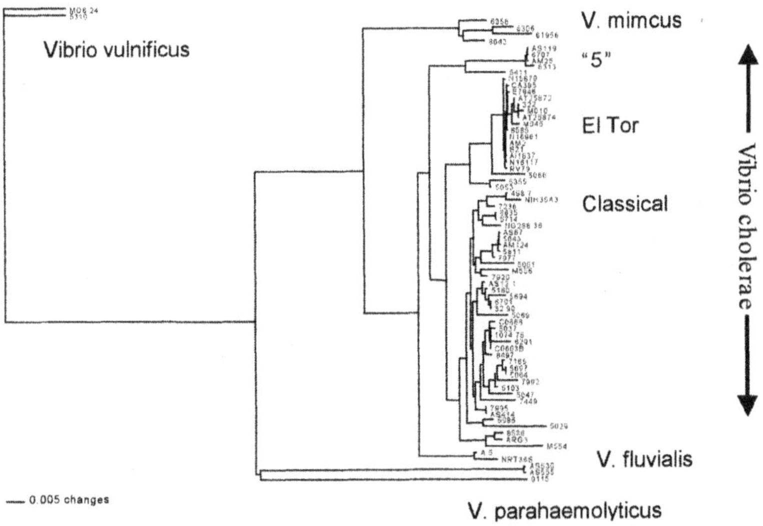

Figure 1. Neighbor joining tree demonstrating the genetic relatedness based on sequence from *recA* of selected species of *Vibrios*. Each of the 76 strains is identified by its number (complete table is in Stine *et al.*, 2000). Within the species *V. cholerae*, the Classical and El Tor clusters designate strains from the 6th and 7th pandemics. The group "5" identifies strains that are outliers by *recA* sequence, but are within the species as defined by biochemical tests. The horizontal bars are proportional to the genetic distance in changes in base pairs over the entire fragment length (705 bp) between strains (see scale bar), the vertical distances are for convenience.

The traditional biochemical tests exploit the known environmental source and the presence and absence of specific enzymes in particular genomes to determine the species of bacteria. Biochemical assays are limited to culturable species and are very labor intensive. One labor saving modification to this methodology was the advent of species specific probes. These probes permit rapid screening of many cultured colonies for a single species. This method is also labor intensive, particularly, if more than one species is to be surveyed. In addition, strain specific probes have not been developed except for virulence genes in some pathogens.

4. Microarrays for Monitoring Microbes

Microarrays can combine both DNA sequence diversity and the presence and absence of specific genes into a single test. Each method can be reduced to whether or not the DNA hybridizes to a probe. For homologous genes, the greater the diversity between the sequences the less the hybridization between those two sequences. If the difference reaches approximately 70%, then the sequences will not hybridize even under conditions of low stringency. Thus, careful selection of the gene sequences used for probes on a microarray will permit two distinct bacterial species to be resolved. For genes that are present in one genome but not the other, any fragment of those genes is likely to distinguish between the species. For example, genes can be used to distinguish between serotypes of *V. cholerae*. The two common epidemic types O1 and O139 can be identified by several distinct genes that are not found in the other strain. However, other strains may complicate the interpretation of the results. Thus, if *V. cholerae* of the serotype O22 is included, in the test the set of genes used as probes would need to be expanded to distinguish this third type that has many genes that are similar to the O139 genes (Yamasaki *et al.*, 1999). However, the solution, using several probes to accommodate each strain, is easy to incorporate into a microarray.

Microarrays for detecting bacterial species should be designed to contain multiple genes to detect each desired taxa. Multiple genes or probes will serve as internal controls. Each probe should provide the same information about the frequency of the taxa. If not, the culprit in the community DNA sample would be similar DNA that cross-hybridizes, as in the *V. cholerae* O139 example. The simplest response is to ignore the outlier and use the data from the other probes. Our plan is to measure the accuracy of identification and quantification by the microarray using culturable bacteria such as *Aeromonas* species, *E. coli*, and *Vibrio vulnificus* as sentinels.

Microarray technology is not yet fully developed. Difficulties exist when trying to quantify from one sample to the next. Most microarray experiments use relative values from one sample to the next rather than absolute quantities (Schena *et al.*, 1996). A second problem is the nature of the fragments to be placed in the array. The two competing methods are PCR amplified DNA and oligonucleotide probes of varying length. The principal advantages for the former are the availability, price and ease. The latter method may provide more specific resolution and better quantification, but the method of choice for a particular experiment is not obvious. A third problem is the cost of performing a microarray analysis and the equipment necessary to evaluate the data. The current technology was developed to be sold to large pharmaceutical companies and price was not limiting. However, if microarrays with fewer capture DNAs are used and simpler optical reading systems are adapted and developed, the cost could be decreased substantially. These and other problems are likely to be solved as more applications of the technology are explored.

The promise of the microarray can only be as good as the database used to build it. Environmental samples need to be characterized so that microarrays can be designed appropriately. The genetic diversity within and between species needs to be explored, at a more detailed level than simply the complete genomic sequence. Distinguishing between species and strains will require known sequence variation at several loci from representatives of the community, as well as species that are of potential concern if found in a particular sample. For example, if the *E. coli* in the Chesapeake Bay were serotype O157, it would be a greater concern than just finding *E. coli*. Thus, microarrays for monitoring microbial communities should include not only all the expected members of the community, but also an assortment of potential pathogens.

Microarrays have the potential to be used for monitoring bacterial populations. Microarrays can confirm whether specific bacteria are present in drinking water, harvest areas for shellfish, recreational waters, and foods. Microarrays can measure both culturable and non-culturable bacteria providing a more comprehensive picture of the changes in microbial communities. Through the use of microarrays to quantify the culturable and non-culturable bacterial species in the Bay, an understanding of the annual fluctuations and the response to environmental change of bacterial species, and evaluating the risk of human illness associated with exposure to and ingestion of water and shellfish will be obtained. Thus, microarrays developed to assess microbial communities should provide a new method to assess and monitor the "health" of the Chesapeake Bay.

References

Altschul S.F., Madden T.L., Schäffer A.A., Zhang J., Zhang Z., Miller W. and Lipman D.J.: 1997, 'Gapped BLAST and PSI-BLAST: a new generation of protein database search programs'. *Nucleic Acids Research* **25**, 3389–3402.

Barns S.M., Fundyga R.E., Jeffries M.W. and Pace N.R.: 1994, 'Remarkable archaeal diversity detected in a Yellowstone National Park hot spring environment', *Proceedings National Academy of Sciences, U.S.A.* **91**(5), 1609–1613.

Bavykin, S.G., Akowski, J.P., Zakhariev, V.M., Barsky, V.E., Perov, A.N. and Mirazabekov, A.D.: 2001, 'Portable system for microbial sample preparation and oligonucleotide microarray analysis', *Applied Environmental Microbiology* **67**, 922–928.

Carnahan, A., 2001, 'The application of DNA sequence diversity to the taxonomy of the genus *Aeromonas*.' Dissertation. University of Maryland.

Carnahan, A.M., Stine, O.C., Morris, J.G., Waddington, M. and Joseph, S.W.: 2001, 'MLST (Multilocus strain typing) of a diverse group of *Aeromonas* strains.' 101st General Meeting, American Society for Microbiology, Orlando, Florida, May 20–24, 2001.

Colwell, R.R., Kaper, J. and Joseph, S.W.: 1997, '*Vibrio cholerae*, *Vibrio parahaemolyticus*, and other vibrios: occurrence and distribution in Chesapeake Bay', *Science* **198**(4315), 394–396.

Dahllof, I., Baillie, H. and Kjelleberg, S.: 2000, '*rpoB*-based microbial community analysis avoids limitations inherent in 16S rRNA gene intraspecies heterogeneity', *Applied Environmental Microbiology* **66**(8), 3376–3380.

Edman, C.F., Mehta, P., Press, R., Spargo, C.A., Walker, G.T. and Nerenberg, M.: 2000, 'Pathogen analysis and genetic predisposition testing using microelectronic arrays and isothermal amplification', Journal of Investigative Medicine 48, 93–101.

Eilers, H., Pernthaler, J., Glockner, F.O. and Amann, R.: 2000, 'Culturability and In situ abundance of pelagic bacteria from the North Sea', *Applied Environmental Microbiology* **66**(7), 3044–3051.

Ewing, B. and Greene, P: 1998, 'Base-Calling of Automated Sequencer Traces Using Phred. II. Error Probabilities', *Genome Research* **8**, 186–194.

Ewing, B., Hillier, L., Wendl, M. and Greene, P.: 1998, 'Base-calling of Automated sequencer Traces Using Phred. I. Accuracy Assessment', *Genome Research* **8**, 175–185.

Field, K.G., Gordon, D., Wright, T., Rappe, M., Urback, E., Vergin, K. and Giovannoni, S.J.: 1997, 'Diversity and depth-specific distribution of SAR11 cluster *rRNA* genes from marine planktonic bacteria', *Applied Environmental Microbiology* **63**(1), 63–70.

Gordon, D., Abajian, C. and Green, P.: 1998, 'Consed: A Graphical Tool for Sequence Finishing', *Genome Research* **8**, 195–202

Green, P.: 1998. 'Phrap, SWAT and CrossMatch', Available with permission from the author at http://www.washington.edu.

Hugenholtz, P., Pitulle, C., Hershberger, K.L. and Pace, N.R.: 1998, 'Novel division level bacterial diversity in a Yellowstone hot spring', *Journal of Bacteriology* **180**(2), 366–376.

Jeanmougin, F., Thompson, J.D., Gouy, M., Higgins, D.G. and Gibson, T.J.: 1998, ' Multiple sequence alignment with Clustal X', *Trends in Biochemical Science* **23**, 403–405.

Kaper, J.B., Lockman, H., Colwell, R.R. and Joseph, S.W.: 1981, '*Aeromonas hydrophila*: ecology and toxigenicity of isolates from an estuary', *Journal of Applied Bacteriology* **50**(2), 359–377.

Knight, I.T., Shults, S., Kaspar, C.W. and Colwell, R.R.: 1990, 'Direct detection of *Salmonella* spp. in estuaries by using a DNA probe', *Applied Environmental Microbiology* **56**(4), 1059–1066.

Mallory, L.M., Austin, B. and Colwell, R.R.: 1997, 'Numerical taxonomy and ecology of oligotrophic bacteria isolated from the esturine environment', *Canadian Journal of Microbiology* **23**, 733–750.

Maidek, B.L., Cole, J.R., Parker, C.T., Garrity, G.M., Larsen, N., Li, B., Lilburn, T.G., McCaughey, M.J., Olsen, G.J., Overbeek, R., Pramanik, S., Schmidt, T.M., Tiedje, J.M. and Woese, C.R.: 1999, 'A new version of the RDP (Ribosomal Database Project). *Nucleic Acid Research* **27**, 171–173.

Maiden, M.C.J., Bygraves, J.A., Feil, E., Morelli, G., Russell, J.E., Urwin, R., Zhang, Q., Zhou, J., Zurth, K., Caugant, D.A., Feavers, I.M., Achtman, M., Spratt, B.G.: 1998, ' Multilocus sequence typing: A portable approach to the identification of clones within populations of pathogenic microorganisms', *Proceedings National Academy of Sciences, U.S.A.* **95**, 3140–3145.

Mollet, C., Drancourt, M. and Raoult, D.: 1997, '*rpoB* sequence analysis as a novel basis for bacterial identification', *Molecular Microbiology* **26**(5), 1005–1011.

Perna, N.T., Plunkett, G., Burland, V., Mau, B., Glasner, J.D., Rose, D.J., Mayhew, G.F., Evans, P.S., Gregor, J., Kirkpatrick, H.A., Posfai, G., Hackett, J., Klink, S., Boutin, A., Shao, Y., Miller, L., Grotbeck, E.J., Davis, N.W., Lim, A., Dimalanta, E.T., Potamousis, K.D., Apodaca, J., Anatharaman, T.S., Lin, J., Yen, G., Schwartz, D.C., Welch, R.A. and Blattner, F.R.: 2001, 'Genomic sequence of entorhaemorrhagic *Escherichia coli* O157:H7', *Nature* **409**, 529–533.

Reid, G.A. and Gordon, E.H.: 1999, 'Phylogeny of marine and freshwater *Shewanella*: reclassification of *Shewanella putrefaciens* NCIMB 400 as *Shewanella frigidimarina*', *International Journal of Systematic Bacteriology* **49**(1), 189–191.

Schena, M., Shalon, D., Heller, R., Chai, A., Brown, P.O. and Davis, R.W.: 1996, 'Parallel human genome analysis: microarray-based expression monitoring of 1000 genes', *Proceedings National Academy of Sciences, U.S.A.* **93**(20), 10614–10619.

Stine, O.C.: 2001, 'Utilizing microarrays of bacterial genes', *SIGBIO Newsletter* **21**, 23–25.

Stine, O.C., Sozhamannan, S., Gou, Q., Zheng, S., Morris, J.G. and Johnson, J.A.: 2000, 'Phylogeny of *Vibrio cholerae* based on *recA* sequence', *Infection and Immunity* **69**, 7180–7185.

Swofford, D.: 2000, 'PAUP*', commercial beta version 4.0b3, Sinauer Assoc. Sunderland, MA.

Williams, H.N., Falkler, W.A. Jr. and Shay, D.E.: 1980, 'Incidence of marine *Bdellovibrios* lytic against *Vibrio parahaemolyticus* in Chesapeake Bay', *Applied Environmental Microbiology* **40**(5), 970–972.

Williams, H.N., Falkler, W.A, Jr. and Shay, D.E.: 1982, 'Seasonal distribution of *Bdellovibrios* at the mouth of the Patuxent River in the Chesapeake Bay', *Canadian Journal of Microbiology* **28**(1), 111–116.

Wright, A.C., Hill, R.T., Johnson, J.A., Roghman, M.C., Colwell, R.R. and Morris, J.G. Jr.: 1996, 'Distribution of *Vibrio vulnificus* in the Chesapeake Bay', *Applied Environmental Microbiology* **62**(2), 717–724.

Wright, A.C., Miceli, G.A., Landry, W.L., Christy, J.B., Watkins, W.D. and Morris, J.G, Jr.: 1993, 'Rapid identification of *Vibrio vulnificus* on nonselective media with an alkaline phosphatase-labeled oligonucleotide probe', *Applied Environmental Microbiology* **62**(2), 717–24.

Yamasaki, S., Shimizu, T., Hoshino, K., Ho, S.T., Shimada, T., Nair, G.B. and Takeda, Y.: 1999, 'The genes responsible for O-antigen synthesis of *Vibrio cholerae* O139 are closely related to those of Vibrio cholerae O22. *Gene* **237**(2), 321–332.

Ziemke, F., Hofle, M.G., Lalucat, J. and Rossello-Mora, R.: 1998, 'Reclassification of *Shewanella putrefaciens* Owen's genomic group II as *Shewanella baltica* sp. nov.', *International Journal of Systematic Bacteriology* **48**(1), 179–186.

USING MULTIPLE ANTIBIOTIC RESISTANCE AND LAND USE CHARACTERISTICS TO DETERMINE SOURCES OF FECAL COLIFORM BACTERIAL POLLUTIION

R. HEATH KELSEY[1,2*], GEOFFREY I. SCOTT[3,4], DWAYNE E. PORTER[1,4], BRIAN THOMPSON[3], AND LAURA WEBSTER[3]

[1]Belle W. Baruch Institute for Marine Biology and Coastal Research, University of South Carolina, Columbia, SC; [2]University of South Carolina, Institute for Public Service and Policy Research, Center for Environmental Policy, Columbia, SC; [3]National Oceanic and Atmospheric Administration Center for Coastal Environmental Health and Biomolecular Research, Charleston, SC; [4]Department of Environmental Health Sciences Norman J. Arnold School of Public Health University of South Carolina, Columbia, SC

**(author for correspondence, email: heath@inlet.geol.sc.edu)*

Abstract. Multiple Antibiotic Resistance (MAR) analysis and regression modeling techniques were used to identify surface water areas impacted by fecal pollution from human sources, and to determine the effects of land use on fecal pollution in Murrells Inlet, a small, urbanized, high-salinity estuary located between Myrtle Beach and Georgetown, South Carolina. MAR analysis was performed to identify areas in the estuary that are impacted by human-source fecal pollution. Additionally, regression analysis was performed to determine if an association exists between land use and fecal coliform densities over the ten-year period from 1989 to 1998. Land-use variables were derived using Geographic Information System (GIS) techniques and were used in the regression analysis.

MAR analyses were conducted by comparing the frequency and patterns of antibiotic resistance found in *Escherichia coli* isolates derived from surface water samples and from sewage sources in the Murrells Inlet sewage collection system. The MAR results suggest that the majority of the fecal pollution detected in the Murrells Inlet estuary may be from non-human sources, including fecal coliforms isolated from areas in close proximity to high densities of active septic tanks.

A MAR Index, which measures the frequency of antibiotic resistance, was calculated for each of twenty-three water samples and nine sewage samples. The antibiotic resistance pattern comparisons were performed using cluster analysis. Although the MAR indices indicated that several surface water sites had potential human-source contamination, the cluster analysis suggests that only one sampling site had MAR patterns that were similar to those found in the sewage samples. This site was in close proximity to several large pleasure boats as well as a sewage collection system lift station, but was not near areas with active septic tanks. The results of the regression analysis also suggest that sewage sources and rainfall runoff from urbanized areas may contribute to fecal pollution in the estuary.

Keywords: Multiple Antibiotic Resistance, *E. coli*, Fecal Pollution, Non-point Source Pollution, Land Use.

1. Introduction

Bacterial contamination in surface waters of coastal areas is a problem affecting recreational and commercial uses of bays, inlets, estuaries and rivers throughout the United States. Water quality degradation from fecal contamination may result in increased health hazards to recreational users (Gersberg *et al.*, 1995) and shellfish consumers, and often results in the closure of shellfish harvesting areas (ISSC, 1997).

The current standard for detection of fecal pollution in surface waters is the determination of fecal coliform bacteria density (Dufour, 1976). Of the fecal coliform group, one species, *Escherichia coli (E. coli)*, is most closely associated with feces from warm-blooded animals (Dufour, 1976). Despite this association, there remains a critical need for the differentiation of fecal pollution from human

Environmental Monitoring and Assessment **81**: 337–348, 2003.

and non-human sources (Pancorbo and Barnhart, 1992). To effectively manage fecal pollution in receiving waters, the sources of the pollution must be known. Strategies to reduce human and non-human source fecal pollution are likely to be different.

Several studies have indicated that relationships exist between urbanized land use and fecal coliform densities in run-off and receiving waters (Young and Thackston, 1999; Watershed Protection Techniques, 1999).

The objectives of the research presented here were to examine the effect of land use on fecal coliform densities, and to differentiate between fecal coliform bacteria from human and non-human sources, in Murrells Inlet, South Carolina. Murrells Inlet is a small, high-salinity estuary located on the South Carolina coast, 90 km north of Charleston, and covers portions of both Horry and Georgetown Counties (Figure 1). The northern two-thirds of the estuary is characterized by residential,

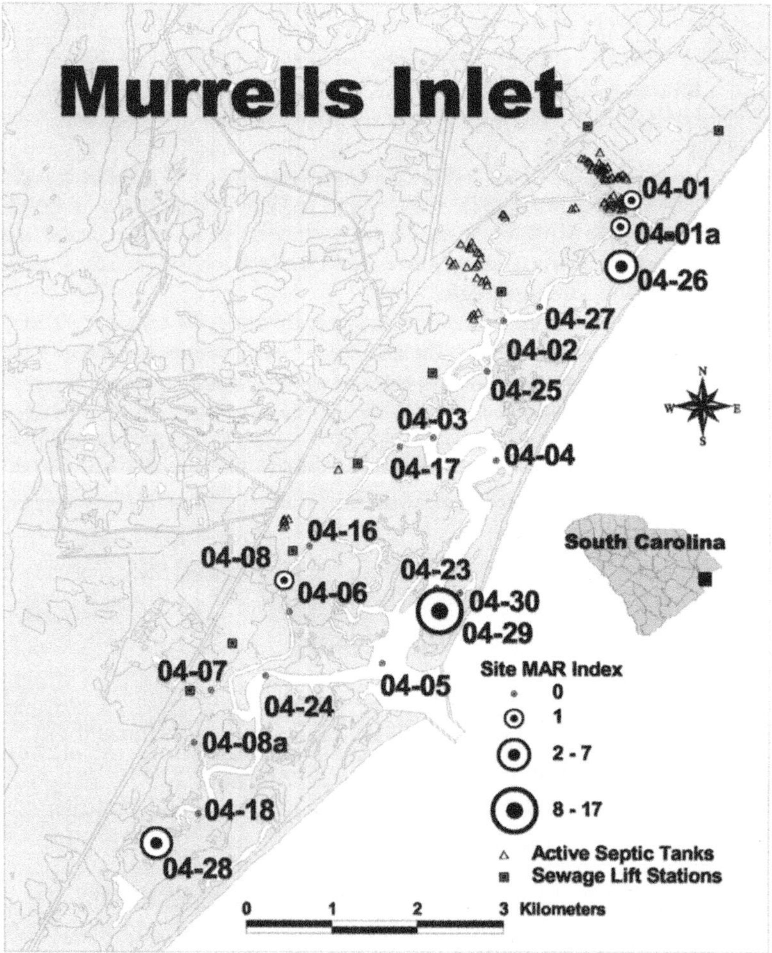

Figure 1. Sampling site locations and MAR Indices in Murrells Inlet, South Carolina.

commercial and tourism related land uses, while the southern third is dominated by an undeveloped state park. There are no wastewater treatment facilities discharging into the surface waters of Murrells Inlet (Newell, 1997).

2. Methods

MAR analysis was used to differentiate fecal coliform bacteria from human and non-human sources. By applying the MAR technique to *E. coli* obtained at surface water sites and sewage system lift stations, the *E. coli* may be grouped into those of probable human origin and those of probable non-human origin. The MAR technique is based on the relative exposures of bacteria from human and other animal sources to antibiotics. Humans, having greater exposure to antibiotics, will likely have enteric bacteria that have developed more resistance to a greater number of antibiotics than other animals, which have less exposure to antibiotics. Several studies based on this exposure difference have demonstrated the potential utility of MAR analyses to examine the sources of fecal pollution. Kaspar *et al.* (1990) have shown that antibiotic indexing of *E. coli* can be used to determine the likely sources of fecal pollution in surface waters. Similarly, Parveen *et al.* (1998) have shown that the patterns of antibiotic resistance in *E. coli* can be used to determine if the bacteria is likely to be from human, non-human, point source or non-point source pollution.

2.1 MAR ANALYSIS

The MAR profiling in this research involved sample collection from both surface water and sewage system sources, analysis of fecal coliform density in the samples, isolation of *E. coli* from the samples, testing of the *E. coli* isolates for antibiotic resistance and statistical analysis of the results. Each of these processes is discussed below.

To perform the MAR analysis, 321 *E. coli* isolates were obtained from both surface waters (198) and from sewage sources (123) in Murrells Inlet, and were examined for antibiotic resistance. Samples were obtained from twenty-three surface water sites and nine sewage system lift stations (Figure 1). Antibiotic resistance patterns exhibited by *E. coli* isolated from the surface water were compared to those from sewage. Sampling procedures were in accordance with field sampling methods as described in the American Public Health Association Methods for Examining Water and Wastewater (APHA, 1989).

2.1.1 Laboratory Analysis for Fecal Coliform Density

The analysis of fecal coliform density was performed by South Carolina (SC) Department of Health and Environmental Control (DHEC) laboratory personnel, at the DHEC Environmental Laboratory in Charleston, SC using the multiple-tube fermentation technique (A-1 media) as described in APHA (1989). Tubes positive for fecal coliforms were then transferred to the National Oceanic and Atmospheric Administration National Ocean Service Center for Coastal Environmental Health and Biomolecular Research Laboratory (NOAA NOS CCEHBR) at James Island, SC for further microbiological analyses.

2.1.2 *Isolation of E. coli from Water Samples*

E. coli was isolated from test tubes which were positive for fecal coliforms in the fecal coliform density analysis described above. The isolates were obtained by streaking for isolation onto Violet Red Bile Agar (VRBA, Difco) plates, transferring and streaking suspected *E. coli* colonies onto Plate Count Agar (PCA, Difco), and confirming the identity of the species as *E. coli* using API 20 E test kits (bioMérieux Vitek, Inc).

2.1.3 *Antibiotic Resistance Testing*

Each confirmed *E. coli* isolate was tested for resistance to ten different antibiotics based on the research protocols established by Parveen *et al.* (1998), as adapted by Webster *et al.* (1999).

2.1.4 *Analysis of MAR Testing Results*

The results of the tests were analyzed by antibiotic resistance index (Kaspar *et al.*, 1990), and by antibiotic resistance pattern cluster analysis (Parveen *et al.*, 1997).

The results from the MAR testing created a matrix consisting of a series of ten test results for each isolate obtained from the sampling sites and sewage system lift stations. To compare the resulting patterns of antibiotic resistance from the water and sewage samples, cluster analysis in SAS/STAT(r) was used. The procedure produces a visualization of the clusters called a "dendrogram" (Parveen *et al.*, 1997). The clusters were examined for patterns from the surface water isolates that were grouped with those from the sewage isolates.

2.2 REGRESSION MODELING

To determine the effect of land use on fecal coliform densities observed in Murrells Inlet, an explanatory regression model was developed, using environmental and land use variables to explain fecal coliform density variability at the DHEC surface water sites from 1989 to 1998. DHEC collects fecal coliform and environmental data at twenty-one surface water sampling sites monthly, as part of a routine survey for shellfish growing area sanitation.

Environmental variables obtained or calculated for the regression modeling included:

- Fecal coliform density, (transformed to [LOG (fecal coliform density + 1)]),
- Tidal stage at sampling time,
- Water temperature,

- Salinity, and
- Rainfall in the 24-, 48-, and 72-hour, and 14-day periods prior to the sampling dates.

GIS techniques were used to calculate values for several "distance or proximity land-use variables" which were included in the regression modeling. These variables included those relating distance from the sampling sites to land use characteristics such as population density, housing unit density, total population, total number of housing units, area of developed (urbanized) and undeveloped (non-urbanized) land, number of septic tanks, sewage system lift stations, roads, boating intensity, boat landings and marinas. Each of these parameters was considered to have a potential impact on fecal pollution in receiving waters (Gannon and Busse, 1989; Young and Thackston, 1999; Watershed Protection Techniques, 1999). Variables that were significant in the regression model may indicate both the effect of land use on, and the origin of, the fecal pollution observed at the surface water sampling sites.

Land use variables calculated include:

- Distances to nearest urban and rural land uses,
- Weighted distances to areas of urbanized land in nearest sub-watersheds,
- Weighted distances to number of housing units and population, and housing and population density in nearest watersheds,
- Weighted distances to septic density in nearest watersheds,
- Distance to nearest sewer system lift station, road, marina, and boat landing,
- Boating intensity at the sampling sites, and
- Waterway depth, width and distance to estuary mouth.

Weighted distance variables were calculated by dividing the area of the selected feature within the nearest watershed by the distance of the monitoring site to the outlet of that watershed.

3. Results and Discussion

3.1 MAR Analysis

Results are presented here for the fecal coliform density analysis at the surface water sampling and sewage system lift stations, the antibiotic resistance testing of the confirmed *E. coli* isolates, MAR indices calculation, cluster analysis of MAR patterns and comparisons to patterns from previous studies.

3.1.1 Fecal Coliform Densities

At the surface water sampling sites, three sites had non-detectable fecal coliform densities (MPN < 2), and twenty sites had positive detection of fecal coliform densities ranging from "2" to "95" MPN per 100 milliliters.

The sewage system lift stations all were positive for fecal coliform densities quantified as > "160,000" MPN per 100 milliliters.

3.1.2 Antibiotic Resistance Testing of Confirmed E. coli Isolates

For the surface water sampling sites, a total of 198 isolates from twenty sites were tested for resistance to the ten antibiotics, creating 1980 individual tests. Fifty-one (2.58%) of the tests were positive for antibiotic resistance. Of the isolates exhibiting resistance, twenty-three (66%) of the isolates were resistant to one antibiotic, eleven (33%) were resistant to two antibiotics and one (3%) isolate was resistant to three or more (seven) antibiotics.

For the sewage system lift stations, a total of 123 isolates from nine lift stations were tested for resistance to the ten antibiotics, creating a total of 1230 individual tests. Eighty-seven (7.07%) of the tests were positive. Of the isolates exhibiting resistance, three (13%) of the isolates were resistant to one antibiotic, two (9%) were resistant to two antibiotics, and eighteen (78%) were resistant to three or more antibiotics. Table 1 presents the number of isolates resistant to antibiotics at the sampling sites and sewage system lift stations.

3.1.3 Calculation of MAR Indices

Table 1 presents the Site MAR Indices calculated for each of the surface water sampling sites and sewage lift stations. Figure 1 presents the MAR Index at each surface water sampling site. The MAR Indices show that average Site MAR Indices for the surface water sampling sites are low when compared to the sewage system lift stations. However, several sampling sites had relatively high MAR Indices, suggesting that the fecal pollution at these sites may be from human sources. Examination of the cluster analysis results however (see below), suggests that most of the fecal coliforms detected on the sampling date were from non-human sources. Significantly, the sites closest to the areas with high septic tank use (Site Nos. 04-01, 04-01A, and 04-02) have low MAR indices.

3.1.4 MAR Pattern Analysis

Pattern analysis was performed using PROC CLUSTER (average linkage) in SAS. The results of the MAR testing created a matrix consisting of a series of ten test results for each isolate obtained from the sampling sites and sewage system lift stations. The cluster procedure uses the matrix to determine "distance" between the individual patterns observed (SAS Institute, Inc., 1999). The dendrogram produced from the procedure revealed that the MAR patterns found in the *E. coli* isolates from the surface waters generally were not similar to the patterns found in the isolates from the sewage samples (Figure 2). Three general groups of clusters

Table 1. Number of isolates resistant to one, two, or three or more antibiotics and MAR Indices for each surface water sampling site.

	Number of Isolates Resistant to:			
Site	*1 Antibiotic*	*2 Antibiotics*	*3 or More Antibiotics*	*Site MAR Index*
Surface Water Sampling Sites				
04-01	1	0	0	0.0111
04-01A	1	0	0	0.0091
04-02	0	0	0	0.0
04-03	0	0	0	0.0
04-04	0	0	0	0.0
04-05	—	—	—	—
04-06	0	0	0	0.0
04-07	0	0	0	0.0
04-08	1	0	0	0.0100
04-08A	0	0	0	0.0
04-16	0	0	0	0.0
04-17	0	0	0	0.0
04-18	0	0	0	0.0
04-23	—	—	—	—
04-24	0	0	0	0.0
04-26	3	2	0	0.0700
04-27	0	0	0	0.0
04-28	8	0	0	0.0727
04-29	4	3	1	0.1700
04-30	0	0	0	0.0
GC-1	3	5	0	0.1300
MC-5	2	1	0	0.0571
Total	**23**	**11**	**1**	**0.0265**
	(66%)	**(31%)**	**(3%)**	**(Av)**
Sewage System Lift Stations				
EPA 302	0	0	3	0.0714
EPA 303	0	0	3	0.1000
EPA 304	0	0	1	0.0214
EPA 305	0	1	4	0.0933
EPA 306	0	1	1	0.0909
EPA 307	0	0	0	0.0
LS 2	2	0	2	0.1333
LS 21	1	0	2	0.0571
LS 72	0	0	2	0.0714
Total	**3**	**2**	**18**	**0.0710**
	(13%)	**(9%)**	**(78%)**	**(Av)**

appear in the dendrogram. The MAR patterns from the surface water sample and sewage sample isolates were grouped separately, and an additional group appeared to contain several unique MAR patterns. An exception was found in the surface water isolates obtained from Site No. 04-29, located in a small embayment in a subdivision at the end of the Garden City Peninsula (Figure 1). All of the isolate MAR patterns at Site No. 04-29 were clustered with the patterns from the sewage samples. No isolate MAR pattern at this site was clustered with the isolate MAR patterns from other surface water sampling sites.

A comparison of specific patterns of resistance found in the *E. coli* isolated from the surface water samples to the patterns found in the *E. coli* isolated in the sewage samples reveals obvious differences. The majority (97%) of the surface water isolates were resistant to one or two antibiotics (Table 1). Only one isolate, from Site No. 04-29, was resistant to more than two antibiotics. Conversely, the majority (78%) of the resistant isolates from the sewage samples exhibited resistance to three or more antibiotics. These differences suggest that the fecal pollution in most of the surface water samples, except Site No. 04–29, may not be from the same source as that in the sewage samples.

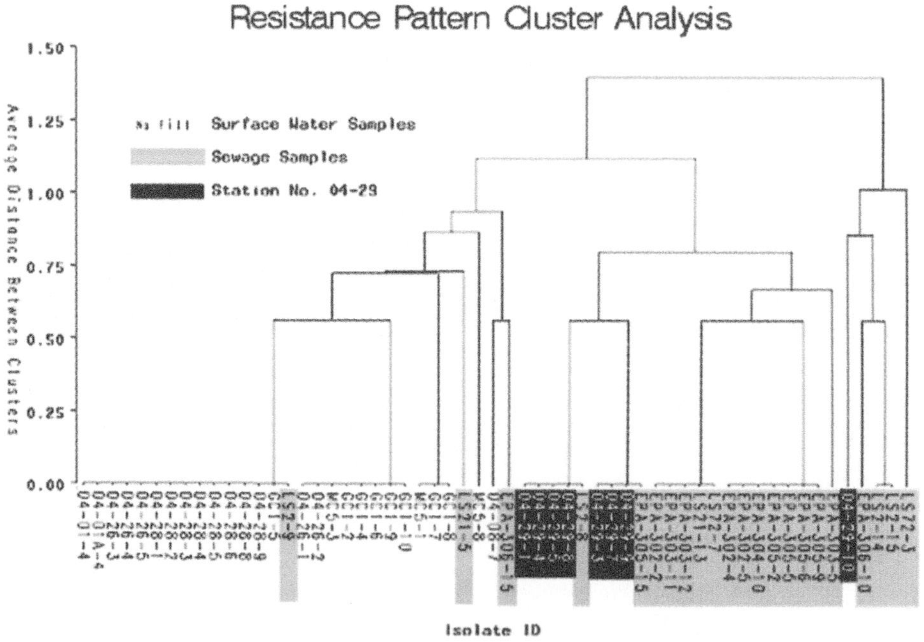

Figure 2. Cluster Analysis Dendrogram Results. Modified output from SAS (shading added).

3.2 REGRESSION ANALYSIS

The regression modeling indicated that, in addition to variables for tide, salinity and month, predictors of fecal coliform densities included rainfall (48 hours and 14 days prior to the sampling dates), proximity to septic tanks, and proximity to urbanized areas. The R^2 for the model was 0.4401.

Following manual model selection procedures, the following land-use variables were retained in the model (a significance level of 0.05 was used to determine the importance of variables as predictors of fecal coliform densities):

- Distances to nearest urban land uses,
- Weighted distances to areas of urbanized land in nearest sub-watersheds,
- Weighted distances to septic density in nearest watersheds,
- Distance to nearest sewer system lift station, and marina, and
- Waterway depth, width and distance to estuary mouth.

The regression model retained variables indicating that areas in proximity to high septic tank usage (Site Nos. 04-01, 04-01a, and 04-02) tended to have relatively high fecal coliform densities, which intuitively suggests that the fecal pollution in these areas is from human sources. However, the MAR analysis clearly shows that the frequencies of antibiotic resistance and the MAR patterns found at these sites and at the sewage samples are not similar. This indicates that the fecal pollution in the surface water at these sites is probably not from sewage sources.

It is important to note that the R^2 for the model indicated that only 44% of the variability of fecal coliform densities was explained using the variables retained in the model. The remaining 56% of the variability of fecal coliform densities observed was not explained.

4. Conclusions

When examined together, the results of the regression modeling and MAR analysis suggest that the majority of fecal pollution in Murrells Inlet is not from human sources. Although septic tank use appears to be an important factor in fecal pollution in Murrells Inlet, as suggested by the regression modeling, the MAR analysis suggests that the fecal pollution in those areas near septic tanks is not from human sources.

The major source of fecal pollution in Murrells Inlet appears to be stormwater runoff, particularly from urban areas, as indicated by the regression model predictors of rainfall and proximity to urbanized areas. Additionally, the results of the MAR testing indicate that, in localized areas such as at Site No. 04-29, fecal pollution from human sources may originate from boats or malfunctioning sewage collection system components.

The results of this research indicate that MAR analysis can be used to identify areas impacted by human sources of fecal pollution and that land-use characteristics can be used to predict fecal coliform densities in surface waters. The water quality management implications of this research include the identification of strategies to reduce or intercept urban stormwater runoff, reduction of waste dumping from boats, and reduction of pet waste. Additionally, the results of the regression analysis indicate that a measure of rainfall in the 48-hours prior to the sampling date is a better predictor of fecal coliform densities in the surface waters of Murrells Inlet than 24-hour rainfall. DHEC currently uses a measure of 1.12 inches of rainfall in one 24-hour period as an indication of fecal coliform densities that would necessitate closure of the shellfish harvesting areas in Murrells Inlet for 14 days (Newell, 1997). The results from this study show that a rainfall measure incorporating 48-hour rainfall may be appropriate for use in regulating closure of the shellfish harvesting areas in Murrells Inlet.

This study clearly shows the impacts of human activities on fecal pollution in Murrells Inlet, South Carolina. Increases in development appear to be correlated with increases in fecal pollution, and the majority of the pollution appears to be from non-human sources. The utility of MAR analysis to differentiate human and non-human source fecal pollution is also demonstrated in this study. By comparing MAR indices and patterns from surface water and local sewage sources, it may be possible to identify areas impacted by human source fecal pollution sources and suggest remedies for this pollution. MAR analyses could provide a useful tool for this differentiation in Murrells Inlet as well as in other estuaries. With additional research, it may become possible to define the source of a bacterium based solely on the antibiotics to which it is resistant.

As Murrells Inlet is typical of the rapidly developing estuaries of the southeastern United States, the results of this study could indicate that processes similar to those occurring in Murrells Inlet are likely occurring in other estuaries. Management strategies such as those described for Murrells Inlet, South Carolina could be applicable to management of fecal pollution in other developing estuaries in the United States and worldwide.

Acknowledgements

The authors would like to acknowledge SC DHEC, the SC Department of Natural Resources, and the Georgetown and Horry County Water and Sewer Authorities for their help and cooperation in obtaining information and data crucial to this study. Additionally, acknowledgement is necessary for various researchers and personnel at the Norman J. Arnold School of Public Health, the National Oceanic and Atmospheric Administration, and the Baruch Institute for Marine Biology and Coastal Research for their assistance with this research.

The National Ocean Service (NOS) and the University of South Carolina (USC) do not approve, recommend, or endorse any proprietary product or material mentioned in this publication. No reference shall be made to NOS or USC, or to this publication furnished by NOS or USC, in any advertising or sales promotion which would indicate or imply that NOS or USC approves, recommends, or endorses any proprietary product or proprietary material mentioned herein or which has as its purpose any intent to cause directly or indirectly the advertised product to be used or purchased because of NOS or USC publication.

This publication represents manuscript number 1295 of the Baruch Marine Science Library. The research described within was supported, in part, by a Grant from the National Oceanic and Atmospheric Administration (Grant Number NA96060P0358) awarded to the University of South Carolina.

References

APHA: 1989, Standard Methods for the Examination of Water and Wastewater. 17th Edition. American Public Health Association, American Waterworks Association, Water Pollution Control Federation.

Dufour, A.P.: 1977, 'Escherichia coli: The fecal coliform', in: Hoadley, A. W. and Dutka, B. J. (eds), Bacterial indicators/ health hazards associated with water. ASTM STP635. American Society for Testing and Materials.

Gannon, J.J., and Busse, M. K.: 1989, '*E. coli* and enterococci levels in urban stormwater, river water and chlorinated treatment plant effluent', *Water Research* **23**(9), 1167–1176.

Gersberg, R., Matkovits, M., Dodge, D., McPherson, T., and Boland, J.: 1995, 'Experimental opening of a coastal California lagoon: effect on bacteriological quality of ocean waters', *Journal of Environmental Health* **58**(2), 24.

ISSC: 1997, *Interstate Shellfish Sanitation Conference Guide for the Control of Molluscan Shellfish*, U.S. Department of Health and Human Services, Public Health Service, Food and Drug Administration. 1997 Revision.

Kaspar, C. W., Burgess, J. L., Knight, I. T., and Colwell, R. R.: 1990, 'Antibiotic resistance indexing of *Escherichia coli* to identify sources of fecal contamination in water', *Canadian Journal of Microbiology* **36**, 891–894.

Newell, C.: 1997, Shellfish management area IV; 1997 annual update. South Carolina Department of Health and Environmental Control. Columbia, South Carolina.

Pancorbo, O. C., and Barnhart, H. M.: 1992, 'Microbial pathogens and indicators in estuarine environments and shellfish: critical need for better indicators of human-specific fecal pollution', *Journal of Environmental Health* **54**(5), 57–63.

Parveen, S., Murphree, R. L., Edminston, L., Kaspar, C. W., Portier, K. M., and Tamplin, M. L.: 1997, 'Association of multiple-antibiotic resistance profiles with point and nonpoint sources of *Escherichia coli* in Appalalchicola Bay', *Applied and Environmental Microbiology* **63**(7), 2607–2612.

SAS Institute, Inc.: 1999, SAS Release 8.01. System Help. SAS Institute, Cary, North Carolina.

Center for Watershed Protection: 1999, 'Concentrations sources and pathways', *Watershed Protection Techniques* **3**(1), 544–565.

Webster, L. F., Thompson, B. C., Gilbert, S., Hagan, J., Payne, D., Woodley, C. M., and Scott, G. I.: 1999, 'Identification of human and animal sources of *Escherichia coli* in South Carolina estuaries', abstract Q-49, in: Abstracts of the 99th General Meeting of the American Society for Microbiology 1999. American Society for Microbiology, Washington, D.C., p.542.

Young, K. D., and Thackston, E. L.: 1999, 'Housing density and bacterial loading in urban streams', *Journal of Environmental Engineering* **125**(12), 1177–1180.

LONG-TERM PHYTOPLANKTON TRENDS AND RELATED WATER QUALITY TRENDS IN THE LOWER CHESAPEAKE BAY, VIRGINIA, U.S.A.

HAROLD G. MARSHALL*, MICHAEL F. LANE, AND KNEELAND K. NESIUS

Department of Biological Sciences, Old Dominion University, Norfolk, VA, U.S.A 23529-0266
**(author for correspondence, e-mail: hmarshal@odu.edu)*

Abstract. Long-term trends (i.e., 1985 through 1999; 14 1/2 yrs) of the phytoplankton community in Chesapeake Bay indicated patterns of increasing phytoplankton abundance and biomass associated with mainly diatoms and chlorophytes, and to a lesser degree dinoflagellates. Decreasing trends in productivity rates above the pycnocline were present over a shorter time period (10 1/2 yrs.), with evidence for increasing nitrogen limitation is indicated. Reduced light availability is inferred due to decreasing trends of Secchi depths and increased suspended solids trends, which were associated with decreasing trends in productivity rates.

Keywords: Chesapeake Bay, phytoplankton, trends, water quality parameters, monitoring

1. Introduction

Phytoplankton composition studies for the Chesapeake Bay are reviewed by Marshall (1994), with the highest phytoplankton biomass associated with spring in the northern Bay (Malone *et al.*, 1991), and in the southern Bay occurring during late winter and early spring (Marshall and Lacouture, 1986; Harding, 1994). The contribution of the autotrophic picoplankton to productivity and biomass has also been stressed by Malone *et al.* (1991), Affronti and Marshall (1994), and Marshall and Nesius (1996). This category is defined as cells 0.2 to 2.0 μ in size, with the presence of the autotrophic picoplankton considered separately from the larger phytoplankton cells within this study (e.g., Marshall, 1995). In general, the lower Bay is dominated by a diatom flora throughout the year, in addition to various phytoflagellates and cyanobacteria as sub-dominant flora. Harding (1994), using data collected over a 40-year period (1950–1990), evaluated phytoplankton trends based on chlorophyll and phaeophytin values. These indicated seasonal and inter-annual fluctuations among the phytoplankton that were associated with regional differences in dissolved inorganic nitrogen and orthophosphates. Other assessments of the status and trends in phytoplankton and water quality in the lower Chesapeake Bay and its tributaries have been conducted by the Chesapeake Bay Monitoring Program (e.g., Alden *et al.*, 1992; Dauer *et al.*, 1999), where results of shorter length studies were reported. The purpose of this study is to identify long term trends in phytoplankton community composition and relate them to long-term trends in water quality in the southern (lower) region of the Chesapeake Bay from 1985 through 1999. Information regarding the distribution, abundance, and seasonal patterns for phytoplankton in this data set have been presented by Marshall (1994; 1995), Marshall and Alden (1990), and Marshall and Nesius (1996).

Environmental Monitoring and Assessment **81**: 349–360, 2003.

2. Materials and Methods

2.1 SAMPLING PROCEDURES

Water quality parameters and phytoplankton community composition were monitored above and below the pycnocline at seven stations in the Chesapeake Bay (Figure 1). From July 1985 through September 1990, collections were taken twice a month from March through October, and monthly November through February. Beginning in October 1990, monthly samples were taken at all stations. The autotrophic picoplankton and primary productivity (measured above the pycnocline) were added in July 1989. Field sampling procedures for the water quality variables, phytoplankton, and picoplankton are described by Alden *et al.* (1992) and Marshall and Alden (1990).

2.2 LABORATORY SAMPLE PROCESSING

Water quality variables were measured following procedures given in USEPA (1983), Aspila *et al.* (1976), D' Elia *et al.* (1977), and APHA (1985) [see Alden *et al.*, 1992]. These included total nitrogen (TN), dissolved inorganic nitrogen (DIN), total phosphorus (TP), dissolved inorganic phosphorus (DIP), total suspended solids (TSS), dissolved oxygen (DO), chlorophyll a (CHLa), Secchi depths, plus the total nitrogen to total phosphorus ratio. These variables were selected due to their relationships to the abundance and composition of phytoplankton in this Bay system (Alden *et al.*, 1992; Malone *et al.*, 1996; Marshall and Alden, 1990; among others). Samples for phytoplankton analyses were passed through a series of settling and siphoning steps to produce a concentrate that was examined using a modified Utermöhl method with an inverted plankton microscope (Marshall and Alden, 1990). The autotrophic picoplankton were observed using epifluorescence protocols under oil at 1000X magnification (Marshall, 1995). Cell volumes for the various taxa were calculated and used in determining the cellular biomass (cell carbon) according to Strathmann (1967). The productivity protocols used are described in Marshall and Nesius (1996).

2.3 DATA ANALYSIS

The seasonal Kendall test was used to identify monotonic long-term trends at each station, and the Van Belle and Hughes test to confirm the homogeneity of trends between seasons at each station (Gilbert, 1987). The seasonal divisions consisted of the following monthly categories: January-February (Winter); March-May (Spring), June-September (Summer), and October-December (Fall). A $p = 0.05$ value was the statistical test criterion for all trend analyses according to Dauer *et al.* (1999). Analysis for trends within the various communities were conducted on: total phytoplankton abundance (excluding picoplankton), total phytoplankton biomass, species diversity (diversity index according to Margalef, 1958), and produc-

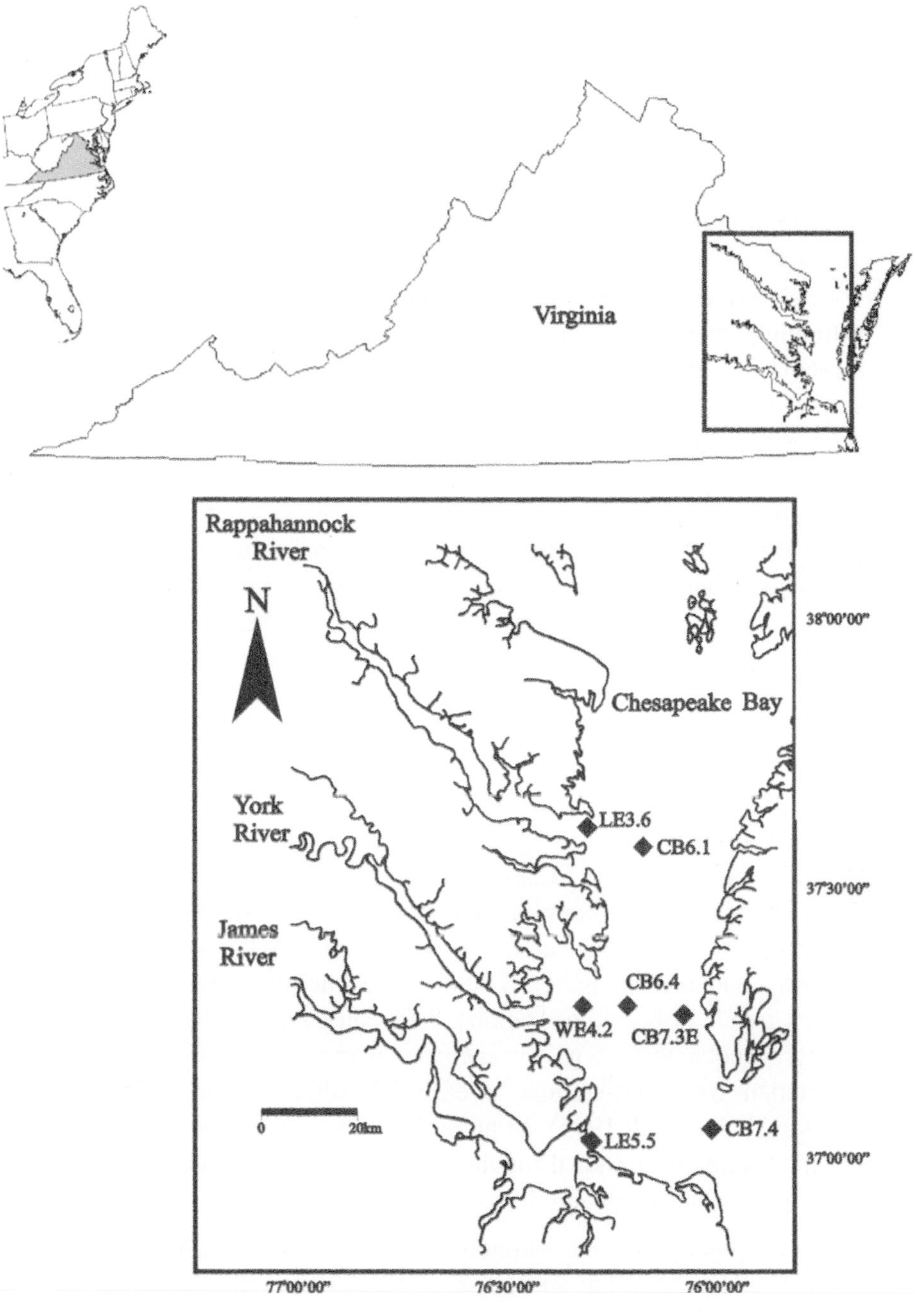

Figure 1. Map of study area showing the locations of the phytoplankton and water quality monitoring stations for the Lower Chesapeake Bay main stem, and sites at mouths of the James, York, and Rappahannock Rivers. The station name prefixes are as follows: CB = Chesapeake Bay main stem; LE = lower estuarine; and WE = western embayment.

tivity, plus the biomass trends for autotrophic picoplankton, chlorophytes, dinoflagellates, cyanobacteria, cryptophytes, and diatoms.

3. Results

3.1 PHYTOPLANKTON TRENDS

There were 26 trends above and 32 trends below the pycnocline at the seven Bay stations (Table 1). A common pattern above the pycnocline and throughout the lower Bay was a decreasing trend in phytoplankton productivity rates. The highest mean monthly rates occurred in spring along the Bay's western coastal region, and are located at the mouths of the Rappahannock, York, and James Rivers (e.g. mean April rates of 73.5, 84.6, and 100.8 mgC m^3 hr^{-1} were at stations LE3.6, WE4.2, and LE5.5, respectively). For each monthly mean throughout the year, LE5.5 had the highest productivity of the seven stations. The mean main stem station rates ranged from 71.7 mgC m^3 hr^{-1} at CB6.1 southward to 42.5 mgC m^3 hr^{-1} at CB7.4. In contrast to these decreasing rates, there were common trends of increased total phytoplankton abundance at six of the seven stations. This increase was accompanied by increased total biomass throughout the water column at the mouths of the James and York Rivers (LE5.5, WE4.2), and below the pycnocline at the Rappahannock River mouth (LE3.6). Increased biomass of diatoms, dinoflagellates, and chlorophytes were associated with these increases. The concentrations were least developed in early winter, with peak abundance during the spring months (March, April, May) when mean concentrations were 7-8 x 106 cells l^{-1} . This was also the period when highest biomass concentrations occurred, and were mainly the product of diatoms (e.g. *Skeletonema costatum*). The lowest mean values for both cell abundance and biomass were in December. The only trend in species diversity (decreasing) was at the Bay entrance (CB7.4). The autotrophic picoplankton had decreasing trends below the pycnocline at three stations, with no significant trends at the other stations. This category was predominantly represented by cyanobacteria (e.g., *Synechococcus* spp.) which were ubiquitous, but most abundant from mid-summer to early fall, often reaching concentrations of 106 cells ml^{-1} (Marshall,1995). In an earlier report, these picoplankters showed decreasing abundance trends throughout the lower Bay (Dauer *et al.*, 1999). This overall pattern was not present in this analysis.

Below the pycnocline the common trends were increased total phytoplankton abundance and biomass. The species composition was a mixture of neritic species entering the Bay and existing estuarine taxa. These were associated with increased biomass of diatoms, chlorophytes, and dinoflagellates at several stations. With the exception of Station LE3.6, there were no other significant trends associated with the cyanobacteria biomass in the lower Bay. The ubiquitous cryptophytes had only one trend which was increasing biomass below the pycnocline at Station LE3.6.

Table 1. Long-term trends in phytoplankton community composition and productivity at stations within the lower Chesapeake Bay for the period of July 1985 through December 1999. A "▲" indicates a significant increasing trend was detected while a "▼" indicates a significant decreasing trend was detected. The number presented is the p value for the seasonal Kendall test. An "NS" indicates that the trend analysis result was considered not significant. An "NS*" indicates that although a significant trend was detected ($p = 0.05$) the seasonal Kendall slope was equal to zero and the trend was not considered to be ecologically significant. The station name prefixes are as follows: CB = Chesapeake Bay main stem; LE = lower estuarine; and WE = western embayment.

Above Pycnocline

	Stations						
Parameter	*LE3.6*	*WE4.2*	*LE5.5*	*CB6.1*	*CB6.4*	*CB7.4*	*CB7.3E*
Total Abundance	▲ 0.0226	NS 0.1199	▲ 0.0044	NS 0.0700	▲ 0.0091	▲ 0.0005	▲ 0.0428
Total Biomass	NS 0.4969	▲ 0.0309	▲ 0.0108	NS 0.8670	NS 0.1191	▲ 0.0309	NS 0.17530
Species Diversity	NS 0.9639	NS 0.7964	NS 0.6315	NS 0.7723	NS 0.8105	▼ 0.0165	NS 0.4556
Productivity	▼ 0.0009	▼ 0.0002	▼ < 0.0001	▼ 0.0067	▼ 0.0013	NS 0.1484	▼ 0.0116
Diatom Biomass	NS 0.5560	▲ 0.0008	▲ < 0.0001	NS 0.8670	▲ 0.0386	▲ 0.0179	NS 0.2063
Dinoflagellate Biomass	NS 0.0591	NS 0.5560	NS 0.1679	NS 0.5526	NS 0.3226	NS 0.3685	NS 0.4932
Chlorophyte Biomass	▲ < 0.0001	▲ < 0.0001	▲ < 0.0001	▲ < 0.0001	▲ < 0.0001	▲ < 0.0001	▲ < 0.0001
Cyanobacteria Biomass	NS 0.2942	NS 0.8264	NS 0.1623	NS 0.5683	NS 0.4505	NS 0.6761	NS 0.8683
Cryptophyte Biomass	NS 0.4412	NS 0.4873	NS 0.9880	NS 0.6339	NS 0.6315	NS 0.1055	NS 0.4836
Picoplankton Biomass	NS 0.1114	NS 0.3153	NS 0.1627	NS 0.0784	NS 0.0057	NS 0.0653	NS 0.0502

Table 1—*Continued.* A "▲" indicates a significant increasing trend was detected while a "▼" indicates a significant decreasing trend was detected. The number presented is the p value for the seasonal Kendall test. An "NS" indicates that the trend analysis result was considered not significant. An "NS*" indicates that although a significant trend was detected ($p = 0.05$) the seasonal Kendall slope was equal to zero and the trend was not considered to be ecologically significant. The station name prefixes are as follows: CB = Chesapeake Bay main stem; LE = lower estuarine; and WE = western embayment.

Below Pycnocline

	Stations						
Parameter	*LE3.6*	*WE4.2*	*LE5.5*	*CB6.1*	*CB6.4*	*CB7.4*	*CB7.3E*
Total Abundance	▲ 0.0002	▲ 0.0164	▲ 0.0004	NS 0.3529	▲ 0.0010	▲ 0.0040	▲ 0.0033
Total Biomass	▲ 0.0415	▲ 0.0052	▲ 0.0002	NS 0.2290	NS 0.0631	NS 0.0514	NS 0.0015
Species Diversity	NS 0.6615	NS 0.4413	NS 0.1191	NS 0.9636	NS 0.7416	▼ 0.0932	NS 0.1392
Diatom Biomass	NS 0.0880	▲ 0.0024	▲ 0.0033	NS 0.8431	NS 0.1774	▲ 0.0118	▲ 0.0012
Dinoflagellate Biomass	▲ 0.0386	NS 0.7742	▲ < 0.0001	NS 0.4374	NS 0.5488	▼ 0.0287	NS 0.2063
Chlorophyte Biomass	▲ < 0.0001	▲ < 0.0001	▲ < 0.0001	▲ < 0.0001	▲ < 0.0001	▲ < 0.0001	▲ < 0.0001
Cyanobacteria Biomass	▲ 0.0135	NS 0.3371	NS 0.7233	NS 0.3941	NS 0.0042	NS 0.8710	NS 0.0743
Cryptophyte Biomass	▲ 0.0497	NS 0.2389	▲ 0.0002	NS 0.6369	▲ 0.0064	NS 0.2673	NS 0.0590
Picoplankton Biomass	▼ 0.0101	NS 0.1484	NS 0.1744	▼ 0.0032	▼ 0.0370	NS 0.7342	NS 0.5685

3.2 WATER QUALITY TRENDS

There were 25 trends above and 20 trends below the pycnocline among the water quality parameters at these stations (Table 2). Above the pycnocline, decreasing trends in Secchi depth were at all 7 stations, along with increasing TSS trends at the three southern most stations. Decreasing trends in TN were at the mouths of the Rappahannock River (LE3.6) and James River (LE5.5), and the Bay main stem station CB6.1. There were increasing trends in TN at CB6.4 and CB7.3E. No significant general trends for DIN were detected, except for a decreasing trend at LE5.5. Decreasing trends in TP and DIP were also present at LE5.5, which had a long term pattern of decreasing nutrient levels for each of the nutrient fractions measured. Increasing trends of the TN:TP ratio were at LE5.5 and the three stations nearer the Bay entrance while a decreasing trend in the ratio was at CB6.1. Mean monthly TN:TP ratios decreased moving along the Bay mainstem toward the Bay entrance, and along the western Bay coastline. At five stations the mean monthly TN:TP ratios were less than the Redfield ratio (=16), and ratios above this value were at LE3.6 and CB6.1. These locations represented the most northern of the seven stations in the lower Bay. No significant trends in chlorophyll a concentrations were detected.

Below the pycnocline, decreasing trends in TN were at the mouths of the three tributaries (LE3.6, WE4.2, and LE5.5) and at the northernmost lower Bay station (CB6.1). A decreasing trend in DIN were detected at LE3.6. Decreasing trends in TP were at the mouth of James River (LE5.5) and at the three southernmost stations in the Bay main stem. A single decreasing trend in DIP was present at station LE5.5. Increasing trends in TSS were at the three southernmost stations. Decreasing trends in nutrients below the pycnocline were coupled with increasing trends in DO at LE3.6 and LE5.5, and a decreasing trend in chlorophyll a at CB6.1. An increasing trend in chlorophyll a was at LE5.5. The TN:TP ratios also decreased moving down the Bay, and indicated a similar relationships to the Redfield ratios as above the pycnocline.

4. Discussion

This appraisal of existing trends within the phytoplankton community and certain water quality parameters indicate these populations are subject to light and nutrient limitation regarding their development and productivity in the lower Chesapeake Bay. A major trend associated with the phytoplankton community at each of the lower Bay stations above the pycnocline was decreasing productivity rates. This occurred with several trends of increasing phytoplankton biomass within the diatom, dinoflagellate, and chlorophyte assemblages, plus increasing phytoplankton abundance. This reduction in production can be related to the increase in TSS and diatom biomass at several of the stations, both of which would influence the amount of light entering the water column that would be available for photosyn-

Table 2. Long-term trends in water quality characteristics at stations within the lower Chesapeake Bay for the period of July 1985 through December 1999. A "▲" indicates a significant increasing trend was detected while a "▼" indicates a significant decreasing trend was detected. An "NS" indicates that no significant trend was detected. An "NS*" indicates that although a significant trend was detected ($p = 0.05$) the seasonal Kendall slope was equal to zero and the trend was not considered to be ecologically significant. Water quality parameters are abbreviated as follows: TN = Total Nitrogen; DIN = Dissolved Inorganic Nitrogen; TP = Total Phosphorus; DIP = Dissolved Inorganic Phosphorus; TSS = Total Suspended Solids; DO = Dissolved Oxygen; CHL a = Chlorophyll a and; TN:TP = Ratio of Total Nitrogen to Total Phosphorus. The station name prefixes are as follows: CB = Chesapeake Bay main stem; LE = lower estuarine; and WE = western embayment.

Above Pycnocline

	Stations						
Parameter	*LE3.6*	*WE4.2*	*LE5.5*	*CB6.1*	*CB6.4*	*CB7.4*	*CB7.3E*
TN	▼ < 0.001	NS 0.072	▼ < 0.001	▼ 0.003	▲ 0.046	NS 0.413	▲ 0.024
DIN	NS* 0.004	NS* 0.001	▼ 0.001	NS* 0.012	NS* 0.009	NS 0.325	NS* 0.034
TP	NS* 0.007	NS 0.059	▼ < 0.001	NS* 0.026	NS* < 0.001	▼ < 0.001	NS* < 0.001
DIP	NS 0.710	NS 0.665	▼ < 0.001	NS 0.861	NS* < 0.001	NS* 0.001	NS* < 0.001
TSS	NS 0.586	NS 0.536	▼ < 0.001	NS 0.432	▲ < 0.001	▲ < 0.001	▲ < 0.001
Secchi	▼ < 0.0001	▼ < 0.0001	▼ < 0.0001	▼ < 0.0001	▼ < 0.0001	▼ < 0.0001	▼ < 0.0001
CHL *a*	NS 0.324	NS 0.131	NS 0.635	NS 0.786	NS 0.310	NS 0.424	NS 0.795
TN:TP Ratio	NS 0.328	NS 0.407	▲ 0.0001	▼ < 0.0001	▲ < 0.0001	▲ < 0.0001	▲ < 0.0001
TN:TP Ratio Mean	23.8	12	11.4	20.8	13.8	10.7	10.1

Table 2—*Continued.* A "▲" indicates a significant increasing trend was detected while a "▼" indicates a significant decreasing trend was detected. An "NS" indicates that no significant trend was detected. An "NS*" indicates that although a significant trend was detected ($p = 0.05$) the seasonal Kendall slope was equal to zero and the trend was not considered to be ecologically significant. Water quality parameters are abbreviated as follows: TN = Total Nitrogen; DIN = Dissolved Inorganic Nitrogen; TP = Total Phosphorus; DIP = Dissolved Inorganic Phosphorus; TSS = Total Suspended Solids; DO = Dissolved Oxygen; CHL a = Chlorophyll a and; TN:TP = Ratio of Total Nitrogen to Total Phosphorus. The station name prefixes are as follows: CB = Chesapeake Bay main stem; LE = lower estuarine; and WE = western embayment.

Below Pycnocline

	Stations						
Parameter	*LE3.6*	*WE4.2*	*LE5.5*	*CB6.1*	*CB6.4*	*CB7.4*	*CB7.3E*
TN	▼ 0.006	▼ < 0.001	▼ 0.008	▼ 0.029	NS 0.203	NS 0.825	NS 0.316
DIN	▼ 0.027	NS 0.053	NS 0.057	NS 0.159	NS 0.082	NS 0.354	NS* 0.010
TP	NS* 0.025	NS* 0.021	▼ < 0.001	NS 0.055	▼ < 0.001	▼ < 0.001	▼ < 0.001
DIP	NS 0.260	NS 0.619	▼ < 0.001	NS 0.733	NS* < 0.001	NS* < 0.001	NS* < 0.001
TSS	NS 0.909	NS 0.432	NS 0.177	NS 0.081	▲ < 0.001	▲ < 0.001	▲ < 0.001
D.O. (Summer)	NS 0.181	NS 0.457	▲ 0.015	NS 0.577	NS 0.620	NS 0.655	NS 0.143
CHL *a*	NS 0.149	NS 0.932	▲ 0.008	▼ 0.047	NS 0.853	NS 0.484	NS 0.110
TN:TP Ratio	NS 0.698	NS 0.962	▲ < 0.001	NS 0.117	▲ < 0.001	▲ < 0.001	▲ < 0.001
TN:TP Ratio Mean	20.7	15.5	8.3	19.0	13.2	9.1	11.5

thesis. Another influencing factor to this association is the difference in starting dates and length of the data sets for productivity and the floral data. Productivity measurements began in 1989, and the phytoplankton analyses in 1985. This resulted in a shorter data base for productivity values and produced a temporal disconnect between the two, with comparative results possibly not as conclusive. As a result, the primary production trend may be less than indicated.

Specific long term trends determined from the monthly water quality data obtained at these stations provide measures to characterize conditions and trends for this region. A major pattern within the lower Chesapeake Bay were trends of reduced Secchi depths at six of the seven stations. This condition was accompanied by trends of increased total suspended solids at CB6.4, CB7.4, and CB7.3E. These conditions infer reduced light conditions prevailing in the water column that were associated with the transport of suspended solids in the Bay (Harding, 1994), which would influence, or be a limiting factor to phytoplankton development and productivity rates (Cloern, 1987). There were several, but varied long term trends associated with TN, DIN, TP, and DIP at these stations. However, when comparisons are made between T and P values in reference to Redfield's ratio (=16), two patterns emerge. The two most northern lower Bay stations (CB6.1, LE3.6) have monthly mean ratios greater than Redfield, whereas the other stations (CB6.4, CB7.4, CB7.3e, WE4.2, LE5.5), had monthly mean ratios less than Redfield. This pattern is also similar below the pycnocline. Fisher *et al.* (1992), when considering the mesohaline and polyhaline regions of Chesapeake Bay, associated nitrogen limitation conditions with low Redfield ratios (common in summer and fall), and to be phosphorus limited with the higher ratios in spring. These results support the previous findings by Fisher *et al.* (1992) and Malone *et al.* (1996) that the lower Bay algal development and productivity is under the seasonal influence of nitrogen and phosphorus limitation and that this condition is prevalent in the lower Chesapeake Bay (mesohaline/polyhaline) region.

Nitrogen limitation is more likely to occur in waters near the Bay entrance and along its eastern shore line. These conditions would be less pronounced at the more northern stations in the lower Bay region, where phosphorus limitation appears more likely during the period of the spring bloom. The increased trend in the abundance of suspended solids, and the pattern of reduced Secchi readings infers that reduced light passage within the water column would accompany this pattern, with less light made available to the present phytoplankton populations for photosynthesis. As a result, the reduced primary production above the pycnocline could be a result of increases in the total suspended solids and decreasing Secchi depth readings in this region of the Bay. If these trends continue, they may initiate a transition to different seasonal development patterns among the phytoplankton, including those that would be less valuable as a food resource within this estuarine system (e.g. increased abundance of cyanobacteria).

Acknowledgments

This study was supported by the Virginia Department of Environmental Quality. Appreciation is given to the contributions of numerous graduate research assistants throughout this study who assisted in field collections, laboratory analysis, and data entry.

References

Affronti, L. and Marshall, H.G.: 1994, 'Using frequency of dividing cells in estimating autotrophic picoplankton growth and productivity in Chesapeake Bay', *Hydrobiologia* **284**,193–203.

Alden, R.W. III., Birdsong, R.S., Dauer, D.M., Marshall, H.G. and Ewing, R.M.: 1992, Virginia Chesapeake Bay water quality and living resources monitoring programs: Comprehensive technical report, 1985–1989. Applied Marine Research Laboratory Technical Report No. 848, Norfolk VA.Final Report to the Virginia State Water Control Board, Richmond,Virginia.

Aspila, I., Agemian, H., and Chau, A.:1976, 'A semi-automated method for the determination of inorganic, organic and total phosphate in sediments', *Analyst* **101**, 187–197.

APHA (American Public Health Association), 1985, Standard Methods for examination of Water and Wastewater. 16th ed., American Public Health Association, Washington, D.C.

Cloern, J.: 1987, ' Turbidity as a control on phytoplankton biomass and productivity in estuaries', *Continental Shelf Research* **7**, (11/12), 1367–1381.

Dauer, D. M., Lane, M., Marshall, H., Carpenter, K. and Diaz, R.: 1999, Status and trends in water quality and living resources in the Virginia Chesapeake Bay: 1985–1998. Old Dominion University Applied Marine Research Laboratory, Norfolk, Va. Tech. Rpt. 40, 244 pp.

D'Ella, C.F., Steudler, P.A. and Corwin, N.: 1977 ''Determination of Total Nitrogen in Aqueous Samples using persulfate digestion', *Limnology and Oceanography* **22**, 760–764.

Fisher, T.R., Peele, E., Ammerman, J. and Harding, L.:1992, 'Nutrient limitation of phytoplankton in Chesapeake Bay', *Marine Ecological Progress Series* **82**, 51–63.

Gilbert, R. O.: 1987, Statistical Methods for Environmental Pollution Monitoring. Van Nostrand Reinhold Co. New York, N.Y., 320 pp.

Harding, L.W.: 1994, ' Long-term trends in the distribution of phytoplankton in Chesapeake Bay: roles of light, nutrients and streamflow', *Marine Ecological Progress Series* **104**, 267–291.

Malone, T., Ducklow, H., Peele, E. and Pike, S.: 1991, 'Picoplankton carbon flux in Chesapeake Bay', *Marine Ecological Progress Series* **78**, 11–22.

Malone, T., Conley, D., Fisher,T., Glibert, P., Harding, L., and Sellner, K.: 1996, 'Scales of nutrient-limited phytoplankton productivity in Chesapeake Bay', *Estuaries* **19**, 371–385.

Margalef, R.: 1958, 'Information theory in ecology', General Systematics 3, 36- 71.

Marshall, H.G.: 1994, 'Chesapeake Bay phytoplankton: I. Composition' *Proceedings of the Biological Society of Washington* **107**, 573–585.

Marshall, H.G.: 1995, 'Autotrophic picoplankton distribution and abundance in the Chesapeake Bay, U.S.A.' *Marine Nature* **4**, 33–42.

Marshall, H.G. and Alden, R.W.: 1990, 'A comparison of phytoplankton assemblages and environmental relationships in three estuarine rivers of the lower Chesapeake Bay' *Estuaries* **13**, 287–300.

Marshall, H.G. and Nesius, K.K:. 1996, 'Phytoplankton composition in relation to primary production in Chesapeake Bay', *Marine Biology* **125**, 611–617.

published record of a HAB in SC estuarine or marine waters was a *Karenia brevis* (formerly *Gymnodinium breve*) bloom transported from the Gulf of Mexico in 1988 (Tester *et al.*, 1991). The historical lack of HAB reports in SC may be related to the relatively low impact of anthropogenic nutrient loading along the SC coast, the relatively high flushing characteristics of SC estuaries, and/or the historical lack of HAB monitoring efforts (Lewitus *et al.*, 1999, In Press). Bricker *et al.* (1999) characterized SC estuaries as having low to moderate symptoms of eutrophication, but predicted that the exceptionally high coastal population growth in the state would lead to future problems related to eutrophic conditions, such as HABs. The authors also noted that "The South Atlantic is generally a poorly studied region. The confidence level of the assessment of overall eutrophic condition was low for 12 systems, and an additional six were specifically identified as requiring improvements in basic monitoring. The need for data in this region is critical, given the projected increases in population and the susceptibility of many of the systems". If the lack of HAB problems in SC is related to low nutrient levels, then the prevalence of HABs in SC waters may increase as nutrient inputs from coastal development continue to escalate.

In late 1997, the SC Task Group on Harmful Algae (originally called "SC Task Group on Toxic Algae") was formed to develop a coordinated state strategy to respond to possible *Pfiesteria*-related events. *Pfiesteria* is a toxic dinoflagellate that has caused numerous fish kills in North Carolina waters, and whose toxin has been linked to human neurological illnesses (Burkholder and Glasgow, 1997). The Task Group is an ongoing collaboration between representatives from SC state agencies, SC academic institutes, and SC-based federal institutes, whose efforts have established several programs to increase surveillance and understanding of HABs in SC. With the help of funding through ECOHAB, CDC, and EPA (see "Acknowledgements"), the Task Group: 1) established HAB monitoring efforts in several sites, 2) established a fish event (lesion or kill) sampling program, 3) expanded SCDNR fish survey program to include fish pathology assessment, 4) established a human health surveillance program for potential *Pfiesteria*-related effects, and 5) developed an outreach program to increase public awareness of the HAB issue.

In October 2000, the mission and accomplishments of the SC Task Group on Harmful Algae were extended with the formation of the South Carolina Harmful Algal Bloom Program (SCHABP), the first comprehensive statewide effort to assess the distribution and potential adverse effects of HABs in SC waters. This cooperative multi-institutional program has the following objectives: 1) to determine the present distribution of harmful algae in SC estuaries, 2) to determine the environmental factors that favor HAB formation in SC estuaries so that future effects on ecosystem and human health can be predicted, 3) to establish a statewide surveillance system for HAB detection, 4) to expand upon current communication and education efforts, and implement new efforts to increase awareness of the HAB issue among resource managers, community officials, coastal users, and the general public, and 5) to insure the continued operation of the SC Task Group on Harmful Algae.

Accomplishment of the objectives required a balance of the following approaches:

a) Sufficient sampling frequency to optimize the chance of detecting HAB events;

b) Sufficient spatial characterization to evaluate HAB distribution statewide;

c) Flexibility in research design to allow rapid and adequate response to HAB events;

d) Adequate information from field data and laboratory experiments to understand the relationship between environmental factors and HAB formation;

e) Rapid entry of results into a meta-data compatible database (Microsoft Access), and

f) A strong effort to communicate these findings and general HAB issues to coastal zone managers and the public.

The program design consists of four components: monitoring, event response, nutrient response bioassays, and outreach. Monitoring of physical, chemical, and biological parameters (including harmful algae distribution) includes 3 elements: 1) intensive temporal sampling at "hot spots" (areas characterized by prevalent red tides and/or lesioned fish), 2) extensive statewide spatial sampling on annual to monthly intervals, and 3) a surveillance program conducted by a citizen volunteer network (The South Carolina Phytoplankton Monitoring Network, SCPMN). Event Response to potentially harmful algae detected by the monitoring component consists of 2 elements: 1) identification, isolation, and culturing of harmful algae, and assays for biotoxin production (Analytical Response Team), and 2) ecophysiological assessment of the effects of environmental conditions on HAB formation. Nutrient Response Bioassays are used to determine the potential role of nutrient loading in stimulating HABs, and Outreach efforts include dissemination of HAB information by: 1) hands-on teaching and training (SCPMN, SC Sea Grant Extension Program in-service training program), 2) circulating printed material ("SC Task Group on Harmful Algae Newsletter;" and "Algae: A Sourcebook for Teaching about Harmful Algal Blooms" by N. Anderson), and 3) developing a web site.

Here, we summarize the documentation of HABs in SC waters over the last 3 years. The very recent establishment of concerted HAB monitoring efforts in SC limits our ability to determine whether newly detected HABs are novel to SC waters or a function of increased surveillance. Accordingly, the effects, if any, of coastal development in promoting HAB formation, will be better understood as monitoring progresses long-term and hypotheses on the ecophysiological bases for these blooms are experimentally tested. Anecdotal accounts suggest that the occurrence of at least one of the newly reported SC HABs, *Kryptoperidinium* sp. "red tides", may be unprecedented for this state (Lewitus *et al.,* In Press). However, in several cases in Spring 2001, HABs were detected in systems not previously surveyed; i.e., brackish ponds associated with housing or golf courses. Whether these pond-associated blooms are a typical springtime phenomenon or a product of anomalous conditions (e.g., the severe drought in this area) is a pressing question, especially because most of the documented HABs were potentially or measurably toxic.

2. Monitoring Sampling Design and Methodology

Routine monitoring efforts include intensive temporal sampling and extensive spatial sampling components. The former component involves weekly to biweekly monitoring at "hot spots" where either fish lesions or *Kryptoperidinium* sp. red tides had been noted or, based on (sub)urban sprawl, HABs may be expected to occur (Figure 1, Table 1).

The extensive statewide spatial monitoring component expands the following ongoing monitoring programs (Figure 1): 1) the South Carolina Estuarine and Coastal Assessment Program (SCECAP), a new initiative by SC Department of Natural Resources (SCDNR) and SC Department of Health and Environmental Control (SCDHEC) to evaluate overall habitat quality and the condition of biological resources throughout SC's coastal zone, and 2) the SCDHEC Water Quality Monitoring Program. The SCHABP receives samples annually from 60 SCECAP sites (randomly selected annually) that represent open water and tidal creek habitats, and monthly from 30 SCDHEC fixed sites. The combination of annually fixed sites ("hot spots" and SCDHEC Water Quality sites) and annually randomized sites (SCECAP) strikes a complimentary balance in spatial assessment of measured variables. Over the long term, application to predictive models evaluating the association between HABs and environmental conditions should be optimized by this complimentary approach.

Table 1. SC Harmful Algal Bloom Program Year 1 monitoring sample schedule. Variables measured: nutrients (ammonium, nitrate/nitrite, phosphate, total nitrogen, total phosphorus, dissolved organic nitrogen, dissolved organic phosphorus, dissolved organic carbon, silicate, dissolved inorganic carbon, urea), chlorophyll *a*, high performance liquid chromatographic pigment profiles, harmful algae and microzooplankton counts, polymerase chain reaction based detection of *Pfiesteria piscicida*, *P. shumwayae*, and *Cryptoperidiniopsis* sp., and atrazine (pesticide) concentration.

	Samplings Per Month											
Component	*Feb*	*Mar*	*Apr*	*May*	*Jun*	*Jul*	*Aug*	*Sep*	*Oct*	*Nov*	*Dec*	*Jan*
Hot Spots	16	16	32	32	38	38	41	38	36	16	16	16
SCECAP					24	24	12					
SCDHEC	30	30	30	30	30	30	30	30	30	30	30	30

3. Distribution of HABs

3.1 *PFIESTERIA* spp. AND *CRYPTOPERIDINIOPSIS*

Prior to 2001, *Pfiesteria piscicida*, *P. shumwayae*, and *Cryptoperidiniopsis* sp. had been documented in several SC estuaries, and the former species shown to be potentially very toxic under laboratory conditions (Lewitus *et al.*, 1999, unpublished data). However, although widespread in SC, *Pfiesteria* had never been found in high abundance (compared to North Carolina), nor was there any evidence linking it to SC fish kills or lesion problems. It was hypothesized that the reason why *Pfiesteria* is not abundant or apparently causing problems in SC estuaries is due to the relatively low nutrient levels and high flushing rates that characterize these systems (Lewitus *et al.*, 1999).

On 30 March 2001, a fish kill (600–800 menhaden and mullet) occurred at a 7-acre Hilton Head brackish pond surrounded by a condominium complex. Water collected during the fish kill contained *Pfiesteria*-like organisms (PLOs) at an abundance of 890 cells ml^{-1}, nearly an order of magnitude higher than any previously estimated abundance in SC waters, and comparable to abundances linked to

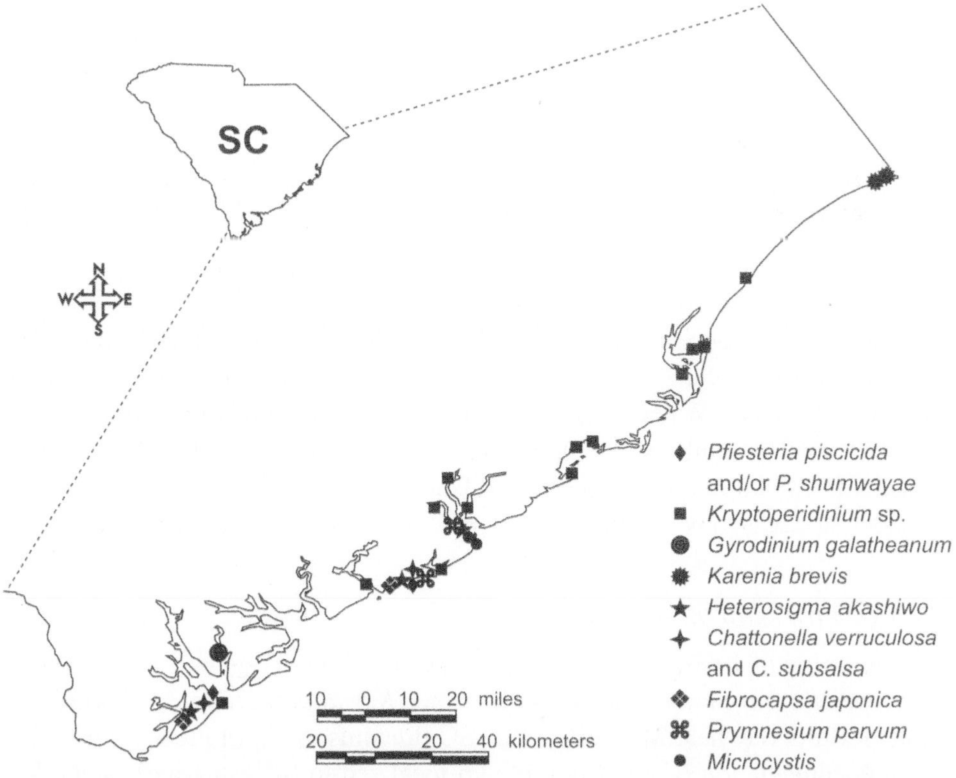

Figure 1. Sampling sites for Year 1 of SC Harmful Algal Bloom Program.

fish kills in the Neuse River (i.e., in the 9 *Pfiesteria*-associated fish kills from 1991–1995, PLO abundance averaged 940 ± 130 cells ml^{-1}; Burkholder *et al.*, 1995). Polymerase chain reaction (PCR) analyses conducted as part of the SCHABP were positive for *Pfiesteria shumwayae* and *Cryptoperidiniopsis* sp. (J.W. Kempton, SCDNR, unpublished data). These results were confirmed by DNA sequencing analyses conducted by an independent laboratory (AMPLICON express, Pullman, Washington, USA). Water and sediment samples were collected from the same pond 12 days after the fish kill event. PLO abundances ranged from 30–80 cells ml^{-1}, again with confirmation of *P. shumwayae* and *Cryptoperidiniopsis* sp. by PCR analyses and DNA sequencing. On April 11, the original sample collected during the fish kill, and sediment and water from the samples collected 12 days after the event, were sent to North Carolina State University's Center for Applied Aquatic Ecology for fish mortality bioassay tests. In separate bioassay experiments, fish death occurred 41 days (original sample) and 25–28 days (sediments and water from latter sampling date) after inoculation, concurrent with an increase in *P. shumwayae* and *P. piscicida*, as confirmed by PCR analyses (J.M. Burkholder and H.B. Glasgow, Jr., unpublished data). The *Pfiesteria* event prompted regulators to post warning signs for fishing and swimming in this pond. Prior to the present findings, the combined observations of high PLO abundance, PCR or scanning electron micrographic confirmation of *Pfiesteria*, and positive results from fish mortality bioassays in samples collected during a fish kill event had only been demonstrated in North Carolina (several times over the last decade) and in Chesapeake Bay (once, summer 1997).

On 25 April, two brackish water ponds were sampled in Kiawah Island as part of the NOAA-funded Urbanization in Southeastern Estuarine Systems (USES) program. One of the ponds was surrounded by a golf course and the other by a housing development. A tidal creek that fed the residential pond was also sampled. In the golf course pond, PLO abundance was estimated at 450 cells ml^{-1}, again potentially toxic population densities but this time not associated with a fish kill. PCR analysis indicated a strong positive for *Pfiesteria piscicida* (J.W. Kempton, unpublished data). Water from the residential pond and tidal creek contained relatively low PLO numbers, but high abundances of other HABs (see below). These results from Hilton Head and Kiawah Island indicate that brackish ponds in SC residential areas may be vulnerable to *Pfiesteria*'s toxic activity.

3.2 *KRYPTOPERIDINIUM* sp.

In the spring of 1998, a potentially new species of dinoflagellate (initially named *Scrippsiella carolinium*, and currently named *Kryptoperidinium* sp.) formed a red tide in Bulls Bay near McClellanville, SC (Lewitus *et al.*, In Press). This was the first documentation of a red tide bloom localized to SC estuarine waters. Since that first observation, blooms of this dinoflagellate have been seen with increased

Table 2. Estuarine systems where *Kryptoperidinium* sp. formed monospecific blooms. The highest abundance (cells ml^{-1}) observed in each system is listed, along with the date observed.

Estuary	*Nearest City*	*Maximum Abundance*	*Date*
Murrells Inlet	Myrtle Beach	5.0×10^3	3 April 2001
North Inlet	Georgetown	2.9×10^5	26 May 1999
Bulls Bay	McClellanville	3.5×10^5	5 May 1999
Cooper River	Charleston	1.0×10^5	18 April 2001
Kiawah River	Charleston	5.3×10^3	25 April 2001
Edisto River	Charleston	3.7×10^3	5 April 2001
Broad River	Hilton Head	2.8×10^3	15 July 1998

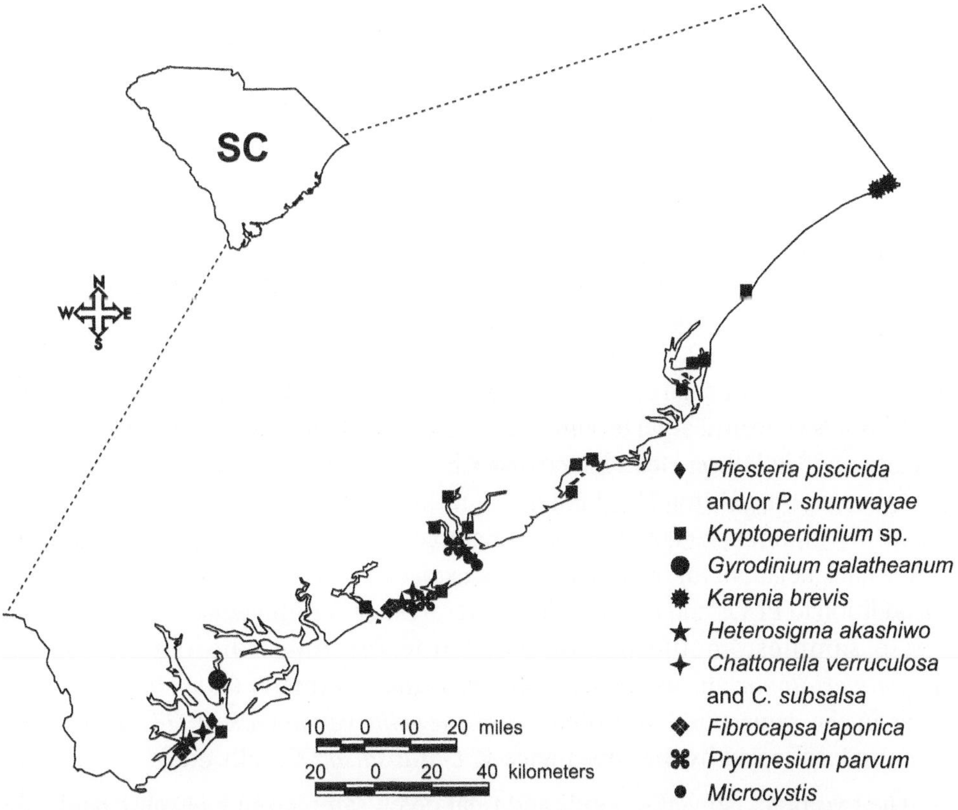

Figure 2. Harmful algal blooms documented in South Carolina as of May 2001 (symbols indicate where species was dominant and, in some cases, monospecific component of plankton community).

frequency from year to year, both spatially and temporally. In the spring of 2001, monospecific populations (> 99% of total phytoplankton biomass) of *Kryptoperidinium* were observed in several SC estuaries where it was not previously known to occur (Table 2, Figure 2). High concentrations of this red tide have been documented from Murrells Inlet southward to Hilton Head Island with cell densities reaching as high as 350,000 cells ml^{-1}.

Our research on *Kryptoperidinium* emphasizes its potential harmful effects, ecology, physiology, evolutionary biology, and taxonomy (e.g. Lewitus *et al.*, In Press). Samples sent to A. Place (Center of Marine Biotechnology, MD) were negative for hemolytic activity (unpublished data). However, oyster bioassays showed a significant increase in lysosomal destabilization when exposed to water samples collected from Kiawah Island containing a monospecific population of *Kryptoperidinium* sp. (A. Ringwood and C. Keppler, SCDNR, unpublished data). The increase in percent lysosomal destabilization indicates increased stress to the organism, and because other feeder algae were supplied during the bioassays, the stress could not be attributed to starvation. These results are in accordance with previous observations that growth of juvenile hard clams was greatly reduced in the presence of the red tide dinoflagellate (A. Ringwood, unpublished data). Also, in 1998, a *Kryptoperidinium* bloom was found in a blue crab shedding house, associated with a blue crab 30% mortality event (A.F. Holland, SCDNR, unpublished data). Based on the widespread occurrence of monospecific *Kryptoperidinium* blooms and evidence for harmful effects on shellfish, its potential ecological and economical impacts are obvious concerns.

3.3 POND-ASSOCIATED HABS

Samples collected in summer 2000 from a Charleston freshwater subdivision pond associated with a cyanobacterial bloom containing *Microcystis* were positive for microcystin (Figure 2, F. Van Dolah and T. Leighfield, National Ocean Service-Charleston, unpublished data). Also, from April-to-May 2001, a succession of toxic or otherwise harmful algal blooms were observed in the abovementioned brackish water ponds in Hilton Head Island and Kiawah Island (see section 3.1). The high PLO event at the Hilton Head pond was followed 12 days later by a *Heterosigma akashiwo* bloom (199,000 cells ml^{-1}). Samples were sent to NOS-Charleston, and came back negative for brevetoxin (M. Busman and S. Morton, unpublished data). In addition to PLOs and *H. akashiwo*, several other harmful algae were detected in these samples, including some red tide-forming dinoflagellates; e.g. *Kryptoperidinium* sp. and *Gyrodinium instriatum* (formerly *G. uncatenum*). Days later in the same pond, a bloom of *Chattonella verruculosa* (97,232 cell ml^{-1}) occurred, followed by the appearance of *C. subsalsa* (27 cells ml^{-1}).

The two brackish water ponds and tidal creek sampled in Kiawah Island were also marked by a succession of HAB types. On the first sampling date (25 April 2001), a bloom of *Chattonella verruculosa* (5840 cells ml^{-1}) was observed in the pond associated with residential housing, and a monospecific population of *Kryptoperidinium* sp. (5273 cells ml^{-1}) was found in the tidal creek. Water containing the *Chattonella* bloom was screened for brevetoxin based on an ELISA assay and tested positive (C. Tomas and J. Naar, University of North Carolina-Wilmington, unpublished data). The ponds were re-sampled on 3 May, and *C. subsalsa* was estimated at 440 cells ml^{-1} in the residential pond. The water again tested positive based on the brevetoxin ELISA assay (K. Steidinger, unpublished data). On 8 May, *C. subsalsa* was estimated at 330 cells ml^{-1} in the same pond, and ELISA assay results on the water were positive, but oysters from a creek that exchanges with the pond tested negative (C. Tomas, unpublished data).

Following the *Chattonella subsalsa* blooms, another potentially toxic raphidophyte, *Fibrocapsa japonica*, reached high concentrations (14,133 to 90,400 cells ml^{-1}) in the residential ponds from Hilton Head and Kiawah on 21 May and 23 May, respectively. Samples from the latter event tested positive based on the brevetoxin ELISA assay (C. Tomas, unpublished data). Finally, on 23 May, another HAB species, *Prymnesium parvum*, was found in high concentrations (609,760 cells ml^{-1}) in samples collected from the golf course pond on Kiawah Island (S. Wilde and J. Wolny, unpublished data).

It is clear from these recent findings that the potential for HAB formation in SC brackish ponds associated with residential development or golf courses may be high. We believe that blooms in these systems are favored by high nutrient and relatively low flow conditions, but information is needed on the specific factors that lead to dominance of a particular bloom type and successional transitions. Ecophysiological analyses (associations with environmental conditions; laboratory experiments with bloom isolates) are ongoing. It should be stressed that 2 out of the 3 SC freshwater residential ponds that were tested for microcystin in summer 2000 (all in Charleston) proved positive (F. Van Dolah and T. Leighfield, unpublished data), and 3 out of 3 brackish ponds sampled contained a number of potentially or measurably toxic HABs. There is a great need to expand the sampling coverage of the SCHABP to assure that routine monitoring and other research efforts include a representative subset (e.g. salinity, age, size) of the numerous ponds associated with housing or golf course development along the SC coast.

4. Discussion

The SCHABP is the first statewide program to assess the status of SC estuaries with respect to the distribution and potential environmental and human impact of HABs. A multi-institutional program involving cooperative efforts from university, state, and federal investigators was deemed necessary to gain information on HAB distribution on adequate spatial and temporal scales, understanding how these blooms form and how human activities could contribute, and to educate the SC public on the HAB issue. The need for such an extensive effort was emphasized by the lack of previous HAB surveillance systems in all but a handful of SC estuaries, a potential link between HAB prevalence and anthropogenic nutrient loading, and the exceptionally high rates of population growth along the SC coast. This report documented our progress towards the objectives of Year 1's research plan. The SCHABP already has significantly extended our understanding of the distribution and ecology of HABs in SC estuaries and freshwater systems, and discovered several potentially toxic blooms (including *Pfiesteria*) never before reported in SC. In the short term, this project has provided the infrastructure for rapid detection and analytical evaluation of these recent HABs. In the long term, the research strategy of the SCHABP monitoring component promises to develop an ecological framework for understanding environmental regulation of HABs in SC waters.

Acknowledgements

We thank the following people for their contributions to the SCHABP and/or this study: SCDNR (C. Johnson, J. Kempton, M. Levisen, P. Maier, G. Riekirk, A. Shurtleff, B. Van Dolah, P. Weinbach, J. Williams, J. Wolny), USC (J. Keesee, R. Tymowski, S. Wilde, B. Willis), SCDHEC (D. Baize, R. Ball, R. Berry, J. Bickley, D. Chestnut, K. Hayes, L. McCormick, W. McDermott, N. Scruggs, B. Younginer), NOS-Charleston (M. DeLorenzo, T. Leighfield, S. Morton, P. Pennington, J. Ramsdell, G. Scott, T. Siewicki, F. Van Dolah), SC Sea Grant Consortium (P. Alford, R. DeVoe, J. Tibbetts), and Medical University of South Carolina (E. Lacy, B. Ribner). This research was funded by the U.S. ECOHAB Program, sponsored by NOAA/NSF/EPA/NASA/ONR, grant NA86OP0493, NOAA grants NA90AA-D-SG672 and NA06OA0675, CDC and EPA grants to SCDHEC, and EPA grant R826944-01-0. Contribution 1344 of the Belle W. Baruch Institute for Marine Biology and Coastal Research, Contribution 480 of SCDNR's Marine Resources Research Institute, ECOHAB Contribution 41.

References

Anderson, D. M. and Garrison, D. .J. (eds.): 1997, 'The Ecology and Oceanography of Harmful Algal Blooms', *Special Issue of Limnology and Oceanography* **42**, 1305 pp.

Anderson, D. M., Hoagland, P., Kaoru, Y. and White, A. W.: 2000, Estimated Annual Economic Impacts from Harmful Algal Blooms (HABs) in the United States. Woods Hole Oceanographic Institute Technical Report WHOI-2000-11, Woods Hole, MA.

Boesch, D. F., Anderson, D. M., Horner, R. A., Shumway, S. E., Tester, P. A. and Whitledge, T. E.: 1997, Harmful algal blooms in coastal waters: Options for prevention, control and mitigation. NOAA Coastal Ocean Program Decision Analysis Series No. 10. NOAA Coastal Ocean Office, Silver Spring, MD, 46 pp. + appendix.

Bricker, S. B., Clement, C. G., Pirhalla, D. E., Orlando, S. P. and Farrow, D. R. G.: 1999, National Estuarine Eutrophication Assessment: Effects of Nutrient Enrichment in the Nation's Estuaries. NOAA, National Ocean Service, Special Projects Office and the National Centers for Coastal Ocean Science. Silver Spring, Maryland.

Burkholder, J. M. and Glasgow Jr., H. B.: 1997, 'The ichthyotoxic dinoflagellate, *Pfiesteria piscicida*: Behavior, impacts, and environmental controls', *Limnology and Oceanography* **42**, 1052–1075.

Burkholder, J. M., Glasgow Jr., H. B. and Hobbs, C. W.: 1995, 'Fish kills linked to a toxic ambush-predator dinoflagellate: distribution and environmental conditions', *Marine Ecology Progress Series* **124**, 43–61.

Lewitus, A. J., Burkholder, J. M., Glasgow Jr., H. B., Glibert, P. M., Willis, B. M., Hayes, K. C. and Burke, M. K.: 1999, 'Mixotrophy and nutrient uptake by *Pfiesteria piscicida*. (Dinophyceae)', *Journal of Phycology* **35**, 1430–1437.

Lewitus, A. J., Hayes, K. C., Gransden, S. G., Glasgow Jr., H. B., Burkholder, J. M., Glibert, P. M. and Morton, S. L.: in press, 'Ecological characterization of a widespread red tide in South Carolina estuaries: a newly observed phenomenon' in: Hallegraeff, G. M., Blackburn, S., Bolch, C. and Lewis, R. (eds.). Proceedings of the Ninth International Conference on Harmful Algal Blooms, IOC UNESCO, Paris.

Tester, P. A., Stumpf, R. P., Vukovich, F. M., Fowler, P. K. and Turner, J. T.: 1991, 'An expatriate red tide bloom: Transport, distribution, and persistence', *Limnology and Oceanography* **36**, 1053–1061.

A PILOT PROJECT TO DETECT AND FORECAST HARMFUL ALGAL BLOOMS IN THE NORTHERN GULF OF MEXICO

WILLIAM S. FISHER[1*], THOMAS C. MALONE[2], AND JAMES D. GIATTINA[3]

[1]*U.S. Environmental Protection Agency, Office of Research and Development, National Health and Environmental Effects Research Laboratory, Gulf Ecology Division, Gulf Breeze, FL 32561;* [2]*University of Maryland Center for Environmental Studies, Horn Point Laboratory, P. O. Box 775, Cambridge, MD 21613;* [3]*U.S. Environmental Protection Agency, Gulf of Mexico Program Office, Stennis Space Center, MS 39529*

**(author for correspondence, email: fisher.william@epa.gov)*

Abstract. More timely access to data and information on the initiation, evolution and effects of harmful algal blooms can reduce adverse impacts on valued natural resources and human health. To achieve this in the northern Gulf of Mexico, a pilot project was initiated to develop a user-driven, end-to-end (measurements to applications) observing system. A key strategy of the project is to coordinate existing state, federal and academic programs at an unprecedented level of collaboration and partnership. Resource managers charged with protection of public health and aquatic resources require immediate notice of algal events and a forecast of when, where and what adverse effects will likely occur. Further, managers require integrated analyses and interpretations, rather than raw data, to make effective decisions. Consequently, a functional observing system must collect and transform diverse measurements into usable forecasts. Data needed to support development of forecasts will include such properties as sea surface temperature, winds, currents and waves; precipitation and freshwater flows with related discharges of sediment and nutrients; salinity, dissolved oxygen, and chlorophyll concentrations (*in vivo* fluorescence); and remotely-sensed spatial images of sea surface chlorophyll concentrations. These data will be provided via a mixture of discrete and autonomous *in situ* sensing with near real-time data telemetry, and remote sensing from space (SeaWiFS), aircraft (hyperspectral imagery) or land (high-frequency radar). With calibration across these platforms, the project will ultimately provide a 4-dimensional visualization of harmful algae events in a time frame suitable to resource managers.

Keywords: Harmful algal blooms, red tide, observing systems, remote sensing, *in situ* sensing, environmental monitoring

1. Introduction

Approximately 5000 species of microalgae exist world-wide, of which about 100 are toxic. More than half of the toxic species occur in the Gulf of Mexico, each with a potential to create public health risks, cause mortalities of marine organisms, or alter the capacity of coastal ecosystems to support living resources. Toxic algae comprise multiple trophic levels and a variety of taxa (dinoflagellates, diatoms, cyanobacteria), and can create toxic effects at relatively low densities. Nonetheless, such events are recognized by a generic term, harmful algal blooms (HABs), which is widely accepted to represent toxic species as well as species that alter habitat (anoxia, hypoxia, persistent turbidity), create physiological dysfunction (respiratory, reproductive), or perturb community relationships (trophic level domination). HAB events worldwide appear to be increasing in frequency and severity (Smayda, 1989). They may be initiated and sustained by natural environmental conditions (Steidinger and Haddad, 1981; Tester *et al.*, 1991), but have also been

Environmental Monitoring and Assessment **81**: 373–381, 2003.

linked to anthropogenic factors. These include watershed runoff of nutrients that stimulate algal growth (Hallegraeff, 1993; Smayda, 1990) and spread of indigenous and non-indigenous algal species through increased commerce (Hallegraeff and Bolch, 1992; Carlton and Geller, 1993).

Coastal economies are affected by HABs through shellfish closures, fish kills, and beach advisories that reduce tourism and consumer confidence in seafood products. It has been estimated (CENR, 2000) that the annual cost of HABs in the U.S. is $300-700 million. Valued marine resources such as finfish, shellfish, manatees, dolphins, and sea turtles are killed by acute exposure to HAB toxins and may be sublethally affected at lower, chronic exposures. Ecologically important species are also affected, which can lead to degraded water and habitat quality. These effects, coupled with the increasing number, extent and severity of harmful algal blooms in the U.S. (Anderson, 1989; Smayda, 1990; Hallegraeff, 1993; Van Dolah, 2000), mobilized Congress to pass the 1998 Harmful Algal Bloom and Hypoxia Research and Control Act (P.L. 105-383), which specifically calls for the establishment of: (1) an inter-agency task force, (2) a national assessment, and (3) an assessment plan for HABs and hypoxia in the Gulf of Mexico. The latter was in recognition of their high visibility and economic importance to the region.

2. Management of Red Tides in the Gulf of Mexico

Among the most common and deleterious blooms in the Gulf of Mexico are those caused by the dinoflagellate *Gymnodinium breve* (Geesey and Tester, 1993), which has recently been renamed *Karenia brevis* (*K. brevis*) (Daugbjerg *et al.*, 2001). In high concentrations, *K. brevis* imparts a reddish hue to the water, a condition commonly referred to as 'red tide'. Red tides routinely occur along the southwest coast of Florida in the late summer and early fall (Gunter *et al.*, 1948; Gilbes *et al.*, 1996) and may persist for three months (Steidinger and Baden, 1984). Recently, *K. brevis* blooms have become more widely distributed; they have been reported in Texas, Mexico, and as far north as North Carolina (Tester *et al.*, 1991). The first confirmed reports of red tides in Alabama, Mississippi and Louisiana were in 1996, supporting the assertion that red tides are occurring more frequently. In 1986, Texas experienced its first red tide in 30 years. Since then, major events have occurred in 1996, 1997 and 2000.

Karenia brevis produces, under certain (yet unknown) environmental conditions, a neurotoxin (brevetoxin), that paralyzes respiratory function in fish and mammals and can accumulate in edible shellfish. Consequently, blooms of *K. brevis* adversely affect public health, economy, and valued coastal resources. Humans are exposed to brevetoxin by aerosols or through ingestion of shellfish that have accumulated the toxin. Aerosol exposure usually causes mild respiratory irritation that leads to a burning throat, coughing and asthma-like symptoms. Ingestion exposure can cause more severe symptoms, including vomiting, muscle cramps and seizures. The severity is tied to high toxin concentrations that can accumulate in edible bivalves such as eastern oysters, *Crassostrea virginica*. To guard against human ingestion, public health managers close shellfish harvesting when *K. brevis* cell densities in the water reach 5 x 10^3/L and re-open them when oysters contain < 80 μg/100 g shellfish. Consequently, a red tide event may last only a few days but hazardous toxin concentrations can persist for months.

Economic costs of red tide events depend on the severity and location of the bloom. Losses related to public health include wages, workdays and medical treatment. Fisheries incur losses when oyster beds are closed and when fish are killed or impaired by red tide toxins or associated hypoxic conditions. There are costs associated with state responses to fish kills, with cleanup of dead and decaying organisms on beaches, and with rehabilitation of protected species. And, even when events are restricted to only a few locations, media reports can reduce tourism across broad coastal areas. In Florida, a single *K. brevis* event is estimated to cost millions of dollars.

All fish and mammals are susceptible to brevetoxin, but some are less mobile and can be trapped in estuaries. Dolphin can easily elude *K. brevis* in the water column but may eat fish that have accumulated brevetoxin. Manatee appear to get trapped when climatic conditions alter their spring migrations or when the blooms are more extensive along the southwestern Florida coast. Notable manatee mortality events have occurred in 1963 (7), 1982 (39) and 1996 (149 manatee killed), coincident with higher than normal levels of *K. brevis* during those years (Landsberg and Steidinger, 1997).

The health, economic and ecological costs of *K. brevis* blooms compel the application of measures to prevent, control or mitigate their occurrence (CENR, 2000). Any management action will depend upon, or benefit from, an accurate system to detect ongoing events, forecast bloom severity and trajectory, and predict when environmental conditions are susceptible to bloom formation. For example, prior knowledge of an event would allow protection (e.g., baffling) of aquaculture shellfish beds or harvesting of threatened oysters prior to an event. Open seasons and bag limits for commercial fish could be altered. In addition, states and animal rehabilitation organizations could mobilize response and cleanup crews in advance of an event. The tourist industry could better estimate their need for employees and could mitigate losses at the beach by offering alternative activities. Forecasts would be valuable at several time scales. Annual and seasonal forecasts would allow states to better budget resources according to estimated response needs. Forecasts in real-time (nowcasts) would provide local and state resource managers with pre-emptive tools to reduce the effects and costs of an event.

Management of blooms and bloom effects may lie within separate political jurisdictions, but forecasting will be better achieved at a regional scale. Initiation and sustenance of HABs is a consequence of both small and large scale environmental and meteorological conditions; changes in circulation patterns, currents, winds, water temperature, climate, watershed runoff and land-use might all interactively influence *K. brevis* blooms, as might a variety of biological and ecological factors (Tester and Steidinger, 1997). Accurate forecasts will require crossing state and local jurisdictions to collect and assemble data into useful, regionally-integrated information. And, despite the logistical issues, a regional approach offers numerous advantages. Specific expertise or capabilities from different jurisdictions can be applied for the good of all, and redundant efforts can be minimized. Regional information also reaches a broader spectrum of users, and the interactions of these users can establish a framework for greater and more sustained coordination among academic, state and federal programs.

3. Initiation of the Harmful Algal Blooms Observing System

A concept for a regional HAB forecasting system was derived through the U.S. Global Ocean Observing System (GOOS) Program (Malone, 2002). The objective of the international GOOS effort is to develop observing systems to improve the timeliness and accuracy for detecting and predicting global scale weather patterns and climate change. The coastal component of GOOS is under development, but includes goals to: (1) improve local and regional weather forecasts; (2) minimize or eliminate public health risks associated with environmental changes (e.g., coastal flooding) and biological events (e.g., harmful algal blooms); (3) improve predictions of the effects of global weather and climate patterns on coastal populations and the safety and efficiency of marine operations, and (4) improve the efficacy of management actions to control and mitigate effects of human activities and natural hazards on coastal ecosystems and to protect and restore critical habitats and biodiversity.

The Harmful Algal Blooms Observing System (HABSOS) is proposed as a pilot project to formulate and implement a user-driven, end-to-end (measurements-to-applications) observing system for the northern Gulf of Mexico (U.S. portion). It will focus on timely access of data and information on the initiation, evolution, location and effects of *K. brevis* blooms with future potential to include additional bloom types. The project will rely on an unprecedented level of inter-jurisdictional cooperation and collaboration, and will serve as 'proof-of-concept' for data collection, integration, interpretation, distribution and archival for a variety of coastal issues. Red tides were selected to pilot the observing system because of their importance to the region, the relatively strong historical record of red tide blooms and their effects, and the ability to detect red tides from satellite and aerial imaging (chlorophyll *a*). The latter will be integral to the pilot project because a primary goal is to demonstrate integration and management of data from a variety of platforms, including *in situ* and remote monitoring. The design and implementation of a *K. brevis* observing system is a test case for determining the types and quality of data collection and data management infrastructure needed for a single, important coastal issue. Once completed, the subsequent goal is to grow the system into a fully integrated and sustained observing system for the northern Gulf of Mexico. Eventually, similar systems developed for other coastal regions will form a national federation of regional observing systems.

Data relevant to *K. brevis* blooms will vary depending on the status and location of the bloom. Data needed to track and forecast the trajectory of an existing bloom will be different than the data needed to identify new blooms or anticipate new blooms from prevailing environmental conditions. In addition, data requirements will depend on whether the bloom is offshore or nearshore. Even though a major bloom has persisted offshore in southwest Florida for many years, more destructive effects are registered for blooms that originate nearshore and in estuaries, where public health is at greater risk. In all cases, the presence and effects of harmful algae must be documented in the context of changes in the physical and chemical characteristics of the water column (e.g., currents, wind, temperature, salinity, nutrients).

Presence of *K. brevis* is best characterized by direct microscopic identification, sometimes performed in routine monitoring of shellfish beds and intermittently performed during research cruises. Remote sensing of red tide blooms is also possible when *K. brevis* cells reach high densities in the water column. The spectral signature of chlorophyll *a* in high-density blooms can be detected from clear Gulf waters using color sensors in airplanes or satellites (Tester and Stumpf, 1998). The Sea-viewing Wide Field-of-View Sensor (SeaWiFS) satellite commercially launched in 1997 can detect chlorophyll *a* at a 1 km resolution across the entire Gulf of Mexico every 1–2 days (Tester *et al.*, 1998). Accounting for variability, the satellite sensor is valid at *K. brevis* cell densities of 5×10^4/L, which is insufficient to signal shellfish bed closures (5×10^3 cells/L), but quite capable of detecting densities (10^5–10^6/L) that are associated with human respiratory distress and aquatic mortality events. Currently, high chlorophyll measurements must be sampled on site (shipboard) to validate presence and dominance of *K. brevis*. Future remote sensors may use spectral signatures for individual phytoplankton species (Cullen *et al.*, 1997). Also, due to limitations of low resolution satellite imagery, chlorophyll detection is not valid within 2 km of shore where many of the most destructive *K. brevis* events occur.

Until better methods are developed, nearshore detection and tracking of red tides may rely on *in situ* monitoring and responses to mortality events. Much of these data, as well as data to investigate underlying environmental causes of HABs, are collected by different state, federal and academic programs. Yet there is little consistency in the objectives of these programs and no existing strategy for integration of data or effective use of collective resources. Forecasting *K. brevis* will require integration of synoptically-collected remote data (e.g., chlorophyll *a*, sea surface temperatures), data collected *in situ* (e.g., cell counts, nutrients, currents, wind) from monitoring programs and data collected during event responses.

4. Integrating and Delivering Data

A principal component of HABSOS is to demonstrate a capacity to integrate existing data collected from multiple sources at different spatial and temporal scales into a cohesive and useful body of information. Such integration must occur with an issue-relevant objective in mind; in this case, the forecast of *K. brevis* blooms. In early discussions of HABSOS, it was determined that state public health and resource mangers, i.e., those who make decisions to mitigate red tide events, were the end-users of the observing system. In the ensuing workshop, these managers expressed a need for moderately-interpreted information; not raw data (e.g., current vectors, cell counts, nutrients) and not prognostication, but an overview of probabilities similar to the integrated meteorological data provided by the Weather Service. State managers also indicated that the priority products for HABSOS should be: (1) routine early warnings (alerts) of events as they are detected, (2) forecasts for time- and space-dependent trajectories of ongoing events, (3) timely access to information on *K. brevis* events and related environmental variables, and (4) alerts of where and when environmental conditions are favorable for *K. brevis* blooms. Because these priorities rely on an integration of regionally-collected data of different types and from different platforms, their achievement will require a model and a clear vision for iterative development and testing.

In a first iteration, data will be mined from previous events and examined for forecasting utility. Phytoplankton experts from across the Gulf of Mexico have begun to aggregate regional data from 1996 (very high *K. brevis* activity across all five states), 1997 (relatively low activity) and 2000 (high activity) with an initial focus on cell counts, chlorophyll *a* (available for 2000 only), salinity and temperature. A determination will be made as to whether these factors could have been used to project the mortalities and shellfish closures that occurred during those years, and whether they differ between high and low red tide years. More refined models will require additional data and may eventually target data that are not currently collected. Fundamental to this process is continued research to elaborate causes, contributing factors, and consequences of *K. brevis* blooms.

Aggregated information from the case study (described above) and from future observations will be presented through a 'portal' system, similar to that used by the Chesapeake Bay Program (www.chesapeakebay.net/cims/). In this system, user queries result in retrieval of information relevant only to their particular needs. Advantages include flexibility in the type and format of data that are retrieved and the knowledge that information is not only timely, but prepared by a subject area expert.

5. Partners in HABSOS

Implementation of HABSOS will require coordination among numerous state, federal and academic institutions. The Gulf of Mexico Program (GMP) was formed in 1988 to develop and implement voluntary, incentive-based management strategies to protect, restore and maintain the health and productivity of the Gulf of Mexico ecosystem. The program is framed to provide interaction among multiple state, federal, and private entities on issues of environmental importance with clear oversight by all stakeholders. Algal blooms are very important to GMP objectives, playing a role in three of four priority issues. The GMP will provide the organizational infrastructure to transform HABSOS from concept to reality. In so doing, it is anticipated that HABSOS will influence the GMP approach to a variety of other environmental issues that require integration of complex and diverse data.

A number of partners are involved in the collection, integration and presentation of HABSOS data. State agencies and academic researchers across all five coastal states will provide cell count data and *in situ* measurements of salinity, temperature and nutrients. Similarly, state resource managers will provide documentation of responses to mortality events through the GMP's Gulf of Mexico Aquatic Mortality Network (Fisher and Sherman, 1999). Forecasts for ocean and near coastal circulation, waves, tides, sediment transport, and water quality in the coastal regions of all five Gulf States is anticipated through expansion of the Northern Gulf Littoral Initiative of the U.S. Navy, University of Southern Mississippi and the GMP. Integration and interpretation of data will become a project of the LabNet and CastNet programs of the 115-member National Association of Marine Laboratories. These are distributed data communications and management systems that link data bases at participating coastal laboratories via the internet to provide visually integrated environmental data on temporal and geospatial scales. Presentation of information via portals will be a project of the newly formed NOAA National Coastal Data Development Center and archival will be facilitated through coordination with NOAA's National Ocean Data Center. Developing forecast tools depends on timely detection of events and on constructive and timely interactions among research, monitoring, and the users of the data and information generated by research and monitoring programs. It is imperative that HABSOS interacts with existing research programs, such as the national, multi-agency ECOHAB Program, to better understand the types of data and interpretive concepts needed to forecast blooms. In return, researchers are provided information in the context of relatively large scale observations critical to hypothesis-driven research. Environmental research is most valuable when conducted in the context of an integrated observing system that is sustained over sufficiently large areas with sufficient resolution to document patterns of variability and change. The successful observing system will incorporate key environmental variables that span regional and national interests while remaining locally relevant. Development of a strategy that provides this critical perspective will be invaluable to future research on causes of algal blooms and aquatic mortalities in the Gulf of Mexico. Such a system will also contribute to better understanding of other changes in coastal ecosystems such as the development and effects of bottom water hypoxia and anoxia.

Acknowledgments

The authors appreciate the efforts of the HABSOS Workshop Steering Committee and participants at the Workshop and two subsequent planning meetings who worked together to establish the HABSOS framework. This is Gulf Ecology Division Contribution 1141.

References

Anderson, D. M.: 1989, 'Toxic Algal Blooms and Red Tides: A Global Perspective', in: Okaichi, T., Anderson, D. M., and Nemoto, T. (eds.), Red Tides: Biology, Environmental Science, and Toxicology. Elsevier Science, NY, pp. 11–16.

Carlton, J. T. and Geller, J. B.: 1993, 'Ecological Roulette: The Global Transport of Nonindigenous Marine Organisms', *Science* **261**, 78–82.

CENR (Committee on Environment and Natural Resources): 2000, National Assessment of Harmful Algal Blooms in US Waters, National Science and Technology Council Committee on Environment and Natural Resources, Washington, DC, 38 pp.

Cullen, J. J., Ciotti, A. M., Davis, R. F. and Lewis, M. R.: 1997, 'Optical Detection and Assessment of Algal Blooms', *Limnology and Oceanography* **42**(5), 1223–1239.

Daugbjerg, N., Hansen, G., Larsen, J. and Moestrup, O.: 2001, 'Phylogeny of Some of the Major Genera of Dinoflagellates Based on Ultrastructure and Partial LSU rDNA Sequence Data, including the Erection of Three New Genera of Unarmoured Dinoflagellates', *Phycologia* **39**, 302–317.

Fisher, W. S. and Sherman, B. H.: 1999, 'Integrated Monitoring of Marine Disease and Mortality' (Abstract), *Journal of Shellfish Research* **18**, 318.

Geesey, M. and Tester, P. A.: 1993, '*Gymnodinium breve*: Ubiquitous in Gulf of Mexico Waters,' in: Smayda, T. J. and Shimizu, Y. (eds.), Toxic Phytoplankton Blooms in the Sea, Elsevier Science, NY, pp. 251–256.

Gilbes, F., Tomas, C., Walsh, J. J. and Muller-Karger, F. E.: 1996, 'An Episodic Chlorophyll Plume on the West Florida Shelf', *Continental Shelf Research* **16**(9), 1201–1224.

Gunter, G., Williams, R. H., Davis, C. C. and Walton Smith, F. G.: 1948, 'Catastrophic Mass Mortality of Marine Animals and Coincident Phytoplankton Blooms on the West Coast of Florida, November 1946 to August 1947', *Ecological Monographs* **18**, 308–324.

Hallegraeff, G. M.: 1993, 'A Review of Harmful Algal Blooms and their Apparent Global Increase', *Phycologia* **32**, 79–99.

Hallegraeff, G. M. and Bolch, C. J.: 1992, 'Transport of Diatom and Dinoflagellate Resting Spores via Ship's Ballast Water: Implications for Plankton Biogography and Aquaculture', *Journal of Plankton Research* **14**, 1067–1084.

Landsberg, J. H. and Steidinger, K. A.: 1997, 'A Historical Review of Red Tide Events Caused by *Gymnodinium breve* as Related to Mass Mortalities of the Endangered Manatee (*Trichechus manatus latirostris*) in Florida, USA', in: Reguera, B., Blanco, J., Fernandez, M. and Wyatt, T. (eds.), Harmful Microalgae, Xunta de Galicia and IOC of UNESCO Publishers, Paris, France.

Malone, T. C.: 2002, 'The Coastal Component of the U.S. Integrated Ocean Observing System', *Environmental Monitoring and Assessment*, This volume

Smayda, T. J.: 1989, 'Primary Production and the Global Epidemic of Phytoplankton Blooms in the Sea: A Linkage?', in: Cosper, E. M., Bricelj,V. M. and Carpenter E. J. (eds.), Novel Phytoplankton Blooms:Causes and Impacts of Recurrent Brown Tide and Other Unusual Blooms, Springer-Verlag, NY, pp. 213–228.

Smayda, T.: 1990, 'Novel and Nuisance Phytoplankton Blooms in the Sea: Evidence for a Global Epidemic', in: Graneli, E., Sundstrom, B., Edler, L. and Anderson, D. M. (eds.), Toxic Marine Phytoplankton. Elsevier Science, NY, pp. 29–40.

Steidinger, K. A., and Baden, D.G.: 1984, 'Toxic Marine Dinoflagellates', in: Spector, D. L. (ed.), Dinoflagellates, Academic Press, NY, pp. 201–261.

Steidinger, K. A. and Haddad, K.: 1981, 'Biologic and Hydrographic Aspects of Red Tides', *Bioscience* **31**, 814–819.

Tester, P. A. and Steidinger, K. A.: 1997, '*Gymnodinium breve* Red Tide Blooms: Initiation, Transport, and Consequences of Surface Circulation', *Limnology and Oceanography* **42**(5), 1039–1051.

Tester, P. A. and Stumpf, R. P.: 1998, 'Phytoplankton Blooms and Remote Sensing: What is the Potential for Early Warning'?, *Journal of Shellfish Research* **17**(5), 1469–1471.

Tester, P. A., Stumpf, R. P., Vukovich, F. M., Fowler, P. K. and Turner, J. T.: 1991, 'An Expatriate Red Tide Bloom: Transport, Distribution, and Persistence', *Limnology and Oceanography* **36**(5), 1053–1061.

Tester, P. A., Stumpf, R. P. and Steidinger, K.: 1998, 'Ocean color imagery: what is the minimum detection level for *Gymnodium breve* blooms?', in: Reguera, B., Blanco, J., Fernandez, M. and Wyatt, T., (eds.), Harmful Algae, Proceedings of the VIII International Conference on Harmful Algae. Xunta de Galicia and IOC of UNESCO, Paris, France, pp. 149–151.

Van Dolah, F.: 2000, 'Marine Algal Toxins: Origins, Health Effects, and their Apparent Global Increase', *Environmental Health Perspectives* **108**, 133–141.

PRELIMINARY INVESTIGATION OF SUBMERGED AQUATIC VEGETATION MAPPING USING HYPERSPECTRAL REMOTE SENSING

DAVID J. WILLIAMS[1*], NANCY B. RYBICKI[2], ALFONSO V. LOMBANA[2]
TIM M. O'BRIEN[3] AND RICHARD B. GOMEZ[4]

[1]*Landscape Ecology Branch, Environmental Sciences Division, U.S. Environmental Protection Agency, 12201 Sunrise Valley Drive, 555 National Center, Reston, VA 20192;* [2]*National Research Program, Water Resources Division, U.S. Geological Survey, 12201 Sunrise Valley Drive, 555 National Center, Reston, VA 20192;* [3]*Environmental Concern Inc., P.O. Box P, St. Michaels, MD 21663;* [4]*Center for Earth Observing and Space Research, School of Computational Sciences, George Mason University, Fairfax, VA 22030*

**(author for correspondence, e-mail: dxwilliams@usgs.gov)*

Abstract. The use of airborne hyperspectral remote sensing imagery for automated mapping of submerged aquatic vegetation (SAV) in the tidal Potomac River was investigated for near to real-time resource assessment and monitoring. Airborne hyperspectral imagery and field spectrometer measurements were obtained in October of 2000. A spectral library database containing selected ground-based and airborne sensor spectra was developed for use in image processing. The spectral library is used to automate the processing of hyperspectral imagery for potential real-time material identification and mapping. Field based spectra were compared to the airborne imagery using the database to identify and map two species of SAV (*Myriophyllum spicatum* and *Vallisneria americana*). Overall accuracy of the vegetation maps derived from hyperspectral imagery was determined by comparison to a product that combined aerial photography and field based sampling at the end of the SAV growing season. The algorithms and databases developed in this study will be useful with the current and forthcoming space-based hyperspectral remote sensing systems.

Keywords: submerged aquatic vegetation, remote sensing, hyperspectral, species mapping, estuarine ecosystems, epiphyte, reflectance spectroscopy, computational techniques

1. Introduction

Hyperspectral remote sensing systems, or imaging spectrometers, are sensor instruments that obtain image data in many spectral bands. To date, hyperspectral remote sensing has been shown to be useful to geologists as a tool to differentiate between mineral species and lithologic units on the earth's surface (Clark, 1999). The objective of this investigation is to determine if this technology can be applied to aquatic ecology by mapping species composition of submerged aquatic vegetation (SAV) in estuarine environments. Aerial photographs have been used extensively to delineate SAV stands (Orth *et al.*, 2000; Kirkman, 1996) and to discern large changes in SAV coverage and density over time. As a result, SAV coverage information has been combined with water quality information and subsequently been used to determine habitat criteria for restoration of SAV (Batiuk *et al.*, 1992; 2000).

SAV species differ in their tolerance to environmental factors. A shift in conditions of carbon availability, water clarity or salinity, for example could selectively affect the abundance of one species more than another. Emergent and submerged aquatic species presence and absence and diversity have been used to assess stream

Environmental Monitoring and Assessment **81**: 383–392, 2003.

water quality and to rate stream degradation (Small *et al.*, 1996). An adequate species mapping technique is necessary in order to measure change in the distribution of species of SAV and ultimately to determine causal relationships between environmental factors and changes in species coverage and distribution. Although very useful for SAV abundance mapping, the lack of multispectral information in aerial photographs makes this data inadequate for species determination. The objectives of this paper are to: 1) investigate the use of high spatial resolution hyperspectral remote sensing to map SAV distributions and abundance, 2) determine if SAV can be mapped to the species level using this type of data.

2. Materials and Methods

2.1 Study Site

Submerged aquatic vegetation stands were studied in the tidal Potomac River at the mouth of Nanjemoy Creek at Blossom Point, in southern Charles Co. Maryland (Figure 1). This study area is characterized as part of the transition zone between the

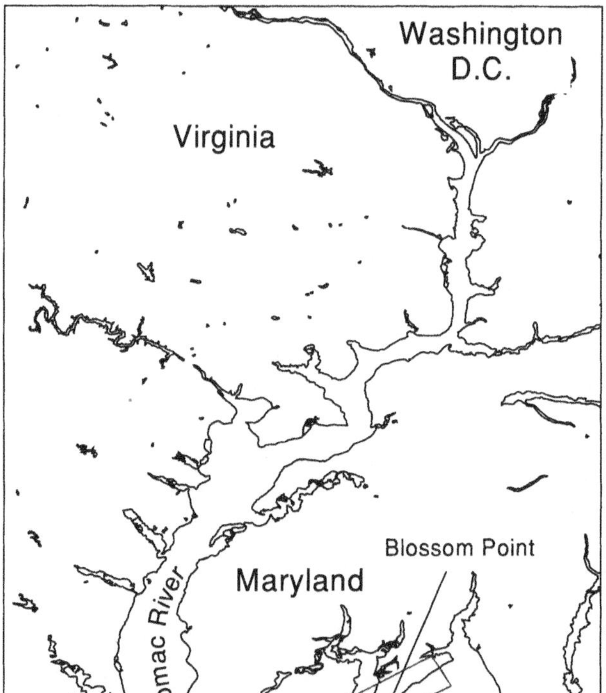

Figure 1. Location of study sites. HyMap flightline denoted as a polygon.

freshwater tidal river and the Chesapeake Bay estuary. The salinity is classified as oligohaline (0.5 to 5.0 ppt). SAV species present at this site are primarily *Vallisneria americana* (wild celery) and *Myriophyllum spicatum* (Eurasian watermilfoil).

2.2 FIELD SPECTROSCOPY

Ground based in-situ spectra were obtained using an Applied Spectral Devices FR porTable field spectrometer. Radiance and reflectance data for sample plots in Nanjemoy Creek (BP) and the Potomac River (PR) were obtained on October 13, 2000, by deploying a fiber optic sensor head over beds of both milfoil and wild celery approximately 1 meter above the water surface. Spectra for several calibration targets, material having uniformly high reflecting spectral response such as beach sand, were used to compare to the airborne reflectance data for quality assurance. Laboratory spectra for field collected milfoil and attached epiphyte colonies were obtained on March 9, 2001. Collected milfoil and wild celery and calibration site spectra were entered into a spectral library database developed in MATLAB. This spectral library has been developed by EPA and George Mason University to enable accurate real-time mapping of target materials in a hyperspectral dataset.

2.3 HYPERSPECTRAL DATA AND IMAGE PROCESSING

Airborne remotely sensed hyperspectral imagery for the site was acquired on October 21, 2000 using the HyMap system (Cocks *et al.*, 1998) (Figure 2). The flightline dimensions were 2.3 x 20 km and the ground sampling distance (pixel size) of the imagery was 4 meters. Sensor radiance data were converted to apparent reflectance using ACORN, an atmospheric correction code based on the MODTRAN 4 radiative transfer model (ImSpec, LLC). Field sample plots were located in the HyMap imagery and spectral signatures of SAV were extracted by averaging over a 50 pixel (200 m^2) area of interest for each plot. A spectral transformation of the reflectance data was accomplished using continuum removal to plot the absorption bands at each wavelength. This procedure isolates the absorption band center and allows for these features to be easily compared with other reflectance spectra (Clark, 1999; Clark and Roush, 1984; Kruse, *et al.*, 1993). The depth of the absorption feature at a specific wavelength is used to identify the two species of SAV.

The first step in identifying species of SAV in the imagery was to suppress the contributions of the optically active components in the ambient water, such as chlorophyll and free floating algae. These components have spectral features that are similar to SAV in certain wavelengths. Using the spectral signatures of the SAV species and the ambient water, a band math algorithm was developed that exploited the spectral differences of SAV versus ambient water at critical wavelengths. The algorithm first processed out the influence of the ambient water by using the continuum removed spectra data. Band differencing was used to set any pixel that did not have absorption features associated with SAV to zero. A band ratio using two SAV absorption bands was then used to map the SAV beds. Because milfoil absorbs more strongly at the 681 nm band than wild celery, the band ratio was set up to take advantage of this difference along with another ratio for the 590 nm band as follows:

(band 1 - band 2) * [(band 1 / band 2) + (band 1 / band 3)]

where band 1= 604 nm, band 2 = 590 nm, band 3 = 681 nm

This equation was used to segment the image and remove potential false positives. This pre-processing step also increased the speed of the next procedure by reducing the amount of data to be processed. Pixels that scored in a set threshold were then passed to a Spectral Feature Fitting (ENVI, 1999) procedure for SAV species identification. Spectral Feature Fitting (SFF) is an algorithm that compares image spectral data to a set of reference spectra, in this case the field-measured spectral library database, by a least-squares fit of the continuum removed spectra (Clark,

Figure 2. HyMap true color composite image of the study area.

1990). The spectral library database of field collected spectra of milfoil and wild celery was compared to each pixel in the hyperspectral image by the SFF procedure. The algorithm produced two images; a scale image measuring the depth of the absorption feature of interest, and root mean square (rms) error image that indicates the degree of match between the reference spectra from the spectral database to the image spectra. Both images were then used to identify SAV by 'best match' to the reference spectra, resulting in a determination of dominant SAV species in each target pixel.

2.4 WATER SAMPLES AND SAV SPECIES CHARACTERIZATION

Water quality parameters were obtained for two sample sites in and outside the SAV beds located at the mouth of Nanjemoy Creek (BP) and on the Potomac River (PR) (Stankelis *et al.*, 2000). Total suspended solids (TSS), total volatile solids (TVS), total chlorophyll-a (CHLA-T), and active chlorophyll-a (CHLA-A) were obtained inside the SAV bed on October 13, 2000. Temperature, salinity, water depth, Secchi depth and Li-Cor underwater light field measurements at stations just below the water surface, midpoint, and at bottom, were obtained October 26, 2000. On October 26, two samples, one in and one outside of the bed, were taken for each site at 100 and 300 meters offshore for BP, and 50 and 200 meters offshore for PR (Bob Wardwell, written communication). The US Geological Survey (USGS) characterized the distribution and abundance of the SAV species for BP and PR on August 31, 2000 by shoreline survey conducted by shallow draft boat (see Ruhl *et al.*, 1999 for methods). Aerial photography (1:24,000 scale) obtained on September 29, 2000, was utilized to map the distribution of SAV for the site (see Orth et al, 2000 for methods).

3. Results and Discussion

3.1 HYPERSPECTRAL IMAGERY INTERPRETATION

SAV beds were present and detecTable in the airborne hyperspectral imagery of Blossom Point (Figure 2). The two species of SAV and water were found to be spectrally separable (Figure 3). The absorption band depths at 0.681 μm, and to a lesser extant 0.574 μm, were more pronounced for milfoil than wild celery (Figure 4). This difference was likely due to the way the SAV plant canopy interacts with light (Asner, 1998) The fully submerged profile of wild celery can be characterized as being meadow forming. Plant stems are mostly vertical. If the remote sensor is oriented at a nadir position, the plant tissue surface available as a reflecting target is negligible. Milfoil is a canopy forming species. Stems are vertical near the bottom, but this species also has numerous small branches that form horizontal surfaces. This type of orientation presents a much larger reflectance target relative to the sensor. Interactions of the plants with epiphyte colonization is another source of spectral variation. These colonies, which can be made up of algae, sediment,

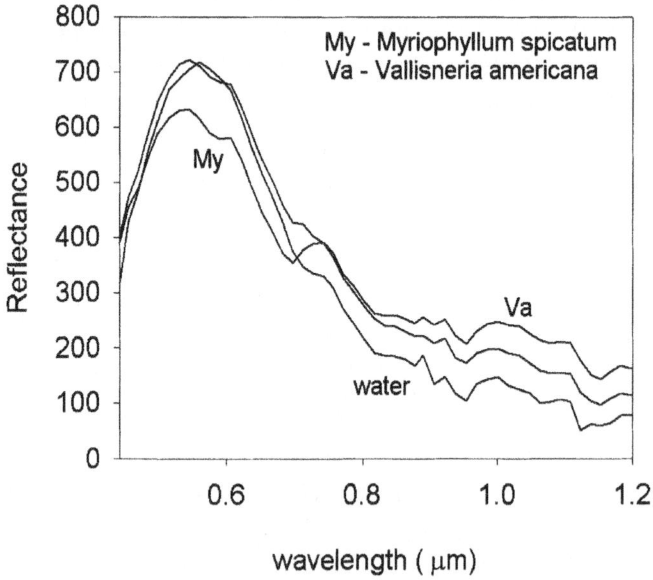

Figure 3. Relative reflectance spectra of two species of SAV and ambient water.

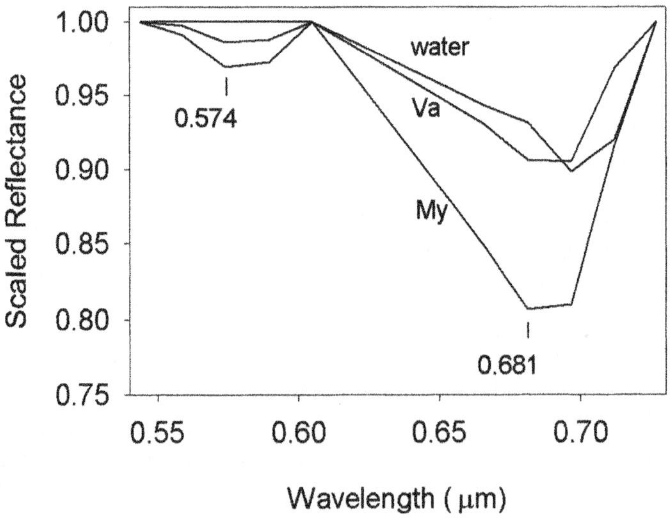

Figure 4. Continuum removed spectra of water and two species of SAV.

bryophytes, and other micro and macro organisms, coat the SAV leaf surface and therefore decrease the amount of light reaching the leaf surface (Orth *et al.*, 1982; Stankelis *et al.*, 2000). The species milfoil, due to its morphology, has more surface area for epiphyte attachment than the species wild celery. Comparison of spectra obtained in the lab for milfoil with and without attached epiphytes indicate that the absorption at approximately 0.574 μm is mostly a result of epiphytes and sediment coating on the SAV.

Two remnant beds of milfoil and wild celery were identified by the hyperspectral technique (Figure 5b). The locations of the beds correspond well to the locations of the bed during the peak growing season determined by the USGS (Figure 5a). The accuracy of the hyperspectral remote sensing derived SAV species map was assessed by comparison to the SAV map produced by the USGS National Research Program using field data and aerial photography. The USGS map shows SAV abundance and distribution in August 31, 2000, and does not reflect the coverage of the SAV at the time of the hyperspectral overflight. Nevertheless, the USGS map reinforces species determinations from the hyperspectral project and can be used to estimate the accuracy of the results.

3.2 WATER AND SAV FIELD DATA

The chemistry and optical properties of the ambient water at the study area are listed for the two sampling dates (Table 1). Irradiance attenuation coefficients (Kd) for the sample sites, BP-100, BP-300, PR-50, PR-200, were 1.96, 1.28, 1.15, and 1.47 (m-1) respectively. The ambient water for these sites can be characterized as moderately transparent (Kd < 2).

Table 1. Water chemistry and optical parameters for the Blossom Point (BP) and Potomac River (PR) sites for October 13 (A) and October 26 (B) of 2000.

(A) October 13

Station	*Date*	*TSS (mg/L)*	*TVS (mg/L)*	*CHLA-T (μg/L)*	*CHLA-A (μg/L)*
BP	10/13/00	8.20	2.60	2.90	2.38
PR	10/13/00	7.60	2.40	1.86	1.35

(B) October 26

Station	*Date*	*Water Temp. (deg. C)*	*Salinity (ppt)*	*Water Depth (m)*	*Secchi Depth (m)*	*LiCor 5cm*	*LiCor Mid*	*LiCor Bottom*
							(μmol s^{-1} m^{-2})	
BP-100	10/26/00	18.2	7.3	0.86	0.5	1,766.6	631	328.5
BP-300	10/26/00	17.7	6.2	1.33	0.74	1,910.5	610	350.5
PR-50	10/26/00	18.1	7.5	0.52	0.52	1,342	1,103	736.7
PR-200	10/26/00	18	7.7	2.16	0.86	1,166.5	270	48.92

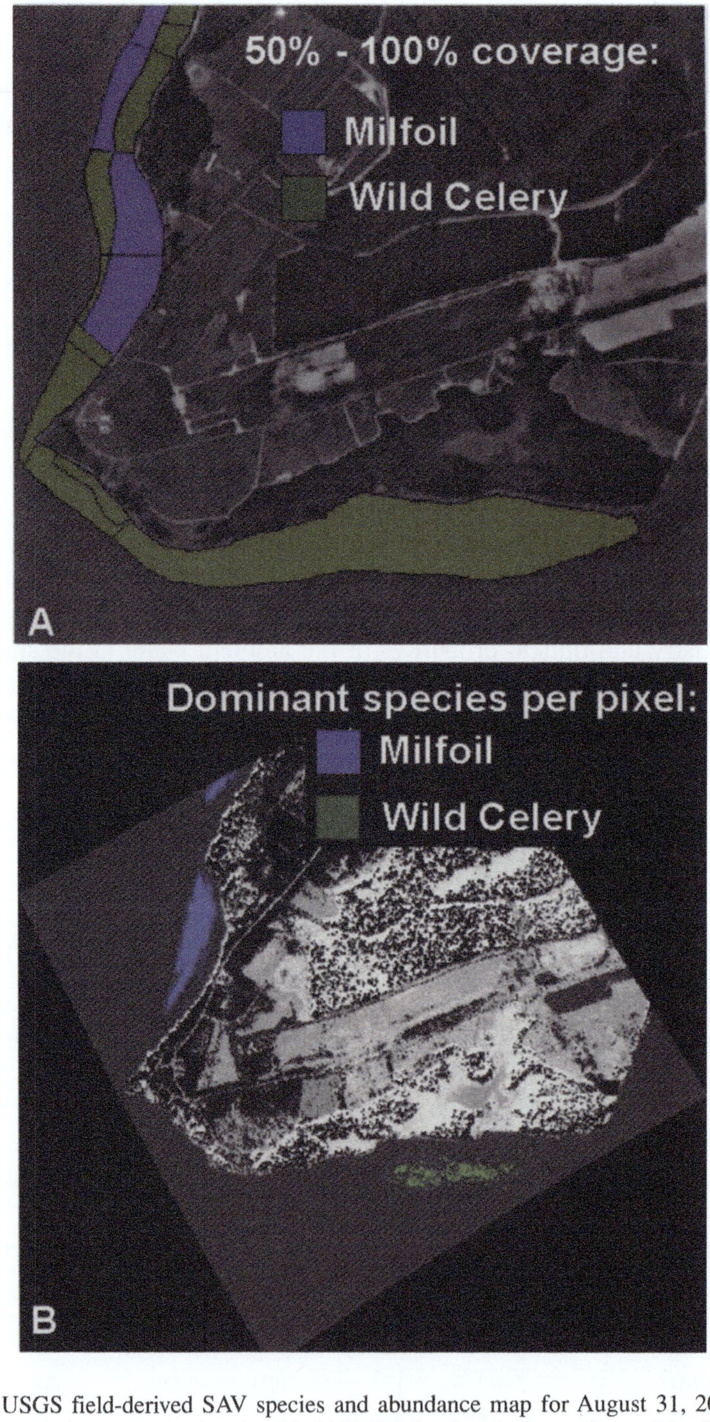

Figure 5. a) USGS field-derived SAV species and abundance map for August 31, 2000; and b) HyMap imagery-derived map for October 21, 2000. By October, large portions of the SAV population has senesed, only a remnant of the August bed is present in October.

Field data showed dominance of milfoil at BP and wild celery at PR (Figure 5a) in August, the peak of the growing season. In October, 2000, each bed was field checked and the dominance of milfoil and wild celery was established, although each bed had mixed composition and the demarcation between species in Figure 5a are often a gradient rather than a sharp separation. The species composition for the BP site were 90 % mifoil and 10% wild celery, while the PR site was 97% wild celery, 3% milfoil, and 1% *Ceratophyllum demersum.*

4. Conclusion

Hyperspectral imagery was used to identify and classify SAV beds in an aquatic environment that can be characterized as optically complex given the significant concentrations of suspended solids and chlorophyll. The primary absorption band for photosynthesis (680 nm) was detecTable in the submerged plant canopies. The differentiation of SAV species was done by exploiting the way light is scattered or absorbed by physically different plant canopies, rather than by some unique biochemical signature. Our data suggests that the presence of epiphytes and sediment coating on the SAV obscure species' biochemical reflectance signatures. This biophysical methodology might be limited in beds of SAV plants with similar canopy profiles, or during low tide when the leaves of meadow forming species become horizontal at the surface. Further research into the spectral signatures of various SAV species with a focus on bio-chemical differences is warranted. Proper timing of the overflight to collect SAV spectra before dense epiphyte colonization might allow for more accurate species identification.

Acknowledgments

The U.S. Environmental Protection Agency through its Office of Research and Development partially funded the research described here. It has been subjected to Agency review and approved for publication. Mention of trade names or commercial products does not constitute an endorsement or recommendation for use. The authors would like to thank the reviewers who have helped improve the quality of this paper.

References

Asner, G.P. 1998, 'Biophysical and biochemical sources of variability in canopy reflectance', *Remote Sens. Environ.* **64**, 234–253.

Batiuk, R., Heasley, Orth, R., Moore, K., Stevenson, J.C., Dennison, W., Staver, .L, Carter, V., Rybicki, N.B., Hickman, R.E., Kollar, S., Bieber, S. and Bergstrom, P.: 1992, 'Chesapeake Bay submerged aquatic vegetation habitat requirements and restoration goals: A Technical Synthesis', USEPA CBP/TRS 83/92, 162 pp.

Batiuk, R., Bergstrom, P., Kemp, M.,. Koch, E., Murray, L., Stevenson, J. C.,. Bartleson, R, Carter, V., Rybicki, N. B., Landwehr, J. M., Gallegos, C., Karrh, L., Naylor, M., Wilcox, D., Moore, K. A., Ailstock, S. and Teichberg, M.: 2000, 'Chesapeake Bay submerged aquatic vegetation water quality and habitat-based requirements and restoration targets: a second technical synthesis', USEPA, CBP/TRS 245/00. web address: http://www.chesapeakebay.net/pubs/sav/index.html

Clark, R.N.: 1999, 'Spectroscopy of rocks and minerals, and principles of spectroscopy', In: A. Rencz,(Ed), The Manual Of Remote Sensing, 3rd Edition (A Series): Volume 3: Remote Sensing for the Earth Sciences. John Wiley and Sons, New York

Clark, R. N., Swayze, G.A., King, T.V.V., Livo, E., Kokaly, R.F., Dalton, B.J., Vance, S. J., Rockwell, B.W. and McDougal, R.R.: 1999, 'Surface reflectance calibration of terrestrial imaging spectroscopy data: a tutorial using AVIRIS', USGS Open File Report 99.

Clark, R.N. and Roush, T.L.: 1984. Reflectance spectroscopy: quantitative analysis techniques for remote sensing applications', *J. Geophysical Res.* **89**(B7), 6329–6340.

Cocks, T., Jenssen, R., Stewart, A., Wilson, I. and Shields, T.: 1998. 'The HyMap airborne hyperspectral sensor: the system, calibration and performance', Proceedings of the 1st EARSEL Workshop on Imaging Spectroscopy, Zurich, October 1998.

ENVI.: 1999, 'ENVI users guide', Research Systems. Boulder Colorado.

Kirkman, H.: 1996. 'Baseline and monitoring methods for seagrass meadows', *J. Environ. Mgmt.* **47**, 191–201.

Kruse, F.A., Lefkoff, A.B. and Dietze, J.B.: 1993, 'Expert system-based mineral mapping in northern Death Valley, California/Nevada using the Airborne Visible/InfraRed Imaging Spectrometer (AVIRIS)', *Remote Sens. Environ.* **44**, 309–336.

Orth, R.J., Wilcox, D.J., Nagey, L.S., Whiting, J.R.,and Fishman, J.R.: 2000, '1999 distribution of submerged aquatic vegetation in the Chesapeake Bay and coastal bays', VIMS Special Scientific Report Number 141. Final report to U.S. EPA, Chesapeake Bay Program, Annapolis, MD. Grant No.CB993777-02-0. web address: http://www.vims.edu/bio/sav/sav99

Orth, R. J., Moore, K. A, and van Montfrans, J.: 1982, 'Submerged aquatic vegetation: Distribution and abundance in the lower Chesapeake Bay and interactive effects of light, epiphytes and grazers', Final Report, USEPA Chesapeake Program Grant No. X003246. 1982. 141 pp.

Ruhl, H.A., Rybicki, N.B., Reel, J.T. and Carter, V.: 1999, 'Distribution and abundance of submersed aquatic vegetation in the tidal Potomac River and upper Potomac Estuary, Maryland, Virginia, and the District of Columbia, 1993-1998', U.S. Geological Survey Open-File Report 99–233.

Small A.M., Roberts D.L., Adey W.H., Lutz S.M. and Reese E.G.: 1996, 'A macrophyte-based rapid biosurvey of stream water quality: restoration at the watershed scale', *Restoration Eco.* **4**(2), 124–145.

Stankelis, R.M., Boynton, W.R. and Frank, J.M.: 2000. 'Submerged aquatic vegetation (SAV) habitat evaluation', In Chesapeake Bay water quality monitoring program ecosystem processes component (EPC) level one report #17 (interpretive). University of Maryland Center for Environmental Science, Technical Report Series No. TS-252-00-CBL, Ref. No. [UMCES]CBL 00-0174.

EFFECT OF EL NIÑO ON DEMOGRAPHIC, MORPHOLOGICAL, AND CHEMICAL PARAMETERS IN TURTLE-GRASS (*Thalassia testudinum*): AN UNEXPECTED TEST OF INDICATORS

PAUL R. CARLSON, JR.[1*], LAURA A. YARBRO[1], KEVIN MADLEY[1], HERMAN ARNOLD[1], MANUEL MERELLO[1], LISA VANDERBLOEMEN[1], GIL MCRAE[1], AND MICHAEL J. DURAKO[2]

[1]*Florida Marine Research Institute, St. Petersburg, FL, U.S.A.;*
[2]*University of North Carolina at Wilmington, Wilmington, NC, U.S.A.*
[*]*(author for correspondence, e-mail: paul.carlson@fwc.state.fl.us)*

Abstract. We examined the response of demographic, morphological, and chemical parameters of turtle grass (*Thalassia testudinum*), to much-higher-than-normal rainfall associated with an El Niño event in the winter of 1997–1998. Up to 20 inches of added rain fell between December 1997 and March 1998, triggering widespread and persistent phytoplankton blooms along the west coast of Florida. Water-column chlorophyll concentrations estimated from serial SeaWiFS imagery were much higher during the El Niño event than in the previous or following years, although the timing and magnitude of phytoplankton blooms varied among sites. Seagrass samples collected in 1997, 1998, and 1999 provided an excellent opportunity to test the responsiveness of *Thalassia* to decline and subsequent improvement of water quality and clarity in four estuaries. Using a scoring technique based on temporal responsiveness, spatial consistency, and statistical strength of indicators, we found that several morphological parameters (*Thalassia* shoot density, blade width, blade number, and shoot-specific leaf area) were responsive and consistent measures of light stress. Some morphological parameters, such as rhizome apex density, responded to declines and subsequent improvement in water clarity, but lacked the statistical discriminating power necessary to be useful indicators. However, rhizome sugar, starch, and total carbohydrate concentrations also exhibited spatially and temporally consistent variation as well as statistical strength. Because changes in shoot density, as well as water clarity, affect rhizome carbohydrate levels, a composite metric based on *Thalassia* shoot density and rhizome carbohydrate levels together is probably more useful than either parameter alone as an indicator of seagrass health.

Keywords: seagrasses, ecological indicators, rhizome carbohydrates, El Niño, *Thalassia testudinum*, Gulf of Mexico, seagrass health

1. Introduction

Seagrasses are vital components of the nearshore ecosystem, providing food and shelter for many economically important fish and shellfish species (Zieman, 1982; Nelson, 1992). However, drastic declines in the distribution and abundance of seagrass and submerged aquatic vegetation communities have occurred in many estuaries throughout the Gulf of Mexico (Lewis *et al.*, 1982; Pulich and White, 1991), the United States (Orth and Moore, 1983; Dennison *et al.*, 1993), and the world (Cambridge *et al.*, 1986) over the past 50 years.

In most cases, seagrass loss has resulted from declining water quality and water clarity. All seagrasses require light for photosynthesis, and most seagrass species support considerable amounts of non-photosynthetic root and rhizome tissue. As a

Environmental Monitoring and Assessment **81**: 393–408, 2003.

result, seagrasses require more light than most algal species, and their distribution is primarily limited by light availability (Dennison, 1987; Duarte, 1991).

The scale of seagrass loss due to declines in water column transparency is considerable. In Chesapeake Bay, submerged aquatic vegetation (SAV) was virtually eliminated between 1965 and 1985 (Orth and Moore, 1983). Eelgrass, once abundant in Long Island Sound, is now completely absent from the western half of the Sound, due to turbidity caused by eutrophication (Koch and Beer, 1996). Similar declines in seagrass abundance attributed to increases in water column turbidity and resulting declines in transparency have also been reported for several Australian estuaries (Larkum and West, 1990; Walker and McComb, 1992), the Wadden See (Giesen *et al.*, 1990), and the Mediterranean Sea (Bourcier, 1986; 1989). In Florida, seagrass cover in both Tampa Bay (Johannson, 1991) and Sarasota Bay (Tomasko *et al.*, 1996) has declined as the result of anthropogenic nutrient loading.

Seagrass communities in impacted estuaries have also demonstrated a limited recovery potential if water quality improves. Approximately 80% of the seagrass present in Tampa Bay (FL) in the late 1800's was lost by 1981 (Johannson, 1991; Lewis *et al.*, 1999). However, improved water quality resulted in a 23% increase in seagrass cover between 1982 and 1994 (Lewis *et al.*, 1999; Johansson and Greening, 2000). Between 1985 and 1991, SAV cover in Chesapeake Bay also increased from 20,000 hectares to nearly 30,000 hectares (Moore *et al.*, 2000), although Orth *et al.* (1994) estimated that up to 250,000 hectares of shallow bay bottom was capable of supporting SAV at one time.

Because seagrasses are very responsive to decreased water clarity, they are excellent sentinel species for biological monitoring of estuarine and nearshore water quality, and considerable interest has focused on the development of ecological indicators to assess seagrass community health. At a workshop held in 1992 (Neckles, 1994) seagrass scientists listed a number of parameters which might be useful as biological indicators of stress in seagrasses, including many used in this study. In this study, we have evaluated the utility of several demographic, morphological, and chemical parameters as indicators of light stress for one seagrass species, turtle grass (*Thalassia testudinum*), applying criteria based on those developed by Jackson *et al.* (2000). Demographic parameters such as shoot age and rhizome branching frequency were included in this study because recent studies have shown that stress responses of seagrass shoots might be age-dependent (Duarte *et al.*, 1994; Durako, 1994). Turtle grass was selected as our indicator species because it is a long-lived perennial species which, because of a large investment in belowground tissue, is very sensitive to declines in water clarity and to other stressors.

The occurrence of an El Niño event in 1998 allowed us to evaluate the response of these parameters to decline and subsequent improvement in water clarity. Heavy rainfall in winter 1997 and spring 1998 caused heavy nutrient loading to estuaries, resulting in persistent phytoplankton blooms along the west coast of Florida. Our samples, collected in late summer each year, captured the condition of turtle grass during the growing seasons prior to, during, and after the El Niño-induced phytoplankton blooms. The magnitude of the rainfall anomaly also varied along the west coast of Florida, providing a range of nutrient and salinity impacts among our study sites.

2. Methods

We used data from four sampling sites along the West Coast of Florida to evaluate El Niño impacts: Homosassa River, Anclote River, Tampa Bay, and Charlotte Harbor (Figure 1). All four of these sites are impacted to varying degrees by freshwater runoff: Charlotte Harbor is the most strongly river-dominated and Lower Tampa Bay experiences the least freshwater influence. At each of these sites, we sampled 30 spatially-distributed, randomly-selected points distributed over a 250,000 m^2 grid in late summer to early fall in 1997, 1998, and 1999.

At each sampling point, the areal cover of seagrass and macroalgal species was estimated by the Braun-Blanquet technique (Mueller-Dombois and Ellenberg, 1974). Seagrass tissues were collected at each sampling point using a single, 6-inch diameter core, and mid-depth water temperature and salinity were measured. Seagrass tissues were frozen for subsequent morphological and chemical analyses. In the laboratory, numbers of *Thalassia, Halodule* and *Syringodium* shoots in cores were counted to calculate areal density of seagrass shoots. We also counted *Thalassia* rhizome apices (growing tips) as a measure of the rhizome branching frequency, and the potential for vegetative reproduction at each sampling point. The number, length, and width of *Thalassia* leaf blades were measured, and *Thalassia* leaf scars (a measure of plant age) were counted. Non-structural carbohydrates (sugar and starch) in *Thalassia* rhizomes were also determined using the sequential extraction method of Zimmerman *et al.* (1995). The fraction of total, non-structural, rhizome carbohydrate present as ethanol-extractable sugars was also calculated and is described in subsequent text as the sugar fraction.

Chlorophyll-*a* estimates from SeaWiFS satellite were used to determine the spatial extent and timing of phytoplankton blooms in our study areas. Although SeaWiFS chlorophyll estimates are not as accurate as laboratory analyses of discrete water samples, the satellite data were the only internally-consistent, synoptic, and quantitative measure of phytoplankton abundance available for all of our study sites during our study period. Because the SeaWiFS sensor was not operational until fall 1997, we applied fall 1997 chlorophyll-a estimates to the entire 1997 growing season. Subsequent data (1998 and 1999) were grouped for calculation of quarterly means for each sampling site.

SeaWiFS scenes with acceptable levels of cloud cover were downloaded from NASA's DAAC (Distributed Active Archive Center) web site and processed with NASA's SeaDAS (SeaWiFS Data Analysis System) software package. The OC4 chlorophyll algorithm (O'Reilly *et al.*, 1998) was used to estimate chlorophyll-a concentrations for approximately 100 sampling points along the west coast of Florida for 1997, 1998, and 1999. The chlorophyll anomaly for each site was calculated as the ratio between chlorophyll-a concentrations in spring 1998 (El Niño values) and the mean of chlorophyll-a concentrations in the preceding and following spring seasons (1997 and 1999, respectively). In subsequent text, chlorophyll-a is described simply as chlorophyll.

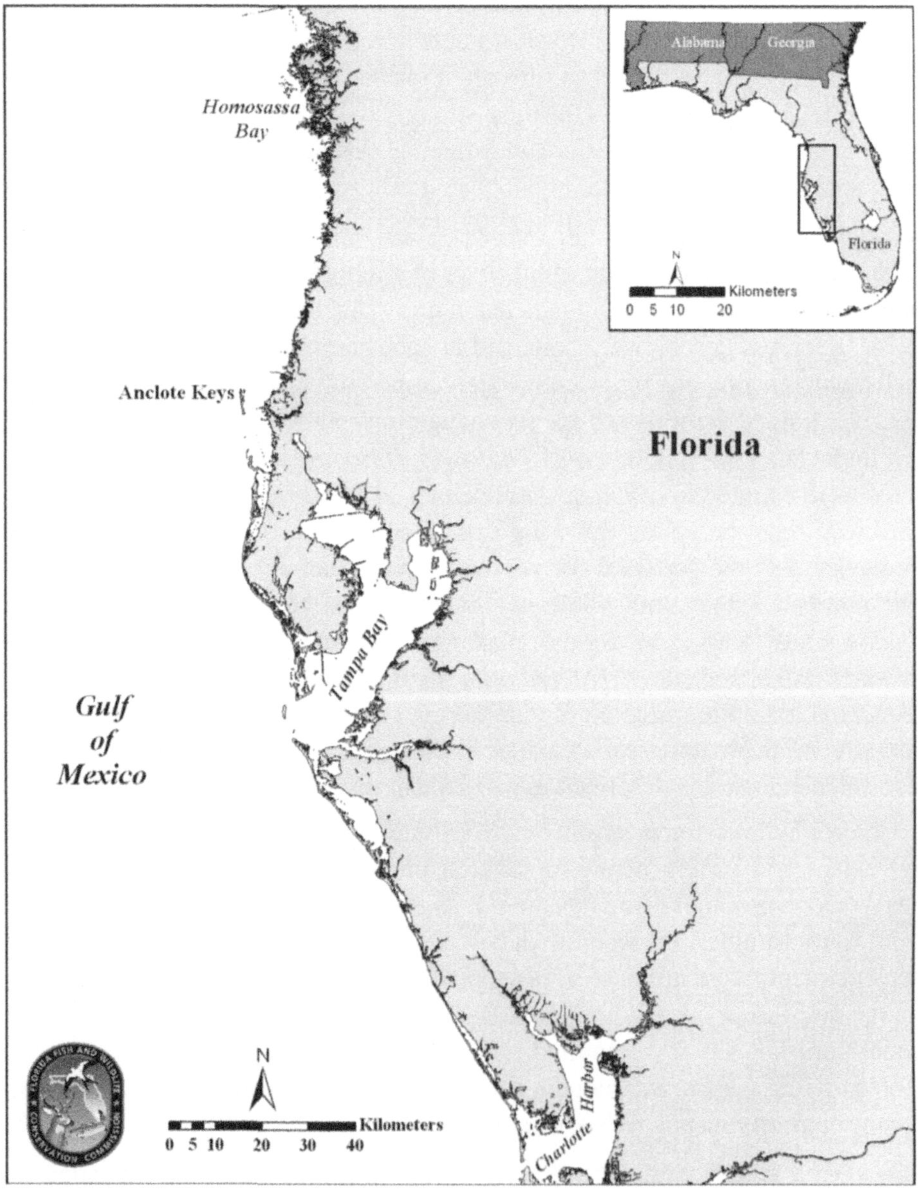

Figure 1. Seagrass study areas along the west coast of Florida, US.

To determine the utility of each parameter as a potential indicator of seagrass health, we compared annual mean values for 1997, 1998, and 1999 at each site using Duncan's Multiple Range Test ($p = 0.05$). We then applied three criteria to calculate a rating of potential usefulness for each parameter. The first criterion-temporal consistency measured the temporal responsiveness of each parameter. Highest values were given to parameters that either increased or decreased between 1997 and 1998 and then exhibited the opposite trend between 1998 and 1999. Our underlying assumption was that, for an indicator to be useful, it must respond to changes in stress, either increases or decreases, within one year. The criterion score was calculated as the number of estuaries where a candidate indicator exhibited the desired temporal pattern. The second criterion was based on the statistical significance of the annual variability at each site. The score for the second criterion is the number of estuaries with significant annual variation ($p = 0.05$) in a parameter without consideration of temporal pattern. The third criterion score, spatial consistency, was calculated as the number of estuaries which exhibited the same temporal pattern. Each criterion score ranged from 0 to 4, so the aggregate indicator performance scores ranged from a minimum score of zero to a maximum score of 12. We also calculated correlation coefficients between seagrass indicators and the El Niño- induced chlorophyll anomalies at each site as a second means of evaluating indicator response to light stress.

3. Results and Discussion

3.1 EL NIÑO RAINFALL AND WATER-COLUMN CHLOROPHYLL CONCENTRATIONS

An average El Niño event delivers 126% of normal rainfall across the state of Florida during winter and spring, and the statewide average of winter-spring rainfall associated with severe events is 153% of normal winter and spring rainfall (NOAA, 2001). The west-central region of the state between Tampa Bay and Charlotte Harbor typically receives the greatest amount of El Niño rainfall. The 1998 El Niño was a severe event, and the spatial pattern of rainfall was consistent with previous events.

The greatest impact of El Niño rainfall was located over Tampa Bay where 20-22 inches of rain in excess of the long-term seasonal average fell between December 1997 and March 1998 (SERCC, 2001). The Homosassa watershed received approximately 14 inches of added rainfall during this period, while the Anclote River and Charlotte Harbor sites received 16-18 inches and 12-18 inches, respectively. Rainfall amounts in the Florida Panhandle, Big Bend region, and South Florida were considerably lower than in Central Florida (Figure 2). Surface water salinity at our four sites varied in response to El Niño rainfall and runoff, but values remained well within tolerance limits for *Thalassia* (3.5 ppt to 60 ppt according to Zieman, 1982). Thus, we attribute the observed changes in seagrass indicators primarily to light stress caused by phytoplankton blooms.

The spatial pattern of elevated water-column chlorophyll concentrations generally coincided with the distribution of El Niño rainfall (Figure 2). The chlorophyll anomaly (the ratio of spring 1998 chlorophyll concentrations to 1997 and 1999 concentrations) was generally low in the Panhandle (PAN), Big Bend (BIG) and Florida Bay (FLB) regions where less than 10 inches of "excess" rain fell during winter 1997 and spring 1998. Highest chlorophyll anomalies occurred at central Florida sites such as Homosassa (HOM), Anclote River (ANC), and Tampa Bay (MTB), where more than 10 inches of additional rain fell .

The magnitude and timing of peaks in water- column chlorophyll varied among Central Florida sites (Figure 3). In the fall of 1997, prior to the onset of El Niño rainfall, SeaWiFS chlorophyll values ranged from 6 µg/L at the Homosassa (HOM) site to 25 µg/L at the Charlotte Harbor (UCH) site (Figures 2, 3). Tampa Bay (LTB), and Anclote River (ANR) sites had values of 9 and 12 µg/L, respectively. Chlorophyll concentrations at the Anclote site rose steadily from 12 µg/L in fall 1997 to over 40 µg/L in fall 1998. After declining during fall and winter, chlorophyll values increased again during summer 1999 to values between 15 and 20 µg/L. Homosassa chlorophyll concentrations peaked near 20 µg/L in spring 1998 and then declined to levels below 10 µg/L for the remainder of the study period.

The response of water column chlorophyll concentrations to El Niño rainfall and runoff also varied within estuaries. For example, chlorophyll concentrations in middle

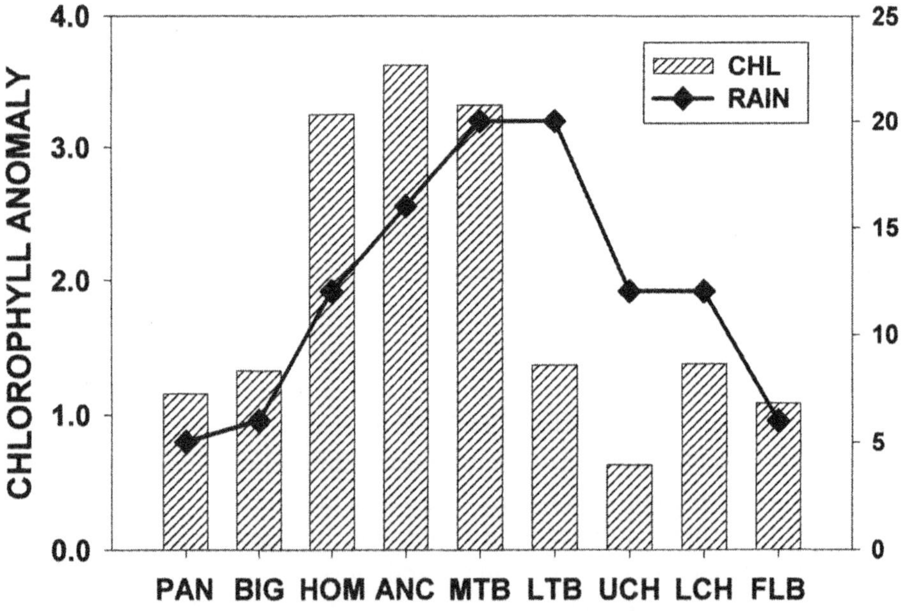

Figure 2. Rainfall anomaly and chlorophyll anomaly along the Florida West Coast. PAN = Florida Panhandle region, BIG = Big Bend region, HOM = Homosassa River, ANC = Anclote River, MTB = Middle Tampa Bay, LTB = Lower Tampa Bay, UCH = Upper Charlotte Harbor, LCH = Lower Charlotte Harbor, and FLB = Florida Bay.

Tampa Bay (MTB) rose dramatically during 1998 (Figure 2), but values in the lower Bay (LTB) varied from 9 μg/L in fall 1997 to 4 μg/L in winter 1999. Chlorophyll concentrations in the vicinity of our Charlotte Harbor site (UCH) were lower in summer 1998 than in the preceding or following year (Figure 3), and similar temporal patterns were seen in color and dissolved organic carbon measurements made on the Peace River, one of the two major tributaries of Charlotte Harbor (U. S. Geological Survey water quality data not shown). However, chlorophyll concentrations in lower Charlotte Harbor (LCH) were higher during 1998 than in 1997 or 1999 (Figure 2), suggesting that high river discharge displaced phytoplankton blooms seaward from our study site in this estuary.

3.2 MORPHOLOGICAL AND DEMOGRAPHIC PARAMETERS

Thalassia shoot density declined in 1998 at three of the four sites, although the declines were only significant at two sites, Homosassa River and Tampa Bay (Table 1). *Halodule* shoot density declined significantly between 1997 and 1998 at the Anclote River and Charlotte Harbor sites, but *Halodule* shoot density increased significantly at the Homosassa River site. *Syringodium* shoot density declined significantly between 1997 and 1998 at the Anclote and Homosassa River sites, but *Syringodium* density increased at the Charlotte Harbor site.

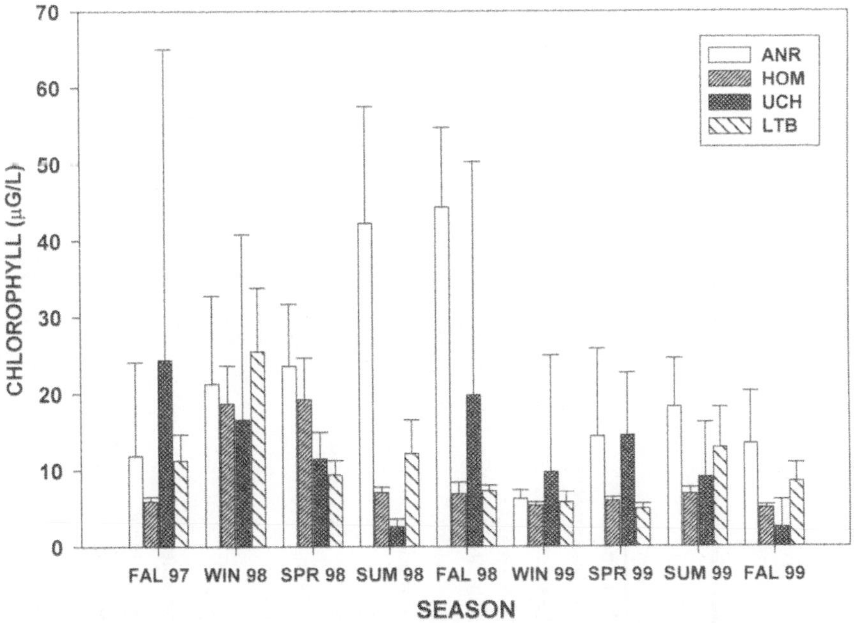

Figure 3. Seasonal variation in SeaWiFS chlorophyll estimates for four West Florida sites between fall 1997 and fall 1999. Bar height and error bars represent mean and 2 SE for five locations at each site and up to six satellite scenes. ANR = Anclote River, HOM = Homosassa River, UCH = Upper Charlotte Harbor, and LTB = Lower Tampa Bay.

Table 1. *Annual Variability of Seagrass Indicators at Four West Florida Sites*. For each site, values of each parameter which have the same letter subscript are not significantly different ($p = 0.05$).

	Anclote River			*Charlotte Harbor*			*Homosassa River*			*Tampa Bay*		
Parameter	*1997*	*1998*	*1999*	*1997*	*1998*	*1999*	*1997*	*1998*	*1999*	*1997*	*1998*	*1999*
Morphological and Demographic Parameters												
Thalassia shoot density	1.00_a	1.00_a	0.42_a	2.83_a	1.50_a	3.91_a	13.90_a	6.57_b	8.23_b	8.37_a	3.88_b	5.13_b
Halodule shoot density	3.13_a	0.02_b	0.08_b	4.43_a	0.25_b	0.81_b	0.13_b	11.70_a	10.20_a	1.17_a	2.80_a	3.60_a
Syringodium shoot density	17.40_a	0.90_b	6.29_b	0.00_b	3.16_b	15.10_a	6.70_a	0.17_b	3.30_{ab}	0.00_a	0.38_a	0.00_a
Thalassia blade width	0.61_a	0.52_a	0.59_a	0.42_a	0.44_a	0.45_a	0.84_a	0.44_b	0.51_b	0.75_a	0.70_b	0.69_b
Thalassia leaf blade number	3.07_b	2.45_b	3.80_a	3.27_a	2.82_b	3.52_a	—	2.51_b	3.13_a	2.63_b	3.38_a	2.62_b
Shoot leaf area	31.50_{ab}	21.50_b	42.70_a	18.10_b	22.40_{ab}	24.60_a	18.10_b	21.70_b	38.20_a	46.80_a	52.50_a	38.90_b
Thalassia rhizome apex density	0.33_a	0.11_a	0.12_a	0.57_a	0.22_a	0.48_a	1.70_a	0.93_a	1.63_a	0.97_a	0.63_a	0.90_a
Rhizome branching frequency	0.27_{ab}	0.09_b	0.61_a	0.23_a	0.11_a	0.14_a	0.11_b	0.16_a	0.18_a	0.11_a	0.16_a	0.26_a
Leaf scar number	13.00_a	20.70_a	20.80_a	14.90_a	14.90_a	13.70_a	23.10_b	22.50_b	27.30_a	29.30_a	33.30_a	28.70_a
Shoot age	17.60_a	24.20_a	25.60_a	19.70_a	18.70_a	18.80_a	26.20_b	25.80_b	32.10_a	33.10_b	38.00_a	32.80_b
Chemical Parameters												
Rhizome sugar content	29.90_b	13.30_c	52.20_a	14.90_c	54.40_a	30.80_b	35.40_b	21.30_c	61.40_a	43.20_a	29.60_b	38.80_a
Rhizome starch content	62.50_a	18.50_b	73.60_a	72.30_a	74.80_a	72.50_a	46.10_b	35.30_c	68.00_a	70.20_a	27.70_c	51.60_b
Total rhizome carbohydrate	92.40_b	31.80_c	125.80_a	87.20_b	129.20_a	103.30_b	81.50_b	56.50_c	129.40_a	113.40_a	57.30_c	90.40_b
Sugar fraction	0.33_b	0.36_b	0.43_a	0.172_c	0.434_a	0.307_b	0.45_b	0.38_c	0.48_a	0.39_b	0.53_a	0.44_b

These results illustrate the effect of external stressors (in this case, turbidity and light attenuation) on competitive interaction among the three seagrass species. While *Thalassia* is generally regarded as the climax species in subtropical seagrass beds, *Halodule* is a pioneer species with the potential for rapid colonization and expansion within gaps created by declines in *Thalassia* abundance. Moderate light stress levels cause shifts in species abundance from *Thalassia* toward *Halodule* or *Syringodium*. Under conditions of extreme light stress (i.e., Anclote River), however, the abundance of all three seagrass species declined.

After the 1998 El Niño event, the response of seagrasses to improved water clarity varied among species and among sites (Table 1). At Homosassa River and Tampa Bay, *Thalassia* shoot densities increased from 1998 to 1999 but remained significantly lower than in 1997, indicating that *Thalassia* recovers slowly from major stress events like the 1998 El Niño. *Thalassia* shoot density also increased between 1998 and 1999 in Charlotte Harbor, but the annual variation at this site was obscured by within-site variation. *Syringodium* shoot density increased sharply and significantly between 1998 and 1999 at Anclote River, Homosassa River, and Charlotte Harbor.

Thalassia blade width, which has been shown to respond to salinity and light stress in other studies (Phillips and Lewis, 1983; Hall *et al.*, 1991; Lee and Dunton, 1997), declined between 1997 and 1998 at three of four sites (Table 1). However, the decline was significant at only two sites—Homosassa River and Tampa Bay. Blade widths at both of those sites remained significantly lower in 1999 than in 1997. *Thalassia* leaf blade number exhibited significant annual variation at all four sites, but values at the Tampa Bay site increased in 1998 while values at Anclote River and Charlotte Harbor decreased. Between 1998 and 1999, mean leaf blade number declined significantly at the Tampa Bay site while values at the Anclote River, Charlotte Harbor, and Homosassa River sites increased significantly.

Thalassia shoot-specific leaf area exhibited significant annual variation at all four sites, but the pattern of variation was not consistent among sites (Table 1). However, the correlation coefficient of shoot leaf area with chlorophyll anomaly was – 0.63 (Table 2), indicating that this parameter is potentially useful as an indicator of light stress in *Thalassia*. Previous studies have also obtained ambiguous results for this parameter. In shading experiments (unpublished data), we have found that shoot-specific leaf area sometimes increases when shoot density declines, and we have attributed the increase to relaxation of competition for light among *Thalassia* shoots in dense stands. However, in the shading study conducted by Hall *et al.* (1991), shoot leaf area did not increase significantly as shoot density declined. Field studies of *Thalassia* morphology along depth gradients are equally ambiguous: Dawes and Tomasko (1988) found that *Thalassia* shoot leaf area was higher and shoot density was lower at the deep, light-limited, grassbed edge than at the shallow edge of the same grassbed for one study site, but the pattern was

Table 2. Evaluation of Indicator Performance by Two Techniques. Scores for first technique are summed scores for each of three criteria (see text for a description of these criteria). Values for second technique are Pearson correlation coefficients for the relationship between parameter response and chlorophyll anomaly values.

	Indicator Performance Criteria				*Pearson Correlation Coefficients*
Indicator	*Temporal Consistency*	*Statistical Strength*	*Spatial Consistency*	*Overall Score*	*Correlation with Chlorophyll Anomaly*
Thalassia shoot density	4	2	3	9	– 0.36
Halodule shoot density	4	2	0	6	0.30
Syringodium shoot density	4	3	0	7	—
Thalassia blade width	4	2	3	9	– 0.61
Blade number	4	4	3	11	– 0.52
Shoot-specific leaf area	2	4	3	9	– 0.63
Thalassia apex density	4	0	4	8	– 0.40
Rhizome branching frequency	4	2	0	6	– 0.24
Leaf scar number	0	0	0	0	– 0.01
Shoot age	0	0	0	0	– 0.07
Rhizome sugar content	4	4	4	12	– 0.73
Rhizome starch content	4	4	3	11	– 0.64
Total Rhizome carbohydrate	4	4	4	12	– 0.77
Sugar fraction	3	4	3	10	– 0.80

reversed at a second site. Dixon and Leverone (1995) found that shoot leaf area at the deep grassbed edge of one Tampa Bay site consistently exceeded values at the shallow edge, but the pattern was reversed at three other sites. The ratio of shoot leaf area in deep and shallow grassbed edges also varied seasonally. The response of shoot leaf area to light stress appears to be site-specific and seasonal, which might limit its broad utility.

Demographic parameters did not exhibit consistent or statistically significant annual variation among sites. The density of *Thalassia* rhizome apices (a measure of the vegetative reproductive potential of a seagrass bed) declined between 1997 and 1998 and then increased between 1998 and 1999 at all four sites, but changes were not statistically significant. Given the relatively low areal density of rhizome apices, this is not surprising. While the numbers of rhizome apices probably did respond to the El Niño event, adequate sampling for this indicator would require unacceptably large (and destructive) samples. Rhizome branching frequency, calculated as the number of rhizome apices divided by the number of shoots in a given area, showed a similar pattern at two of four sites, but only the Anclote River site showed significant declines and recovery in this parameter. Mean leaf scar number and shoot age (calculated from leaf scar number) exhibited no consistent response within or among sites.

3.3 Chemical Parameters

Thalassia rhizome sugar, starch, and total non-structural carbohydrate showed significant annual variation at all four study sites (Table 1). All three parameters declined significantly between 1997 and 1998 at the Anclote River, Homosassa, and Tampa Bay sites. Values of all three parameters rebounded at these same sites between 1998 and 1999, demonstrating the temporal responsiveness necessary for a good indicator.

In Charlotte Harbor, carbohydrate parameters also exhibited temporal responsiveness, but values of starch, sugar, and total non-structural carbohydrate increased between 1997 and 1998 and declined between 1998 and 1999. This pattern, the mirror image of the other three estuaries, is actually consistent with water column chlorophyll data. As shown in Figure 3, chlorophyll concentrations in Charlotte Harbor were lower in summer 1998 than in the preceding or following summers. The chlorophyll data are also supported by unpublished monitoring data which show that water-column color was higher in summer 1997 and summer 1999 than in summer 1998 (U. S. Geological Survey, unpublished). Color and chlorophyll in Charlotte Harbor both had strong peaks in fall 1997, winter 1998, and fall 1998, but values of both parameters declined to very low levels between May and August 1998, at least in the upper Harbor where our sampling sites were located. It is likely that seagrasses in the lower Charlotte Harbor estuary were significantly, and adversely, impacted by phytoplankton blooms, but rapid flushing appears to have removed phytoplankton and color from the upper estuary early in the growing season of 1998. Although Tomasko and Hall (1999) found that freshwater had an adverse impact on seagrasses in Charlotte Harbor, the chlorophyll and color data suggest that major runoff events might be followed by periods of better than normal water clarity.

The sugar fraction (the portion of total rhizome non-structural carbohydrate present as ethanol-extractable sugars) did not exhibit a consistent temporal pattern among estuaries. While annual variation was significant in three of four estuaries, values rose and then declined in two estuaries (Charlotte Harbor and Tampa Bay), but they declined and then rose at the Homosassa site. Although the partitioning of rhizome carbohydrate reserves between sugar and starch is known to be affected by light availability in *Thalassia* (Lee and Dunton, 1997) and in *Zostera marina* (Burke *et al.*, 1996), it might also be affected by salinity, plant phenology, or other processes.

4. Indicator Evaluation

In order for demographic, morphological, and chemical parameters in *Thalassia* to be useful indicators of light stress they must meet several criteria (loosely adapted from Jackson *et al.*, 2000). First, they must exhibit a quantitative, dose-dependent, response to stress. They must also respond to stress rapidly enough that: 1) response can be definitely tied to a particular stress or stressors; and 2) corrective action can be taken before whole seagrass communities are lost. From these guidelines, we created three criteria to evaluate the performance of seagrass indicators: 1) Temporal responsiveness, 2) Statistical strength of response, and 3) Spatial consistency of response among estuaries. We also evaluated correlation coefficients of response variables and the El Niño-induced chlorophyll anomaly as a fourth criterion to evaluate indicators.

For example, a parameter such as *Thalassia* shoot density exhibits temporal responsiveness to El Niño-induced phytoplankton blooms by declining between 1997 and 1998 (Table 2). The decline is consistent with previous studies which show that seagrass shoot density declines along depth gradients (Dixon and Leverone, 1995) and in response to shading (Hall *et al.*, 1991; Lee and Dunton, 1997). Increases in *Thalassia* shoot density between 1998 and 1999 also indicate the parameter responds fairly rapidly to both onset and relaxation of stress. The pattern is spatially consistent, occurring in three of four estuaries. However, this same parameter exhibits a statistically-significant response in only two of four estuaries, and the correlation coefficient between *Thalassia* shoot density and chlorophyll anomaly is only – 0.36 (Table 2).

Weak statistical response alone is not reason for rejection of this parameter as an indicator of seagrass health, however. The lack of statistical power for this parameter is probably due to inadequate sample size because our shoot density estimates are based on cores with an area of 0.02 m^2. While destructive sampling of larger areas of seagrass beds is not acceptable, non-destructive shoot counts of 0.25 m^2 quadrats would probably increase the power of this parameter.

The most promising morphological indicators in this study are *Thalassia* shoot density, blade width, blade number, and shoot-specific leaf area, confirming several of the parameters chosen by 1992 seagrass workshop participants (Neckles, 1994), shading studies (Hall *et al.*, 1991; Neverauskas, 1988), and field studies (Dawes and Tomasko, 1988; Dixon and Leverone, 1995). Rhizome carbohydrate concentrations—sugar, starch, and total non-structural carbohydrate—also exhibit consistent and quantitative variation in response to light stress. However, these parameters are also affected by shoot density, because light limitation in dense seagrass beds often affects the photosynthetic production of individual shoots, and results in "thinning" of shoots (Dawes and Tomasko, 1988; Dixon and Leverone, 1995). In previous shading studies (Carlson *et al.;* unpublished), we have found significant interactions between shoot density and rhizome carbohydrate concentrations: rhizome carbohydrate levels can be unaffected or rise slightly as shoot density declines in response to shading. As a result, a multi-metric parameter based on *Thalassia* shoot density and rhizome carbohydrate levels is probably more useful than either parameter alone in developing indicators of light stress in *Thalassia*.

For most parameters, the correlation coefficients calculated between potential indicators and chlorophyll anomaly values agree with their indicator performance scores (Table 2). For example, carbohydrate parameters had the highest correlation coefficients with chlorophyll anomaly, and three carbohydrate parameters (sugar, starch, and total non-structural carbohydrate) also had the highest indicator performance scores of any parameters. Parameters like shoot age, rhizome branching frequency, and *Halodule* shoot density had low scores and weak correlation with chlorophyll anomaly. One notable exception is *Thalassia* shoot density which had an indicator performance score of 9 but a correlation coefficient of – 0.36. As noted above, larger sample sizes might improve both statistical strength and correlation coefficients for this parameter.

5. Conclusions

The parameters we tested fall into three groups. The first group is composed of parameters such as shoot age which appear to lack the temporal responsiveness or spatial consistency required for an indicator to be applied throughout the eastern Gulf of Mexico. The second group includes parameters which respond to stress and exhibit similar responses at several sites, but these parameters exhibit a large amount of within-site variability which decreases the ability to detect change. This group includes parameters such as *Thalassia* shoot density, rhizome apex density, and rhizome branching frequency. Parameters in the marginal group might be useful indicators if more samples or larger samples are taken to reduce the within-site variability. However, more and larger destructive samples also might have unacceptable impacts on seagrass beds.

Several morphological and chemical parameters (*Thalassia* shoot density, blade width, blade number, and shoot-specific leaf area, as well as rhizome carbohydrate concentrations) fall into the third group: parameters worthy of additional testing for indicator development. These parameters exhibit temporal and spatial consistency as well as statistical significance in their response to light stress. Assessment thresholds and "dose-response" relationships should be determined for these parameters, and additional effort should focus on testing of multi-metric parameters and improving the statistical power of indicators.

Acknowledgements

Funding for this research was provided by the U. S. Environmental Protection Agency STAR Program, and we gratefully acknowledge Ms. Barbara Levinson's help as Project Manager. SeaWiFS data were obtained from NASA through an existing authorized user agreement between NASA and FMRI. We thank Dr. Brian Melzian and three anonymous reviewers for their helpful comments.

References

Bourcier, M.: 1986, 'Change, over five years, of a *Posidonia oceanica* bed and circalittoral macrobenthos: The combined action of human activity and climatic conditions', *Vie Milieu* **36**, 1–8.

Bourcier, M.: 1989, 'The decline of *Posidonia oceanica* seagrass meadows east of Marseilles under the combined impact of human activity and climatic conditions', in: Boudouresque, C. F., Meinesz, A., Fresi, E. and Gravez, V. (eds), The Second International Workshop on Posidonia Beds, Ischia, Italy, Marseille France, pp. 287–292.

Burke, M. K, Dennison, W. C., Moore, K. A.: 1996, 'Non-structural carbohydrate reserves of eelgrass *Zostera marina*', *Marine Ecology Progress Series* **137**, 195–201.

Cambridge, M. L., Chiffings, A. W., Brittan, C., Moore, L. and McComb, A. J.: 1986, 'The loss of seagrass in Cockburn Sound, Western Australia. II. Possible causes of seagrass decline', *Aquatic Botany* **24**, 269–285.

Dawes, C. J. and Tomasko, D. A.: 1988, 'Depth distribution of *Thalassia testudinum* in two meadows on the west coast of Florida: a difference in effect of light availability', PSNZI *Marine Ecology* **9**, 123–130.

Dennison, W. C.: 1987, 'Effects of light on seagrass photosynthesis, growth and depth distribution', *Aquatic Botany* **27**, 15–26.

Dennison, W. C., Orth, R. J., Moore, K. A., Stevenson, J. C., Carter, V., Kollar, S., Bergstrom, P. W., and Batiuk, R. A.: 1993, 'Assessing water quality with submersed aquatic vegetation', *BioScience* **43**, 86–93.

Dixon, L. K. and Leverone, J. R.: 1995, 'Light requirements of *Thalassia testudinum* in Tampa Bay, Florida', Mote Marine Laboratory Technical Report 425, Sarasota, FL, 75 pp. and appendices.

Duarte, C. M.: 1991, 'Seagrass depth limits', *Aquatic Botany* **40**, 363–377.

Duarte, C. M., Marba, N., Agawin, N., Cebrian, J., Enriquez, S., Fortes, M. D., Gallegos, M. E., Merino, M., Olesen, B., Sand-Jensen, K., Uri, J., Vermaat, J.: 1994. 'Reconstruction of seagrass dynamics: age determinations and associated tools for the seagrass ecologist', *Marine Ecology Progress Series* **107**, 195–209.

Durako, M. D.: 1994, 'Seagrass die-off in Florida Bay (USA): changes in shoot demographic characteristics and population dynamics in *Thalassia testudinum*', *Marine Ecology Progress Series* **110**: 59–66.

Giesen, W. B. J. T., van Katwijik, M. M. and den Hartog, C.: 1990, 'Eelgrass condition and turbidity in the Dutch Wadden Sea', *Aquatic Botany* **37**, 71–85.

Hall, M. O., Tomasko, D. A. and Courtney, F. X.: 1991, 'Responses of *Thalassia testudinum* to *in situ* light reduction', in: Kenworthy, J. D. and Haunert, D. (eds), *The Light Requirements of Seagrasses: Proceedings of a Workshop to Examine the Capability of Water Quality Criteria, Standards, and Monitoring Programs to Protect Seagrasses.* NOAA Technical Memorandum NMFS-SEFC-287, Washington, DC, pp. 77–86.

Jackson, L., Kurtz, J. and Fisher, W.: 2000, *Evaluation Guidelines for Ecological Indicators*, EPA/620/R-99/005, Office of Research and Development, U.S. Environmental Protection Agency, Washington, DC, 107 pp.

Johansson, J. O. R.: 1991, 'Long-term trends of nitrogen loading, water quality and biological indicators in Hillsborough Bay, Florida', in: Treat, S. F. and Clark, P. A. (eds), Proceedings, Tampa Bay Area Scientific Information Symposium 2, Tampa, FL, pp. 157–166.

Johansson, J. O. R. and Greening, H. S.: 2000, 'Seagrass restoration in Tampa Bay: a resource-based approach to estuarine management', in: Bortone, S. (ed) *Seagrasses: Monitoring, Ecology, Physiology, and Management*, CRC Press, Boca Raton, FL, pp. 279–293.

Koch, E. W. and Beer, S.: 1996, 'Tides, light and the distribution of *Zostera marina* in Long Island Sound, USA', *Aquatic Botany* **53**, 97–107.

Larkum, A. W. D. and West, R. J.: 1990, 'Long-term changes of seagrass meadows in Botany Bay, Australia', *Aquatic Botany* **37**, 55–70.

Lee, K., and Dunton, K. H.: 1997, 'Effects of in situ light reduction on the maintenance, growth, and partitioning of carbon resources in *Thalassia testudinum* Banks ex Konig', *Journal of Experimental Marine Biology and Ecology* **210**, 53–73.

Lewis, R. R. III, Durako, M. J., Moffler, M. D. and Phillips, R. C.: 1982, 'Seagrass meadows of Tampa Bay- A Review', in Treat, S. F. (ed.) Proceedings of the Tampa Bay Area Scientific Information Symposium (BASIS). *Florida Sea Grant Report* **65**, pp. 210–246.

Lewis, R. R., III, Clark, P. A., Fehring, W. K., Greening, H. S., Johansson, R. O., and Paul, R. T.: 1999, 'The rehabilitation of the Tampa Bay estuary, Florida, USA, as an example of successful integrated coastal management', *Marine Pollution Bulletin* **37**, 468–473.

Moore, K. A., Wilcox, D. J. and Orth, R. J.: 2000, 'Analysis of the abundance of submersed aquatic vegetation communities in the Chesapeake Bay', *Estuaries* **23**, 115–127.

Mueller-Dombois, D. and Ellenberg, H.: 1974, *Aims and Methods of Vegetation Ecology*. Wiley, New York, 547 pp.

National Oceanographic and Atmospheric Administration (NOAA): 2001, El Niño Impacts: Florida. http://www.nnic.noaa.gov/products/analysis_monitoring/ensostuff/dist/fl_bar.htm.

Neckles, H.: 1994, Indicator development: Seagrass monitoring and research in the Gulf of Mexico, EPA/620/R-94/029, Office of Research and Development, U.S. Environmental Protection Agency, Washington, DC.

Nelson, D. M. (ed.): 1992, Distribution and abundance of fishes and invertebrates in Gulf of Mexico estuaries. Vol. 1. Data summaries. ELMR Rep.10. NOAA/NOS Strategic Assessments Division, Rockville, MD, 273 pp.

Neverauskas, V. P.: 1988, 'Response of a *Posidonia* community to prolonged reduction in light', *Aquatic Botany* **31**, 361–366.

O'Reilly, J. E., Maritorena, S., Mitchell, B. G., Siegel, D. A., Carder, K. L., Garver, S. A., Kahru, M. and McClain C.: 1998, 'Ocean color chlorophyll algorithms for SeaWiFS', *Journal of Geophysical Research* **103**, 24, 937–24, 953.

Orth, R. J. and Moore, K. A.: 1983, 'Chesapeake Bay: an unprecedented decline in submerged aquatic vegetation', *Science* **222**, 51–53.

Orth, R. J., Batiuk, R. A. and Nowak, J. F.: 1994, Trends in the distribution and abundance of habitat quality of submerged aquatic vegetation in Chesapeake Bay and its tidal tributaries: 1971 to 1991. U. S. Environmental Protection Agency, Chesapeake Bay Program Office. CBP/TRS 137/95, Annapolis, MD.

Phillips, R. C. and Lewis, R. R.: 1983, 'Influence of environmental gradients in variations in leaf widths and transplant success in North American seagrasses', *Marine Technology Society Journal* **17**, 59–68.

Pulich, W. M. and White, W. A.: 1991, 'Decline of submerged vegetation in the Galveston Bay system: Chronology and relationships to physical processes', *Journal of Coastal Research* **7**, 1125–1138.

Southeast Regional Climate Center (SERCC): 2001, How does El Niño affect the climate of the Southeast United States. http://water.dnr.state.sc.us/climate/sercc/el_nino/ENSO_Impacts.html.

Tomasko, D. A., Hall, M. O. and Dawes, C. J.: 1996, 'The effects of anthropogenic nutrient enrichment on turtle grass (*Thalassia testudinum*) in Sarasota Bay, Florida', *Estuaries* **19**, 448–456.

Tomasko, D. A. and Hall, M. O.: 1999, 'Productivity and biomass of the seagrass *Thalassia testudinum* along a gradient of freshwater influence in Charlotte Harbor, Florida', *Estuaries* **22**, 592–602.

Walker, D. I. and McComb, A.J.: 1992, 'Seagrass degradation in Australian coastal waters', *Marine Pollution Bulletin* **25**, 191–195.

Zieman, J. C.: 1982, The Ecology of the Seagrasses of South Florida: A Community Profile, FWS/OBS-82/25, U.S. Fish and Wildlife Service, Washington, DC, 123 pp.

Zimmerman, R. C., Reguzzoni, J. L. and Alberte, R. S.: 1995, 'Eelgrass (*Zostera marina L.*) transplants in San Francisco Bay: Role of light availability on metabolism, growth, and survival', *Aquatic Botany* **51**, 67–86.

GPSR Compliance

The European Union's (EU) General Product Safety Regulation (GPSR) is a set of rules that requires consumer products to be safe and our obligations to ensure this.

If you have any concerns about our products, you can contact us on ProductSafety@springernature.com

In case Publisher is established outside the EU, the EU authorized representative is:

Springer Nature Customer Service Center GmbH
Europaplatz 3
69115 Heidelberg, Germany

Batch number: 10868073

Printed by Printforce, the Netherlands